Doing Mathematics

with

Scientific WorkPlace®

and

Scientific Notebook®

Users' Guide to Computing

Version 5

Doing Mathematics

with

Scientific WorkPlace®

and

Scientific Notebook®

Users' Guide to Computing
Version 5

Darel W. Hardy
Colorado State University

Carol L. Walker
New Mexico State University

Printed in the United States of America

10   9   8   7   6   5   4   3   2   1

Trademarks

Scientific Word, *Scientific WorkPlace*, *Scientific Notebook*, and EasyMath are registered trademarks of MacKichan Software, Inc. EasyMath is the sophisticated parsing and translating system included in *Scientific Word*, *Scientific WorkPlace*, and *Scientific Notebook* that allows the user to work in standard mathematical notation, request computations from the underlying computational system (MuPAD in this version) based on the implied commands embedded in the mathematical syntax or via menu, and receive the response in typeset standard notation or graphic form in the current document. MuPAD is a registered trademark of SciFace GmbH. Acrobat is the registered trademark of Adobe Systems, Inc. $\mathrm{T_{E}X}$ is a trademark of the American Mathematical Society. True$\mathrm{T_{E}X}$ is a registered trademark of Richard J. Kinch. PDF$\mathrm{T_{E}X}$ is the copyright of Hàn Thế Thành and is available under the GNU public license. Windows is a registered trademark of Microsoft Corporation. MathType is a trademark of Design Science, Inc. All other brand and product names are trademarks of their respective companies. The spelling portion of this product utilizes the Proximity Linguistic Technology.

This document was produced with *Scientific WorkPlace*.

Authors: *Darel Hardy and Carol Walker*
Manuscript Editors: *Susan Bagby and George Pearson*
Compositor: *MacKichan Software Inc.*
Printing and Binding: *Malloy Lithographing, Inc.*

Dedicated

to the memory of our parents

Alice DeVinny Hardy

and

F. Waldo Hardy

Contents

3   Algebra   53

7   Calculus   221

Preface

Scientific WorkPlace and *Scientific Notebook* provide a free-form interface to a computer algebra system that is integrated with a scientific word processor. They are designed to fit the needs of a wide range of users, from the beginning student trying to solve a linear equation to the professional scientist who wants to produce typeset-quality documents with embedded advanced mathematical calculations. The text editors in *Scientific WorkPlace* and *Scientific Notebook* accept mathematical formulas and equations entered in natural notation. The symbolic computation system produces mathematical output inside the document that is formatted in natural notation, can be edited, and can be used directly as input to subsequent mathematical calculations.

Scientific WorkPlace was originally developed as an interface to a computational system, with partial support from a National Science Foundation Small Business Innovation Research (SBIR) grant. The goal of the research conducted under the SBIR grant was to provide a new type of interface to computer algebra systems. The essential components of this interface are *free-form editing* and *natural mathematical notation*. *Scientific WorkPlace* and *Scientific Notebook* satisfy both criteria. They make sense out of as many different forms as possible, rather than requiring the user to adhere to a rigid syntax or just one way of writing an expression.

The computational components of *Scientific WorkPlace* and *Scientific Notebook* use a MuPAD engine. All versions use standard libraries furnished by Sciface Software. *Scientific WorkPlace* and *Scientific Notebook* provide easy, direct access to all the mathematics needed by the many users. For the sophisticated user, they also allow access to the full range of MuPAD functions and to functions programmed in MuPAD. By providing an interface with little or no learning cost, *Scientific WorkPlace* and *Scientific Notebook* make symbolic computation as accessible as any Windows-based word processor.

Scientific WorkPlace and *Scientific Notebook* have great potential in educational settings. In a classroom equipped with appropriate projection equipment, the program's ease of use and its combination of a free-form scientific word processor and computational package make it a natural replacement for the chalkboard. You can use it in the same ways you would a chalkboard and you have the added advantage of the computational system. You do not need to erase as you go along, so previous work can be recalled. Class notes can be edited and printed.

Scientific WorkPlace and *Scientific Notebook* provide a ready laboratory in which students can experiment with mathematics to develop new insights and to solve interesting problems; they also provide a vehicle for students to produce clear, well-written homework. For situations where the array of possibilities is beyond the scope of a course, you can hide some of the higher level options on the **Compute** menu. To accomplish this, from the **Tools** menu, choose **Engine Setup** and check **Display Simplified Compute Menu**.

This document, *Doing Mathematics with Scientific WorkPlace and Scientific Notebook Version 5*, describes the use of the underlying computer algebra systems for doing mathematical calculations. In particular, it explains how to use the built-in computer algebra system MuPAD to do a wide range of mathematics without dealing directly with the syntax of the computer algebra system.

This document is organized around standard topics in the undergraduate mathematics curriculum. Users can find the guidance they need without going to chapters involving mathematics beyond their current level. The first four chapters introduce basic procedures for using the system and cover the content of the standard precalculus courses. Later chapters cover analytic geometry and calculus, linear algebra, vector analysis, differential equations, statistics, and applied modern algebra. Exercises are provided to encourage users to practice the ideas presented and to explore possibilities beyond those covered in this document.

Users with an interest in doing mathematical calculations are advised to read and experiment with the first five chapters—*Basic Techniques for Doing Mathematics*; *Numbers, Functions, and Units*; *Algebra*; *Trigonometry*; and *Function Definitions*—which provide a good foundation for doing mathematical calculations. You may also find it helpful to read parts of the sixth chapter *Plotting Curves and Surfaces* to get started creating plots. You can approach the remaining chapters in any order.

Experienced MuPAD users will find it helpful to read about accessing other MuPAD functions and adding user-defined MuPAD functions in the chapter *Function Definitions*. You will also want to refer to the tables in that chapter that pair MuPAD names with *Scientific WorkPlace* and *Scientific Notebook* names for constants, functions, and operations.

On-Line Help

The first three items on the Help menu—Contents, Search, and Index—provide three routes for obtaining information.

Contents   To reach the Contents page, press F1 or choose Contents from the Help menu. Choose Computing Techniques for help arranged by topic, basically an on-line version of this manual. Once you are in a computing help document, the Next Document links take you sequentially through all of the computing help documents—click the right arrow on the Link Bar or choose Go + Links + Next Document. The Next Document links also take you sequentially through the tables of contents for the chapters of Computing Techniques.

Search   For a discussion on a particular topic, choose Help + Search and enter key words. Search will find topics in the General Information and Reference Library indexes as well as in the Computing Techniques index.

Index   For a discussion on a particular computing topic, choose Help + Index + Computing Techniques. When you open an Index, use the drop-down list on the Navigate bar, click the GoTo Marker button on the Navigate or History toolbar, or choose Go + To Marker, and choose from the drop-down list that appears.

For a quick start in using *Scientific WorkPlace* or *Scientific Notebook* for text editing and computing, press F1 or choose the Contents menu under Help, and try Take a Tour and Learn the Basics. You will also get many useful hints for computing by working quickly through documents provided with your system on the \play subdirectory. If you save copies of the Help documents in *Scientific WorkPlace* or *Scientific Notebook*, you can interact with the mathematics they contain, experimenting with or reworking the included examples.

For information on the document-editing features of your system, the online **Help** describes how to create transportable LaTeX documents without viewing the syntax of LaTeX, how to typeset with LaTeXand PDFLaTeX, and how to create HTML and RTF output—see **General Information** under **Contents** or **Index**, or choose **Search**. These document-editing features are also described in the document, *Creating Documents with Scientific Word and Scientific WorkPlace Version 5.*

Conventions

Understanding the notation and the terms we use in our documentation will help you understand the instructions in this manual. We assume you are familiar with basic Windows procedures and terminology. In this manual, we use the notation and terms listed below.

General Notation

- **Text like this** indicates text you should type exactly as it is shown.
- Text like this indicates the name of a menu, command, or dialog.
- TEXT LIKE THIS indicates the name of a keyboard key.
- `Text like this` indicates the name of a file or directory.
- *Text like this* indicates a term that has special meaning in the context of the program.
- *Text like this* indicates an expression that is typed in mathematics mode.
- The word *choose* means to designate a command for the program to carry out. As with all Windows applications, you can choose a command with the mouse or with the keyboard. Commands may be listed on a menu or shown on a button in a dialog box. For example, the instruction "From the **File** menu, choose **Open**" means you should first choose the **File** menu and then from that menu, choose the **Open** command. This is often abbreviated as **File + Open**. The instruction "choose **OK**" means to click the **OK** button with the mouse or press TAB to move the attention to the **OK** button and then press the ENTER key on the keyboard.
- The word *check* means to turn on an option in a dialog box.
- When **Compute** menu commands are specified, the word **Compute** is usually suppressed. For example, when you see **Evaluate**, choose **Compute + Evaluate**.

Keyboard Conventions

We also use standard Windows conventions to give keyboard instructions.

- The names of keys in the instructions match the names shown on most keyboards. They appear like this: ENTER, F4, SHIFT.
- A plus sign (+) between the names of two keys indicates that you must press the first key and hold it down while you press the second key. For example, CTRL + G means that you press and hold down the CTRL key, press G, and then release both keys.
- The notation CTRL + **word** means that you must hold down the CTRL key, type the word that appears in bold type after the +, then release the CTRL key. Note that if a letter appears capitalized, you should type that letter as a capital.

Obtaining Technical Support

If you can't find the answer to your questions in the manuals or the online Help, you can obtain technical support from the website at

http://www.mackichan.com/techtalk/knowledgebase.html

or at the Web-based Technical Support forum at

http://www.mackichan.com/techtalk/UserForums.htm

You can also contact the Technical Support staff by email, telephone, or fax. We urge you to submit questions by email whenever possible in case the technical staff needs to obtain your file to diagnose and solve the problem.

When you contact Technical Support by email or fax, please provide complete information about the problem you're trying to solve. They must be able to reproduce the problem exactly from your instructions. When you contact them by telephone, you should be sitting at your computer with the program running. Please provide the following information any time you contact Technical Support:

- The MacKichan Software product you have installed.

- The version and build numbers of your installation (see Help / About).

- The serial number of your installation (see Help / System Features).

- The version of the Windows system you're using.

- The type of hardware you're using, including network hardware.

- What happened and what you were doing when the problem occurred.

- The exact wording of any messages that appeared on your computer screen.

▶ **To contact technical support**

- Contact Technical Support by email, fax, or telephone between 8 A.M. and 5 P.M. Pacific Time:

Internet electronic mail address: support@mackichan.com
Fax number: 360-394-6039
Telephone number: 360-394-6033
Toll-free telephone: 877-SCI-WORD (877-724-9673)

You can learn more about *Scientific WorkPlace* and *Scientific Notebook* on the MacKichan web site, which is updated regularly to provide the latest technical information about the program. The site also houses links to other TeX and LaTeX resources. There is also an unmoderated discussion forum and an unmoderated email list so users can share information, discuss common problems, and contribute technical tips and solutions. You can link to these valuable resources from the home page at **http://www.mackichan.com**.

Darel W. Hardy
Carol L. Walker

1 Basic Techniques for Doing Mathematics

In this chapter, we give a brief explanation, with examples, of each of the basic computational features of *Scientific WorkPlace* and *Scientific Notebook*. You are encouraged to open a new document and work the examples as you proceed.

You can begin computing as soon as you have opened a file.

► **To enter and evaluate an expression**

1. From the Insert menu, choose Math. (If this choice does not appear, your insertion point is already in mathematics mode and you are ready for step 2.)

2. Enter a mathematical expression in the document—for example, $2+2$. (It will appear red on your screen.)

3. Leaving the insertion point in the expression, from the Compute menu, choose Evaluate.

The expression $2 + 2$ will be replaced by the evaluation $2 + 2 = 4$.

Mathematics is automatically spaced differently from text as you enter it—for example, "$2 + 2$" rather than "2+2"—so you do not have to make adjustments.

Inserting Text and Mathematics

The blinking vertical line on your screen is referred to as the *insertion point*. You may have heard it called the *insert cursor*, or simply the *cursor*. The insertion point marks the position where characters or symbols are entered when you type or click a symbol. You can change the position of the insertion point with the arrow keys or by clicking a different screen position with the mouse. The position of the mouse is indicated by the *mouse pointer*, which assumes the shape of an I-beam over text and an arrow over mathematics.

Basic Guidelines

You can enter information in a document in either mathematics or text. The mathematics that you enter is recognized by the underlying computing engine as mathematics, and the text is ignored by the computing engine.

- *Text* is entered at the position of the insertion point when the **Toggle Text/Math** button in the **Standard** toolbar shows ⊤.

- *Mathematics* is entered at the position of the insertion point when the **Toggle Math/-Text** button on the **Standard** toolbar shows Ⅿ.

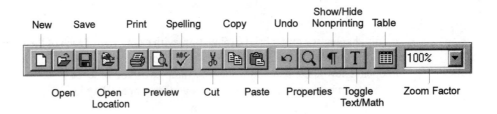

On the screen, mathematics appears in red and text in black. For information on changing this, see **Help + Search + Screen Defaults.**

You can toggle between these two states by clicking the buttons shown earlier or by pressing CTRL + M or CTRL + T on the keyboard. Entering a mathematics symbol by clicking a button on a toolbar automatically puts the state in mathematics at the position in which the symbol is entered. The state remains in mathematics as you enter characters or symbols to the right of existing mathematics, until you either toggle back into text or move the insertion point into text by using the mouse or by pressing RIGHT ARROW, LEFT ARROW, or ENTER. (To customize your system for toggling between mathematics and text, from the **Tools** menu, choose **User Setup** and click the **Math** tab. The choices include toggling with the space bar or the INSERT key. See **Help + Search + User Setup** for further details.)

Choose **Help + Search + toolbars + Customizing the toolbars,** if the **Math Templates** toolbar referred to below—or any other toolbar you would like to use—does not automatically appear on your screen.

▶ **To enter a fraction, radical, exponent, or subscript**

1. Click on ⊟, √◻, ℕˣ, or ℕₓ on the **Math Templates** toolbar.

   or

   From the **Insert** menu, choose **Fraction, Radical, Superscript,** or **Subscript.**

   or

   Press CTRL + F, CTRL + R, CTRL + H (or CTRL + UP ARROW), or CTRL + L (or CTRL + DOWN ARROW).

2. Enter expressions in the input boxes.

The SPACEBAR and ARROW keys move the insertion point through mathematical expressions and the TAB key toggles between input boxes.

▶ **To use various symbols for multiplication and division**

- Click your choice on the Symbol Cache toolbar

or

- Click the Binary Operations button  on the Symbol Panels toolbar

and choose from symbols on the drop-down panel:

You *select* a piece of text with the mouse by holding down the left mouse button while moving the mouse, or from the keyboard by holding down SHIFT and pressing RIGHT ARROW or LEFT ARROW. Your selection appears on the screen in reversed colors. This technique is sometimes referred to as "highlighting" an area of the screen. This is also one of the ways you can select mathematics. See page 9 for a discussion of *automatic* and *user* selections for mathematics.

See the preface (page xxiii) for additional notation and keyboard conventions used in this manual.

Tip   The appearance of the Toggle Text/Math button — $\boxed{M}$ or $\boxed{T}$ — reflects the state at the position of the insertion point.

Displaying Mathematics

Mathematics can be centered on a separate line in a *display*.

$$y = ax + b$$

▶ **To create a display**

1. Click ☰ or, from the Insert menu, choose Display, or press CTRL + D.

2. Type a mathematical expression in the display, or select a piece of mathematics and drag it into the display.

Note   Pressing ENTER immediately before a display will add extra vertical space. If you do not want this space, place the insertion point immediately before the display and press BACKSPACE. (This removes the "new paragraph" symbol.)
Pressing ENTER immediately after a display will add extra vertical space and cause the next line to start a new paragraph. If you do not want this space or indentation, place the insertion point at the start of the next line and press BACKSPACE. (This removes the "new paragraph" symbol.)

You can begin with an existing mathematical expression and put it into a display.

▶ **To put mathematics in a display**

1. Select the mathematics with click and drag or SHIFT+RIGHT ARROW.

2. Click the Display button on the Math Objects toolbar, or from the Insert menu, choose Display.

The default environment in a display is mathematics. You can, however, enter text in a display by toggling to text. Also, you can select mathematics or text in a display and change its state by toggling (see page 2).

Centering Plots, Graphics and Text

If you have text that you wish to center on a separate line, the natural way to do this operation is with Body Center, which you can choose from one of the pop-up menus at the bottom of your screen.

If you have a plot or graphic that you wish to center on a separate line, you should choose the Display setting in the Layout dialog, as discussed in Chapter 6, *Plotting Curves and Surfaces*. To center a group of plots or graphics, choose the In Line setting in the Layout dialog and use Body Center. We generally advise against placing a plot inside the Insert + Display object, as this makes the plot itself a mathematical object, which can sometimes cause difficulties.

Basic Guidelines for Computing

When you respond to the request "place the insertion point in the expression," place the insertion point within, or immediately to the right of, the expression. The position immediately to the left of a mathematical expression is not valid. You can check the state of Toggle Text/Math to verify that your insertion point is in mathematics.

Evaluating Expressions

To enter a mathematics expression for a computation, begin a new line with the mathematics expression or type the expression immediately to the right of text or a text space. If you enter mathematics immediately to the right of other mathematics, the expressions will be combined in ways you may not intend. A safe way to begin is to press ENTER and start on a new line.

1. Click $\boxed{T}$ (or choose Insert + Math, or press CTRL + M) to toggle to mathematics mode, so that the Text/Math button looks like $\boxed{M}$.

2. Type $3 + 8$.

3. Leaving the insertion point in the expression $3 + 8$, do one of the following:

   - Click $\boxed{=?}$ on the Compute toolbar

     or

   - From the Compute menu, choose Evaluate

     or

   - Press CTRL + E.

This sequence of actions inserts $= 11$ to the right of the $3 + 8$, resulting in the equation

$$3 + 8 = 11.$$

Note The contents of the gray boxes (shaded background) display the mathematical expressions you enter, together with the results of the indicated operation. In general, throughout this document, the mathematical contents of gray boxes display both the input for an action and the results. In the case of plots, the input is displayed in the gray box and the results are displayed immediately following the gray box.

By following the *same* procedure, you can carry out the following operations and perform a vast variety of other mathematical computations.

Add

▶ Evaluate

$$235 + 813 = 1048 \qquad 49.3 + 2.87 = 52.17$$

$$\frac{2}{3} + \frac{1}{7} = \frac{17}{21} \qquad (x + 3) + (x - y) = 2x + 3 - y$$

Subtract

▶ Evaluate

$$96 - 27 = 69 \qquad \left(2x^2 - 5\right) - (3x + 4) = 2x^2 - 9 - 3x$$

$$49.3 - 2.87 = 46.43 \qquad \frac{2}{3} - \frac{8}{7} = -\frac{10}{21}$$

Multiply

▶ Evaluate

$$82 \times 37 = 3034 \qquad (936)(-14) = -13104$$

$$14.2 * 83.5 = 1185.7 \qquad \frac{2}{3}\frac{8}{7} = \frac{16}{21}$$

Divide

▶ Evaluate

$$82 \div 37 = \frac{82}{37} \qquad 36/14 = \frac{18}{7}$$

$$\frac{14.2}{83.5} = 0.1700598802 \qquad \frac{-\frac{2}{3}}{\frac{8}{7}} = -\frac{7}{12}$$

Important   Except that it be mathematically correct, there are almost *no rules* about the form for entering a mathematical expression in *Scientific WorkPlace* and *Scientific Notebook*.

For example, the expressions

$$\frac{\frac{2}{3}}{\frac{8}{7}} \qquad \frac{2}{3} \div \frac{8}{7} \qquad \frac{2}{3}/\frac{8}{7} \qquad (2/3)\,/\,(8/7)$$

are equally acceptable ways of entering a quotient of fractions. Also,

$$(936)(14) \qquad 936 \cdot 14 \qquad 936 \times 14 \qquad 936 * 14$$

and many other variations are acceptable for the same product. One of the few exceptions to the claim of "no rules" is that "vertical" notation such as

$$\begin{array}{r} 24 \\ +15 \end{array} \quad \text{and} \quad \begin{array}{r} 235 \\ -47 \end{array} \quad \text{and} \quad 2\overline{)364}$$

used when doing arithmetic by hand is not generally recognized. Write sums, differences, products, and quotients of numbers in natural "linear" or fractional notation, such as $24 + 15$ and $235 - 47$ and 36×14 and $364/2$ or $\frac{364}{2}$ or $364 \div 2$.

Certain constants are recognized in their usual forms—such as π, i, and e—as long as the context is appropriate. On the other hand, they are recognized as arbitrary constants, variables, or indices when appropriate to the context, helping to provide a completely natural way for you to enter and perform mathematical computations.

Note   The number of digits in answers to numerical problems depends on settings that you can change in the **Tools + Computation Setup** dialog discussed on page 29. The examples in this document may differ in this respect from the answers you get with your system, and different examples in this document use different settings.

Interpreting Expressions

If your mathematical notation is ambiguous, it may still be accepted. However, the way it is interpreted may or may not be what you intended. To be safe, remove an ambiguity by placing additional parentheses in the expression.

▶ **To check the interpretation of a mathematical expression**

1. Leave the insertion point in the expression.

2. Press CTRL, and while holding it down, type ?

   or

   Choose **Compute + Interpret**.

▶ CTRL + **?**

$$1/3x + 4 = \tfrac{1}{3}x + 4 \qquad 1/(3x + 4) = \tfrac{1}{(3x+4)}$$
$$1/(3x) + 4 = \tfrac{1}{3x} + 4 \qquad 1/3(x + 4) = \tfrac{1}{3}(x + 4)$$

Tip   Although in most cases different shapes of brackets are interchangeable, as a general rule standard parentheses $(3 + \pi)$ are better for grouping mathematical expressions than other types of brackets, because in a few very special cases, other brackets can be interpreted in a way you don't intend. Also, the *expanding* brackets you enter from the **Insert** menu or the **Math Templates** toolbar or with various keyboard shortcuts are better for grouping mathematical expressions than the single brackets on the keyboard.

The Compute Menu and Toolbar

Click Compute at the top of the screen and a drop-down menu will appear with a number of computing choices, beginning with Evaluate, Evaluate Numerically, Simplify, Combine, Factor, and Expand.

Important Throughout this document, whenever computing commands are specified, the preceding Compute is implied. For example, when you see Evaluate, choose Compute + Evaluate.

The Compute toolbar contains some of the most often used choices from the Compute menu.

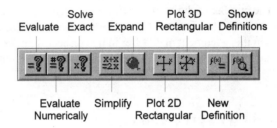

▶ To perform a computation

1. Place the insertion point inside or to the right of the expression on which you want to perform an operation.

2. Click the button or menu item for the operation you want to perform.

Commands on the Compute menu can be executed from the keyboard following standard procedures.

▶ To execute a command on the Compute menu from the keyboard

- Press ALT and, while holding down this key,
  - Press C (for Compute), followed by the command letter underlined on the drop-down menu that appears.
  - If the command is followed by an arrow on the right of the menu, press ENTER followed by another underlined command letter.

Some commands have a shorter keyboard shortcut. (CTRL + KEY is an abbreviation for "Press CTRL and, while holding down this key, press KEY.")

Shortcut	Command
CTRL + E	Compute + Evaluate
CTRL + SHIFT + E	Compute + Evaluate (in place)
CTRL + Y	Compute + Send String (weak evaluation)
CTRL + =	Compute + Definitions + New Definition

There are many other keyboard shortcuts available. For a list of keyboard shortcuts for both mathematics and text, choose Help + Search, keyboard shortcuts and, from the list, choose Keyboard shortcuts.

Selecting Mathematical Expressions

When you perform a mathematical operation, a mathematical expression is automatically selected for the operation, depending on the position of the insertion point and the operation involved. We will refer to these as *automatic selections*. You can also force other selections by selecting mathematics with the mouse. We will refer to the latter as *user selections*.

Understanding Automatic Selections

When you place the insertion point in a mathematical expression and choose an operation from the Compute menu, the automatic selection depends primarily on the command you choose. It also depends on the location of the mathematics, such as in line, in a matrix, or in a display. The following two possibilities occur for mathematical objects that are typed in line.

- Selection of an expression, that part of the mathematics containing the insertion point that is enclosed between a combination of text and the class of symbols—such as $=$,

  $<$, or $\leq$ —known as *binary relations*. (Click $\boxed{\leq\varsigma}$ on the Symbol Panels toolbar to see the full selection of binary relations.)

- Selection of the entire mathematical object, such as an equation or inequality.

Examples in the following two sections illustrate situations where these two types of selections occur.

Operations that Select an Expression   The majority of operations select an expression enclosed between text and binary operations. Place the insertion point anywhere in the *left* side of the equation $2x + 3x = 1 + 4$ except to the left of the 2, and choose Evaluate.

▶ Evaluate

$2x + 3x = 5x = 1 + 4$

The expression $= 5x$ is inserted immediately after the expression $2x + 3x$. This time, only the expression on the left side of the equation was selected for evaluation. Since the result of the evaluation was equal to the original expression, the result was placed next to the expression, preceded by an equals sign. The insertion point is placed at the right end of the result so that you can select another operation to apply to the result without moving the insertion point.

Other commands, including Evaluate Numerically, Simplify, Combine, Factor, and Expand, make similar selections under similar conditions.

Operations that Select an Equation or Inequality   Place the insertion point anywhere within the equation $2x + 3x = 1$ and click the Solve button $\boxed{\text{x?}}$ on the Compute toolbar or, from the Solve submenu, choose Exact.

▶ Solve + Exact

$2x + 3x = 1$, Solution is: $\frac{1}{5}$

$3x + 5 \leq 5x - 3$, Solution is : $4 \leq x$

In these cases, the entire mathematical object—that is, the equation or inequality—was selected. The other choices on the Solve submenu and the operation Check Equality also select the entire mathematical object.

If the mathematics is not appropriate for the operation, no action is taken. For example, applying one of the Solve commands to $x = y = z$ causes a syntax error, because of the pair of equals signs. You see an error message, hear a beep, or see no action, depending on the Error Notification setting. You can change this setting in the Engine Setup dialog on the Tools menu. (See page 13).

Selections Inside Displays and Matrices

You can use a *matrix* to arrange mathematical expressions in a rectangular array. To create a matrix, choose Matrix from the Insert menu or click $\boxed{\vdots}$, set the number of rows and columns, and choose OK. If you see nothing on your screen, choose View and turn on Hidden Lines or Input Boxes. Type a number or mathematical expression in the input boxes of the matrix.

Operations may behave somewhat differently when mathematics is entered in a display or in a matrix. If you place the insertion point inside a display or matrix, the automatic selection is the entire array of entries, for any operation. Some operations apply to a matrix, and others to the entries of a matrix or contents of a display. If the operation is not appropriate for either a matrix or its entries or for all the contents of a display, you may receive a report of a syntax error.

When you click $\boxed{=?}$ or apply Evaluate with the insertion point in the left side of the displayed equation

$$2x + 3x = 3 + 5$$

you get the result : $5x = 8$ in line outside the display, rather than the result $2x + 3x = 5x = 3 + 5$, which would appear if the equation had been placed in line. Inside a display or matrix, the automatic selection is all the mathematics, and the result is generally returned outside the display.

When you click $\boxed{=?}$ or apply Evaluate to a matrix of expressions, all the expressions will be evaluated and the result will be returned as a matrix. Evaluate Numerically, Simplify, Factor, and choices from the Combine submenu behave similarly.

▶ **Evaluate (or Simplify)**

$$\left(\begin{array}{cc} x+x & 5+3 \\ 5/2 & 6^2 \end{array} \right) = \left(\begin{array}{cc} 2x & 8 \\ \frac{5}{2} & 36 \end{array} \right)$$

▶ **Evaluate Numerically**

$$\left(\begin{array}{cc} x+x & 5+3 \\ 5/2 & 6^2 \end{array} \right) = \left(\begin{array}{cc} 2.0x & 8.0 \\ 2.5 & 36.0 \end{array} \right)$$

▶ **Factor**

$$\left(\begin{array}{cc} x+x & 5+3 \\ 5/2 & 6^2 \end{array} \right) = \left(\begin{array}{cc} 2x & 2^3 \\ \frac{5}{2} & 2^2 3^2 \end{array} \right)$$

Selections Inside Tables

If you have mathematics in a table, placing the insertion in the mathematics will automatically select all of the mathematics in the cell that is adjacent to the insertion point. For example, Evaluate will select an equation if one is present, rather than just an expression. The result of the operation will appear in the cell. The table itself is not a mathematical object, and the behavior is somewhat different than for mathematics in a matrix or display.

Understanding User Selections

You can restrict the computation to a selection you have made and so override the automatic choice. Recall that you can *select* a piece of mathematics by holding down the left mouse button while moving the mouse; your *selection* is the information that appears on the screen in reversed colors.

There are two options for applying operations to your selection—"operating on a selection" displays the result of the operation but leaves the selection intact, and "replacing a selection" replaces the selection with the result of the operation. Following are two examples illustrating the behavior of the system when operating on a selection. The option of replacing a selection is referred to as *computing in place*, and examples are shown in the next section.

▶ **To operate on a user selection**

1. Use the mouse or press SHIFT + ARROW to select an expression.

2. Applying a command to the expression.

Example   Use the mouse or press SHIFT + ARROW to select $2 + 3$ in the expression $2 + 3 - x$. From the **Compute** menu, choose **Evaluate**. The answer appears to the right of the expression, following a colon (:).

$$2 + 3 - x : 5$$

Use the mouse or press SHIFT + ARROW to select $(x + y)^3$ within the expression $(x + y)^3 (7x - 13y)^3 + \sin^2 x$. From the **Compute** menu, choose **Expand**. The answer appears to the right of the whole expression, following a colon.

$$(x + y)^3 (7x - 13y)^3 + \sin^2 x : x^3 + 3x^2 y + 3xy^2 + y^3$$

In general, the result of applying an operation to a *user selection* is not equal to the entire original expression, so the result is placed at the end of the mathematics, separated by something in text (in this case, a colon). You can then use the word-processing capabilities of your system to put the result where you want it in your document.

▶ **To replace a selection**

1. Use the mouse or press SHIFT + ARROW to select an expression.

2. Press and hold CTRL while applying a command to the expression.

The system replaces the selected expression with the output of the command. This is an in-place computation, as described in the following section.

Computing in Place

With the help of the CTRL key, you can perform any computation in place; that is, you can replace an expression directly with the results of that computation. This "computing in place" is a key feature. It provides a convenient way for you to manipulate expressions into the forms you desire.

Select with the mouse the expression that you wish to replace, and while holding down the CTRL key, choose the desired operation from the **Compute** menu. The response that replaces the original expression will remain selected, making it convenient to add parentheses around the new expression when needed simply by clicking the parentheses button.

▶ **Select expression and choose** CTRL + Evaluate **or choose** CTRL + SHIFT + E

$2/3$ is replaced by $\frac{2}{3}$ $146 + 529 - 19 + 6$ is replaced by 662

▶ **Select expression and choose** CTRL + Expand

$2345/567$ is replaced by $4\frac{11}{81}$ $(a + b)^3$ is replaced by $a^3 + 3a^2 b + 3ab^2 + b^3$

This feature, combined with copy and paste, allows you to "fill in the steps" in demonstrating a computation.

Example   To replace $(x - 2y)^2$ in the expression $(x - 2y)^2 (7x - 13y) (x^2 + 1)$ with its expansion, select $(x - 2y)^2$, hold down the CTRL key, and click ⬛ or choose **Expand**. Your selection, $(x - 2y)^2$, is replaced by its expansion. The expansion has no parentheses around it, but since it remains selected, you can click ⬛ to add the needed parentheses. Following this procedure,

$$(x - 2y)^2 (7x - 13y) (x^2 + 1)$$

is replaced by

$$(x^2 - 4xy + 4y^2)(7x - 13y)(x^2 + 1)$$

You can return the expression to a factored form by selecting $(x^2 - 4xy + 4y^2)$, holding down the CTRL key, and choosing **Factor**.

Stopping a Computation

Most computations are done more or less instantaneously, but some may take several minutes to complete, and some may take a (much) longer time. So it is convenient to be able to interrupt the computing and regain control of your document.

▶ **To stop a computation**

- Click the stop sign ⬛ on the **Stop** toolbar.
  or

- Press CTRL + BREAK.

Try it out by applying **Factor** to $2^{91} + 3$. This expression will be factored in several minutes, but you can stop the computation if you wish.

Computational Engine

The computational engine embedded in *Scientific WorkPlace* and *Scientific Notebook Version 5* is MuPAD 2.5. To see if this engine is active in your system, or to deactivate the engine, choose **Tools + Computation Setup + Engine Selection**.

For a list of menu commands and a partial list of functions and constants available, and a description of these commands and functions in terms of the native commands of MuPAD, see page 130, or go to **Help + Search + function** and choose "A brief description of commands and functions."

Error Handling

From the **Tools** menu, choose **Engine Setup** and click the **Error Handling** tab. On this page you can specify the default settings for **Error Notification**, **Engine Command Notification**, and **Transaction Logging**.

Under Error Notification you can choose None, choose to be notified with Beep, or choose Message to Status Bar or Message to Dialog Box. These are responses to various syntax errors in the mathematics being sent to the computing engine. If you choose to have messages shown, you will see some information concerning these errors.

Under Engine Command Notification, you can choose None, or choose Show Commands on Status Bar or Show Commands in Dialog Box. If you choose to have commands shown in a dialog box, you will see the syntax of commands being sent to the computing engine.

The factory defaults for these choices are those shown in the graphic.

- Error Notification: Beep
- Engine Command Notification: None.
- Transaction Logging: None

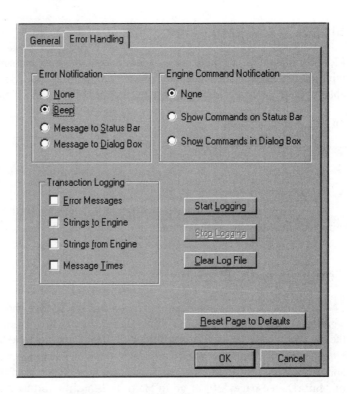

To return to these defaults, choose Reset Page to Defaults and choose OK.

Under Transaction Logging, you can choose to have some, all, or no records of mathematical messages sent to the computing engine (transactions) recorded in an ASCII log file. Transaction Logging always resets to the default of not logging. To accumulate entries in the log file, you must take the following action *each time you open a session.*

▶ **To start logging transactions**

1. From the Tools menu, choose Engine Setup, and click the Error Handling tab.

2. Choose Start Logging.

3. Check any or all of the choices for Transaction Logging: Error Messages, Strings to Engine, Strings from Engine.

4. If you want the time recorded in the log file, check Message Times.

With Start Logging checked, your system will create a file named **engine.log** on the root directory of your *Scientific WorkPlace* or *Scientific Notebook* system, and record all of your transactions in this file for the rest of the session or until you press Stop Logging. The transactions you have logged will be saved in this file until you choose **Clear Log File**. You can read the file **engine.log** with any ASCII editor.

Frequently Asked Questions

Here, in question and answer form, are some situations that might arise when you are working in a document.

Q. My screen has gotten cluttered with lines or marks that don't belong there (or I can't see something on the screen that I know is there). What can I do?

A. Press ESC or choose View + Refresh to refresh your screen.

Q. What can I do if I type an expression in text mode that I meant to have in mathematics mode?

A. Select the expression with the mouse or select the expression from the keyboard by placing the insertion point to the left of the expression and pressing SHIFT + RIGHT ARROW. Then click the Toggle Text/Math button $\boxed{\text{T}}$ to change it to $\boxed{\text{M}}$.

Q. What can I do if I cannot see all of my work on the screen either horizontally or vertically?

A. If a piece of mathematics extends beyond the width of the screen, you can scroll horizontally using the scroll bars at the bottom of the screen. If the mathematics has possible breaking points, add an Allow Break at appropriate places from the Insert + Spacing + Break menu.
To see more of your document on the screen at a time, you can reduce the size of the screen font. Click View and choose Working. Change the percentage in the Working View box and choose OK. The 1x on the View menu gives 100% and the 2x gives 200%. The range for Working View is 50% (very small) to 400% (huge).

Q. Can I change the behavior of the SPACEBAR, ENTER, and TAB keys?

A. A setting on the Edit page of Tools + User Setup will allow you to change these behaviors. The default behavior allows you to enter multiple spaces, horizontal or vertical, by pressing SPACEBAR, ENTER, or TAB. Follow directions on the menu to change spacing behaviors.

For a variety of spacing options, choose Insert + Spacing + Horizontal Space or Vertical Space, and choose an appropriate size space. If you check Custom Space, you can specify the width or height of the space or choose stretchy spaces. Mathematics is automatically spaced appropriately for most situations. *To keep a mathematical expression meaningful for computation, be sure any added space stays in mathematics mode.*

Q. I tried a computation and nothing happened (although my system does carry out other computations). How can I find out what I did wrong?

A. The most common problem is a forgotten definition. Choose Compute + Definitions + Show Definitions and look for a definition that is interfering with your computation. Apply Compute + Definitions + Undefine to the variable or function that is causing the problem.

If that does not solve your problem, click Tools, choose Engine Setup, Error Handling page, and change the setting for Error Notification. With a setting of None, you get no response to errors. With a setting of Beep, you get a warning sound with an error. With other settings, you get messages with information about the error—usually the error message generated by the computing engine. (See page 13 for more detailed information about error handling.)

Q. I tried to take the absolute value of an expression and nothing happened. What is wrong?

A. The symbols for absolute value are the vertical lines from the dialog box under

Brackets [▯] . (The keyboard vertical line will also work, but expanding brackets are less vulnerable to misinterpretation.) Perhaps you used the vertical lines from

the symbol panel under the Binary Relations button [≤⊊] . Although they appear similar, they are not the same symbols and will not be interpreted as absolute-value symbols.

Q. How can I be sure exactly how my mathematical expression is being interpreted?

A. Select the expression and press CTRL + ? or choose COMPUTE + INTERPRET. The expression will be presented in an unambiguous form. For example,

$$\sin a / \sin b = \frac{\sin a}{\sin b} \qquad \sin x/y = \sin \frac{x}{y} \qquad \int xy = \int xy \, d?$$

Add parentheses or change the expression some other way to remove an ambiguity.

Q. The mathematics I entered is being misinterpreted but it looks okay on the screen. What should I change?

A. Is your expression in a display? If you place the insertion point in a display and choose Evaluate, the entire contents of the displayed object will be evaluated, even if part of it is text. To avoid this type of behavior, select the expression or equation with click and drag before choosing a command. If you are using braces, square brackets, or non-expanding parentheses in place of expanding parentheses, try changing all of these to expanding parentheses. **It is advisable, as a general rule, to use expanding parentheses whenever parentheses are called for.** For a variety of reasons, other choices for brackets are vulnerable to misinterpretation—*particularly if the left and right parentheses do not match.* Example: $(2)(3)$ is entered with the outer parentheses "()" expanding brackets and the inner parentheses ")(" non-expanding parentheses. Evaluating this non-matched expression gives $(2)(3) = 2$, which is probably not what is intended! Although the expanding brackets under the Brackets button , the expanding brackets and , and non-expanding brackets from the keyboard are generally interchangeable (when properly matched), there are a few circumstances in which the square brackets or keyboard brackets or even keyboard parentheses do not work properly. In particular, the less-than and greater-than symbols on the keyboard should *not* be used as brackets. These two symbols, as well as the symbols on the panel under , are binary relations and generally will not be interpreted as brackets. Square brackets and braces have some special meanings for the computing engine, and even though the interface is designed to accept as many ordinary mathematics expressions as possible, the use of nonexpanding or unusual brackets can lead to misinterpretations.

Q. My expression will not plot. What can I try?

A. Hold down the CTRL key while giving the plot command. This will cause the Plot Properties dialog to come up before the system makes the plot. Choose the Items Plotted page and try changing the settings for Variables Intervals and/or uncheck the Adjust Plot for Discontinuities option.
Choose Compute + Definitions + Show Definitions, or click the Show Definitions button on the Compute toolbar, and look to see if any of the variables you are using are defined. If so, select the variable and choose Compute + Definitions + Undefine.

Q. My document contains complicated plots, and scrolling through the document is very slow. Can I turn off the plots temporarily to save myself time as I edit the file?

A. You can turn off the computing engine temporarily: Choose Tools + Computation Setup, go to the Engine Selection page, and choose None. Alternately, you can minimize a plot so that it appears as an icon: Select the plot, choose Edit + Properties, and on the Layout page, under Screen Display Attributes, check Iconified. Or, if your plots are in final form, you can rename the plot snapshots and import them as pictures that take much less time to load. See page 208 for details on this option.

2 Numbers, Functions, and Units

Numbers and functions to be used for computing should be entered in mathematics mode and appear red (or gray) on your screen. If that is not the case, select the expression and click $\boxed{\text{T}}$ to change it to mathematics. Units to be used for computing must be entered as a Unit Name (see page 44).

▶ **To enter a mathematics expression for a computation**

- Begin a new line with the mathematics expression.
  or

- Type the expression immediately to the right of text or a text space.

Note   If you enter mathematics immediately to the right of other mathematics, the expressions will be combined in ways you may not intend. A safe way to begin is to press ENTER and start on a new line.

Integers and Fractions

The first examples are centered around rational numbers—that is, integers and fractions. You will find examples of many of the same operations later in this chapter, using real numbers and then complex numbers. Similar operations will be illustrated in later chapters with a variety of different mathematical objects.

Addition and Subtraction

▶ **To add 3, 6, and 14**

1. To put the insertion point in mathematics mode, do one of the following:
   - Choose Insert + Math. (If you see Text on the Insert menu, you are already in mathematics mode.)

   - Click the Text/Math button $\boxed{\text{T}}$ on the Standard toolbar. (If you see $\boxed{\text{M}}$ on the toolbar, you are already in mathematics mode.)
   - Press CTRL + M.

2. Type $3 + 6 + 14$ (This expression should appear red on your screen.)

3. Leaving the insertion point in the expression $3 + 6 + 14$, do one of the following:
   - From the Compute menu, choose Evaluate.
   - Click the Evaluate button ≡? on the Compute toolbar.
   - Press CTRL + E.

This sequence prompts the system to insert $= 23$ to the right of the $3 + 6 + 14$, resulting in the equation $3 + 6 + 14 = 23$.

By following the *same* procedure, you can carry out subtraction and perform a vast variety of other mathematical computations. With the insertion point in the sum (or difference) choose Evaluate

▶ Evaluate

$$235 + 813 = 1048 \qquad \frac{2}{3} - \frac{8}{7} = -\frac{10}{21} \qquad 96 - 27 + 2 = 71$$

Note Following a command, the mathematical contents of gray boxes (the shaded areas on these pages) display the mathematical expressions you enter, together with the results of the indicated command. In general, throughout this document, the mathematical contents of gray boxes display both the input for an action and the results.

▶ **To obtain the fraction template**

   - Place the insertion point in the position where you want the fraction, and
      - Choose Insert + Fraction.
      or
      - Click the Fraction button ▱ on the Math Templates toolbar.
      or
      - Press CTRL + F or CTRL + / or CTRL + 1.

The template will appear with the insertion point in the upper input box, ready for you to begin entering numbers or expressions. Choose View and check Input Boxes to see input boxes on the screen.

Multiplication and Division

Use any standard linear or fractional notation for multiplication and division, and with the insertion point in the product (or quotient), choose Evaluate

▶ Evaluate

$$16 \times 37 = 592 \qquad (84)(-39) = -3276 \qquad \frac{2}{9}\frac{13}{7} = \frac{26}{63}$$

$$103 \div 37 = \frac{103}{37} \qquad 8.2/3.7 = 2.2162 \qquad \frac{-\frac{2}{9}}{\frac{13}{7}} = -\frac{14}{117}$$

Mixed Numbers and Long Division

A number written in the form $14\frac{5}{9}$ is interpreted as the *mixed number* $14 + \frac{5}{9}$. Most commands applied to a mixed number return a fraction. For example, applying Evaluate or Simplify to $14\frac{5}{9}$ gives the result $\frac{131}{9}$. The reverse is accomplished by Expand, which converts a fraction to a mixed number. These commands are also available directly on the Compute toolbar:

- Click ▣ for Evaluate, ▣ for Simplify, and ▣ for Expand.

▶ Evaluate or Simplify

$$1\frac{2}{3} = \frac{5}{3} \qquad 193\frac{87}{94} = \frac{18229}{94} \qquad 1\frac{2}{3} + 2\frac{3}{4} = \frac{53}{12}$$

▶ Expand

$$\frac{18229}{94} = 193\frac{87}{94} \qquad \frac{53}{12} = 4\frac{5}{12}$$

The expansion of a fraction to a mixed number uses the familiar *long-division algorithm*. In the preceding example, 18229 divided by 94 is equal to 193 with remainder 87.

Elementary Number Theory

The arithmetic of positive integers exhibits many interesting properties. Many of these properties are related to integers called *primes*.

Prime Factorization

An integer greater than 1 is a *prime* if it is not evenly divisible by any positive integer except 1 and itself. The list of primes begins with $2, 3, 5, 7, 11, 13, 17, \ldots$. Every positive integer greater than 1 can be factored into a product of powers of primes. You can identify a prime by the fact that it is its own prime factorization.

To factor integers into products of powers of primes, place the insertion point inside the number and choose Factor.

▶ Factor

$$12345 = 3 \times 5 \times 823 \qquad 82723 = 82723$$

$$4733\,64564\,31063\,80000 = 2^5 3^{10} 5^4 7^3 11^2 13 \times 17 \times 19 \times 23$$

Alternately, while in mathematics, type **factor** (it will automatically turn gray), enter the integer (with or without parentheses), and choose Evaluate.

▶ Evaluate

$$\text{factor}\,(12345) = 3 \times 5 \times 823 \qquad\qquad \text{factor}\,(82723) = 82723$$

$$\text{factor}\,(4733\,64564\,31063\,80000) = 2^5 3^{10} 5^4 7^3 11^2 13 \times 17 \times 19 \times 23$$

You can use Simplify or Evaluate to return any of the preceding factorizations to integer form.

Greatest Common Divisor and Least Common Multiple

The *greatest common divisor* of a collection of integers is the largest integer that evenly divides every integer in the collection.

To find the greatest common divisor of a collection of integers,

1. Type gcd in mathematics. (The name gcd should turn gray when you type the d.)

2. Enclose the list of numbers, separated by red commas, in brackets.

3. Leave the insertion point in the list, and click ⬛ or choose Evaluate.

▶ Evaluate

$$\gcd(35, 15, 65) = 5 \quad \gcd\,(910, 2405, 5850, 2665) = 65 \quad \gcd\,(104, 221) = 13$$

Note   If you enter the function gcd from the keyboard while in mathematics mode, the gc appears in red italics until you type the d, then the function name gcd changes to a gray, nonitalic gcd. The function gcd is automatically substituted for the three-letter sequence g, c, and d. You can also choose gcd from the dialog that appears when you click ⬛ or when you choose Insert + Math Name.

The *least common multiple* of a collection of integers is the smallest positive integer that is evenly divisible by every integer in the collection. To find the least common multiple of a collection of integers, evaluate the function lcm applied to the list of numbers enclosed in brackets and separated by commas. Leave the insertion point in the list and choose Evaluate.

▶ Evaluate

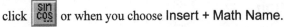

$$\text{lcm}\,(24, 36) = 72 \qquad\qquad \text{lcm}\,(35, 15, 65) = 1365$$

You can enter the function lcm from the keyboard while in mathematics mode. It changes to gray, nonitalic letters on your screen. (If it does not appear on the function list under Insert + Math Name, you can add it to the list by typing it in the Name box and choosing Add.)

You can also determine both the greatest common divisor and least common multiple by inspection after applying Factor to each of the numbers in the list.

Factorials

Factorial is the function of a nonnegative integer n denoted by $n!$ and defined for positive integers n as the product of all positive integers up to and including n

$$n! = 1 \times 2 \times 3 \times 4 \times \cdots \times n$$

and for zero by

$$0! = 1$$

You can compute factorials with Evaluate.

▶ Evaluate

$$3! = 6 \qquad 7! = 5040 \qquad 10! = 3628800$$

Binomial Coefficients

An expression of the form $a + b$ is called a **binomial**. The formula that gives the expansion of $(a + b)^n$ for any natural number n is

$$(a + b)^n = \sum_{k=0}^{n} \frac{n!}{k! \, (n - k)!} a^{n-k} b^k$$

This is the same formula that gives the number of combinations of n things taken k at a time. The coefficients $\frac{n!}{k!(n-k)!}$ that occur in this formula are called *binomial coefficients*. These coefficients are often denoted by the symbols $\binom{n}{k}$ or $C_{n,k}$ or $_nC_k$. Use the symbol $\binom{n}{k}$ to compute these coefficients.

▶ To enter a binomial coefficient $\binom{n}{k}$

1. Click the Binomial button 🔳 on the Math Objects toolbar, or choose Insert + Binomial.

2. Choose None for line and choose OK.

3. Type numbers in the input boxes.

▶ Evaluate

$$\binom{5}{2} = 10 \qquad \binom{35}{7} = 6724\,520$$

You can use the Rewrite command to change a symbolic binomial to a factorial expression.

▶ Rewrite + Factorial

$$\binom{m}{n} = \frac{m!}{n!\,(m-n)!} \qquad \binom{m}{5} = \frac{1}{120}\frac{m!}{(m-5)!}$$

Real Numbers

The real numbers include the integers and fractions (rational numbers), as well as irrational numbers such as $\sqrt{2}$ and π that cannot be expressed as quotients of integers.

Reminder   Numbers used in computing must be entered in mathematics mode and appear red on your screen (or another color that you have specified). If that is not the case, select each number, or combination of numbers, and click ⊤ (or choose Insert + Math) to convert to mathematics.

Basic Operations

The result of applying Evaluate to a combination of integers and fractions will be a fraction. However, if any of the components of a combination of numbers is written in floating point form—that is, with a decimal—the result will be in decimal notation. Symbolic real numbers such as $\sqrt{2}$ and π will retain symbolic form unless evaluated numerically. To evaluate the following expressions, place the insertion point in or immediately to the right of the mathematical expression you want to evaluate, then choose Evaluate or Evaluate Numerically. To evaluate these expressions and replace them with their value, press CTRL + SHIFT + E.

▶ Evaluate (or CTRL + E)

$$9.6\pi - 2.7\pi = 6.9\pi \qquad 42\left(\frac{2}{3}+\frac{1}{7}\right)\sqrt{2} = 34\sqrt{2}$$

▶ Evaluate Numerically

$$9.6\pi - 2.7\pi = 21.677 \qquad 42\left(\frac{2}{3}+\frac{1}{7}\right)\sqrt{2} = 48.083$$

▶ CTRL + SHIFT + E

$235.3 + 813$ will be replaced by 1048.3

$42\left(\frac{2}{3}+\frac{1}{7}\right)\sqrt{2}$ will be replaced by $34\sqrt{2}$

You can control the number of decimal places returned by Evaluate Numerically in the Tools + Computation Setup and Compute + Settings dialogs (see page 29).

To change a floating point number to a rational number, use Rewrite from the Compute menu.

▶ Rewrite + Rational

$$0.125 = \tfrac{1}{8} \qquad 4.72 = \tfrac{118}{25} \qquad 6.9\pi = \tfrac{69}{10}\pi \qquad 3.1416 = \tfrac{3927}{1250}$$

To change a rational number or a symbolic number to a floating point number, use Rewrite from the Compute menu.

▶ Rewrite + Float

$$\tfrac{1}{8} = 0.125 \qquad \tfrac{118}{25} = 4.72 \qquad \tfrac{69}{10}\pi = 21.677 \qquad \tfrac{3927}{1250} = 3.1416$$

Typing **float** while in mathematics gives the grayed function float. Evaluating float at a rational number gives the floating point form of the number.

▶ Evaluate

$$\text{float}\left(\tfrac{1}{8}\right) = 0.125 \qquad \text{float}\left(\tfrac{118}{25}\right) = 4.72$$

Powers and Radicals

To raise numbers to powers, use common notation for powers and apply Evaluate.

▶ **To obtain the radical template**

- Place the insertion point in the position where you want the radical, and
  - Click the Radical button $\sqrt{\Box}$ on the Math Templates toolbar.
    or
  - Choose Insert + Radical.
    or
  - Press CTRL + R or CTRL + 2.

The template will appear with the insertion point in the input box, ready for you to begin entering numbers or expressions.

To see input boxes on the screen, choose View and check Input Boxes.

▶ **To obtain the superscript (subscript) template,**

- Place the insertion point in the position where you want the superscript (subscript), and
  - Choose Insert + Superscript (Subscript).
    or

- Click the **Superscript** button $\boxed{N^X}$ (**Subscript** button $\boxed{N_x}$) on the Math Templates toolbar.

  or

- Press CTRL + H or CTRL + UP ARROW or CTRL + 3 (CTRL + L or CTRL + DOWN ARROW or CTRL + 4).

The template will appear with the insertion point in the upper input box, ready for you to begin entering numbers or expressions.

To see input boxes on the screen, choose **View** and check **Input Boxes**.

▶ **Evaluate**

$$3^4 = 81 \qquad 3^{-4} = \tfrac{1}{81} \qquad \sqrt{2.34} = 1.5297 \qquad (2.5)^{\frac{4}{5}} = 2.081\,4$$

$$\left(\tfrac{2}{5}\right)^{32} = \frac{42949\,67296}{232\,83064\,36538\,69628\,90625} \qquad 0.4^{32} = 1.844\,674\,407 \times 10^{-13}$$

Note that **Evaluate** returns a different answer for $\left(\tfrac{2}{5}\right)^{32}$ and $(0.4)^{32}$. The fraction displayed for $\left(\tfrac{2}{5}\right)^{32}$ is the exact answer, and the number displayed for $(0.4)^{32}$ is the best 10-digit approximation to the exact answer. The exponential and radical notation can evoke different responses for roots of real numbers.

Radical notation for roots

Evaluate and **Simplify** will compute real roots of positive real numbers written in either symbolic or floating point notation, and will compute complex odd roots of floating point numbers (see page 39). The result of either of these operations is presented in symbolic or floating point notation according to the form of the input. **Evaluate** and **Simplify** produce the same result from floating point numbers. Sometimes **Simplify** is useful with symbolic numbers.

▶ **Evaluate**

$$\sqrt[3]{0.008} = 0.2 \qquad \sqrt{24} = 2\sqrt{6} \qquad \sqrt{\tfrac{9}{4}\pi^2} = \tfrac{3}{2}\pi$$

▶ **Simplify**

$$\sqrt[4]{16} = 2 \qquad \sqrt[3]{-8} = 2\sqrt[3]{-1} \qquad \sqrt[3]{\tfrac{16}{27}} = \tfrac{2}{3}\sqrt[3]{2}$$

$$\sqrt[5]{18.234} = 1.787\,2 \qquad \sqrt[5]{-18.234} = 1.445\,9 + 1.050\,5i$$

You can also **Evaluate** the built-in function simplify(expression). (Type **simplify** in mathematics mode and it will automatically turn gray.)

▶ Evaluate

$$\text{simplify}\left(\sqrt[4]{16}\right) = 2 \qquad\qquad \text{simplify}\left(\sqrt[4]{162\pi^6}\right) = 3\pi^{\frac{3}{2}}\sqrt[4]{2}$$

$$\text{simplify}\left(\sqrt[3]{-\tfrac{16}{27}}\right) = -\tfrac{2}{3}\sqrt[3]{2} \qquad \text{simplify}\left(\sqrt[3]{-8}\right) = 2\sqrt[3]{-1}$$

Exponential notation for roots

The exponential notation for roots accepts any real exponent. With the exponential notation for integer roots, Evaluate converts roots of symbolic numbers to radical notation, but otherwise produces a nontrivial response only for floating point numbers. Simplify computes some integer roots of symbolic numbers.

▶ Evaluate

$$(16)^{\frac{1}{4}} = \sqrt[4]{16} \qquad (-8)^{\frac{1}{3}} = \sqrt[3]{(-8)} \qquad (0.008)^{\frac{1}{3}} = 0.2 \qquad (24)^{\frac{1}{2}} = \sqrt{24}$$

$$(0.008)^{\frac{1}{3}} = 0.2 \qquad (3.1416)^{2.7183} = 22.459\,8 \qquad (-0.008)^{\frac{1}{3}} = 0.1 + 0.173\,21i$$

▶ Simplify

$$(16)^{\frac{1}{4}} = 2\sqrt[4]{16} \qquad (-8)^{\frac{1}{3}} = 2\sqrt[3]{-1} \qquad (-0.008)^{\frac{1}{3}} = 0.1 + 0.173\,21i$$

Evaluate Numerically will compute all real roots of positive real numbers. For odd integer roots of negative real numbers, it produces a complex root (see page 39).

▶ Evaluate Numerically

$$8^{\frac{1}{3}} = 2.0 \qquad\qquad \pi^e = 22.459\,2 \qquad\qquad (-8)^{\frac{1}{3}} = 1.0 + 1.732\,1i$$

Rationalizing a Denominator

To rationalize the denominator of a fraction, leave the insertion point in the fraction and from the Compute menu, choose Simplify.

▶ Simplify

$$\frac{1}{\sqrt{2}} = \tfrac{1}{2}\sqrt{2} \qquad\qquad\qquad \frac{1}{\sqrt{2} + \sqrt{3}} = -\sqrt{2} + \sqrt{3}$$

$$\frac{\sqrt{2} + \sqrt{3}}{\sqrt{5} - \sqrt{7}} = -\tfrac{1}{2}\sqrt{2}\sqrt{7} - \tfrac{1}{2}\sqrt{2}\sqrt{5} - \tfrac{1}{2}\sqrt{3}\sqrt{7} - \tfrac{1}{2}\sqrt{3}\sqrt{5}$$

Numerical Approximations

The result of a computation is exact, or symbolic, whenever appropriate and otherwise is a numerical approximation. You can force a numerical result to any evaluation either by choosing Evaluate Numerically from the Compute menu, or by starting with numbers entered in decimal notation. You obtain numerical approximations in response to any operation when you enter numbers with decimals because such a number is interpreted as a floating-point real number and *not* as a rational number (although symbolic numbers such as π, e, or $\sqrt{2}$ retain their symbolic form under Evaluate).

Numerical analysis may be described as the study of errors introduced by using *floating point* arithmetic (*round-off errors*) and by using a finite number of terms when an infinite number of terms is required for exactness (*truncation errors*). Floating point is a data type that is machine dependent. It is important to understand that a floating point number is neither rational nor irrational; indeed, each floating-point number represents an infinite number of possible rational numbers and an infinite number of possible irrational numbers.

Computer algebra systems use what is called *infinite precision* or *extended precision* to represent integers and rationals exactly. Numbers such as $\sin 1$, $\sqrt{2}$, π, and e are examples of numbers that are represented exactly. Evaluate Numerically leads to the following approximations, with Digits Used in Computations and Digits Shown in Results both set to 25 in Engine Setup and Computation Setup, respectively (see page 29).

$$\sin 1 \;=\; 0.\,841\,470\,984\,807\,896\,506\,652\,502\,3$$
$$\sqrt{2} \;=\; 1.\,414\,213\,562\,373\,095\,048\,801\,689$$
$$\pi \;=\; 3.\,141\,592\,653\,589\,793\,238\,462\,643$$
$$e \;=\; 2.\,718\,281\,828\,459\,045\,235\,360\,287$$

Notice that the approximations are broken into blocks of length 3 decimal digits in order to make them more readable. The numbers on the left are exact, while the numbers on the right are merely approximations. In particular,

$$\left(\sqrt{2}\right)^{2} = 2$$

exactly, whereas

$$1.\,414\,213\,562\,373\,095\,048\,801\,689^{2} = 2.\,000\,000\,000\,000\,000\,000\,000\,001$$

with Digits Shown in Results set to 25.

Contrast the results of evaluating the following expressions with Evaluate and Evaluate Numerically.

Evaluate	Evaluate Numerically
$82 \div 37 = \frac{82}{37}$	$82 \div 37 = 2.\,216\,2$
$936/14 = \frac{468}{7}$	$936/14 = 66.\,857$
$93.6/1.4 = 66.\,857$	$936/14.0 = 66.\,857$
$\sqrt{234} = 3\sqrt{26}$	$\sqrt{234} = 15.\,297$

You can change the number of digits displayed in approximations and the threshold for scientific notation by changing settings in a dialog. You can also change the number of digits used in computations in a dialog. These options are discussed in greater detail in the following pages.

Scientific Notation

Any nonzero real number x can be written in the form
$$x = c \times 10^n$$
with $1 \leq c < 10$ and n an integer. A number in this form is in *scientific notation*.

▶ **To write a number in scientific notation**

1. Enter the number c to as many decimal places as appropriate.

2. Choose $\times$ from the **Symbol Cache** toolbar or from the **Binary Operations** symbol panel under $\boxed{\pm \div}$.

3. Enter the number 10.

4. Choose **Insert + Superscript**, or click $\boxed{\text{N}^\text{x}}$, and enter the integer n in the input box.

There is a keyboard shortcut that does most of the work for you.

▶ **To write a number in scientific notation using a keyboard shortcut**

1. Enter the number c (in mathematics mode) to as many decimal places as appropriate.

2. Type *ttt* while still in mathematics mode.

    (This automatically turns into $\times 10$. The superscript box is at the right of the 10—to see it you must have **View + Input Boxes** checked on.)

3. Place the insertion point in the superscript input box and enter the integer n.

Following are some examples of scientific notation:
$$12 = 1.2 \times 10$$
$$1274.9837 = 1.2749837 \times 10^3$$
$$0.000001234 = 1.234 \times 10^{-6}$$

The results of a numerical computation are sometimes returned in scientific notation. This happens when the number of digits exceeds the setting for **Threshold for Scientific Notation**. See page 31 for details on changing this setting.

Computation and Display of Numerical Results

The number of digits used in numerical computations and the display of numerical results are controlled by settings you can change. You can change the defaults that apply generally using Engine Setup and Computation Setup, respectively, on the Tools menu. You can override Computation Setup defaults for individual documents using Settings from the Compute menu. These various settings affect the accuracy and the appearance of numerical results.

Digits Used in Computations

▶ **To set global defaults for the number of digits used in computations**

1. From the Tools menu, choose Engine Setup.

2. Select the General page.

3. Increase or decrease the number in the box.

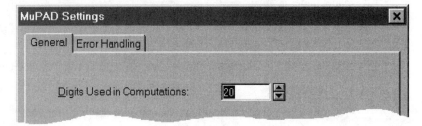

If you choose a large number, you improve accuracy, but computations may be significantly slower. You can try various settings when doing large computations to see how speed and accuracy are affected.

Digits Shown in Results

You can change the defaults for the number of digits shown in the results of a numerical computation, and you can also make a setting that overrides these defaults for an individual document.

▶ **To set global defaults for the number of digits shown in results**

1. From the Tools menu, choose Computation Setup.

2. Select the General page.

3. Increase or decrease the number in the box

4. Choose OK.

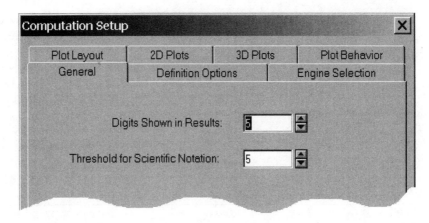

▶ **To set local (document specific) defaults for the number of digits shown in results**

1. From the Compute menu, choose Settings.

2. Select the General page.

3. Check Set Document Values.

4. Increase or decrease the number in the Digits Shown in Results box.

5. Choose OK.

"Digits shown in results" refers to the number of digits put on the screen as the result of a numerical computation. This setting does not affect accuracy in computations, but if you enter a floating-point number in a document using a larger number of digits than the setting under Digits Shown in Results, scientific notation is returned when you choose Evaluate or Evaluate Numerically. Recall that when you enter numbers in decimal (floating-point) notation using no symbolic numbers, Evaluate and Evaluate Numerically have the *same* effect.

Numbers are correctly rounded, depending on the value you have set for Digits Shown in Results. As a general rule, you could reasonably set Digits Shown in Results to the number of places determined by your least accurate measurement.

Threshold for Scientific Notation

If the number of digits to the left of the decimal is greater than the setting under Digits Shown in Results, or if the number of leading zeroes is greater than or equal to that setting, choosing Evaluate Numerically always returns the number in scientific notation. Otherwise, the setting of Threshold for Scientific Notation determines the number of digits that leads to a response in scientific notation, based on the number of digits to the left of the decimal or the number of leading zeroes. The thresholds for a number and its negative are the same. Some of the effects of different settings are illustrated in the following table.

Digits Shown in Results	Threshold for Scientific Notation	Number Entered	Evaluate Numerically
6	3–6	100π	314.159
6	1–2	100π	3.14159×10^2
6	3–6	-100π	-314.159
6	1–2	-100π	-3.14159×10^2
5	5	56789	56789.
5	1–4	56789	5.6789×10^4
4	1–4	56789	5.679×10^4
9	6–9	0.000001111	0.000001111
9	1–5	0.000001111	1.111×10^{-6}
8	1–8	0.000001111	1.111×10^{-6}

▶ **To set global defaults for the threshold for scientific notation**

1. From the Tools menu, choose Computation Setup.

2. Click the General tab.

3. Increase or decrease the number in the Threshold for Scientific Notation box.

4. Choose OK.

▶ **To set local (document specific) defaults for the threshold for scientific notation**

1. From the Compute menu, choose Settings.

2. Click the General tab.

3. Click Set Document Values.

4. Increase or decrease the number in the Threshold for Scientific Notation box.

5. Choose OK.

The dialogs ensure that the threshold for scientific notation cannot exceed the setting for the number of displayed digits. For example, when Digits Shown in Results is set to 5, a setting for Threshold For Scientific Notation greater than 5 is not accepted.

Functions and Relations

Numbers or expressions to be used for computing should be entered in mathematics mode and appear red on your screen. If that is not the case, select the expression and click $\boxed{T}$ or choose Insert + Math to change the expression to mathematics.

Absolute Value

The *absolute value* of a number z is the distance of z from zero. The absolute value of a number or expression is denoted by placing vertical bars around the number or expression, such as $|z|$.

▶ **To put vertical bars around an expression**

1. Select the expression.

2. Click [⬛] and choose the vertical bars, and choose **OK**.

or

Choose **Insert + Brackets**, choose the vertical bars, and choose **OK**.

or

Press CTRL + VERTICAL BAR on the keyboard.

▶ **To find an absolute value**

1. Place the insertion point in an expression enclosed between vertical bars.

2. Choose **Evaluate**.

▶ Evaluate

$$|-7| = 7 \qquad |-11.3| = 11.3 \qquad |43| = 43 \qquad |21 - 13| = 8$$

See page 42 for information on absolute values of complex numbers.

Maximum and Minimum

The functions max and min find the largest and smallest numbers in a list of numbers separated by commas and enclosed in parentheses, or in a vector or matrix. The function names max and min can be chosen from the list provided by [⬛] and **Insert + Math Name**, or they can be entered from the keyboard while in mathematics mode.

You can also find the largest or smallest of several numbers by inserting the binary operations join $\vee$ or meet $\wedge$ between the numbers. You will find these symbols on the **Binary Operations** panel under [⬛]. To compute a join or meet, leave the insertion point in the expression and click [⬛] or choose **Evaluate**.

▶ Evaluate

$$\max(27, \sqrt{236}, \tfrac{65}{2}, -14) = \tfrac{65}{2} \qquad 27 \vee \sqrt{236} \vee \tfrac{65}{2} \vee -14 = \tfrac{65}{2}$$

$$\max\left(27, \min\left(\sqrt{236}, \max\left(\tfrac{65}{2}, -14\right)\right)\right) = 27 \qquad 27 \vee \left(\sqrt{236} \wedge \left(\tfrac{65}{2} \vee -14\right)\right) = 27$$

$$\min(27, \sqrt{236}, \tfrac{65}{2}, -14) = -14 \qquad 27 \wedge \sqrt{236} \wedge \tfrac{65}{2} \wedge -14 = -14$$

$$\max\begin{pmatrix} 1 & 3 \\ -5 & 4 \\ \pi & 6 \end{pmatrix} = 6 \qquad\qquad \max\begin{pmatrix} 1 & 3 & 1/6 & 5 \\ -5 & 4 & \pi & e^2 \end{pmatrix} = e^2$$

To find the maximum or minimum of a finite sequence, enter the limits on the integer variable as a subscript on max or min, either in the form of a double inequality such as $1 \le n \le 10$ or as membership in an interval such as $k \in [1, 10]$.

▶ Evaluate

$$\max_{1 \le n \le 10}(\sin n) = \sin 8 \qquad\qquad \min_{k \in [1,10]}(\cos k) = \cos 3$$

$$\max_{1 \le n \le 10}(\sin 1.5n) = 0.997\,49 \qquad \min_{k \in [1,10]}(\cos 2.6k) = -0.994\,18$$

▶ Evaluate Numerically

$$\max_{1 \le n \le 10}(\sin n) = 0.989\,36 \qquad\qquad \min_{k \in [1,10]}(\cos k) = -0.989\,99$$

$$\max_{-2 \le x \le 2}(x^3 - 6x + 3) = 8.0 \qquad \min_{k \in [1,10]}(\cos 2.6k) = -0.994\,18$$

Note that the functions max and min look only at the sequence of values for *integer* variables. The notations $x \in [-2, 2]$ and $-2 \le x \le 2$ both indicate that x assumes the range of values in the 5-element set $\{-2, -1, 0, 1, 2\}$. In the last example the maximum is picked from among values of $x^3 - 6x + 3$ for $x = -2, -1, 0, 1, 2$. This is *not* the maximum of the continuous polynomial function $x^3 - 6x + 3$.

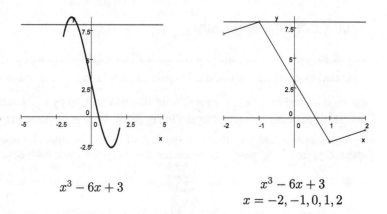

$$x^3 - 6x + 3 \qquad\qquad\qquad x^3 - 6x + 3$$
$$\qquad\qquad\qquad\qquad x = -2, -1, 0, 1, 2$$

Greatest and Smallest Integer Functions

You can find the greatest integer less than or equal to a number by using the floor function, denoted $\lfloor z \rfloor$.

▶ **To put floor brackets around an expression**

1. Select the expression with click and drag.

2. Click ▨ or choose Insert + Brackets, click the left floor bracket ⌊ , and choose OK.

▶ **To find a greatest integer value**

1. Place the insertion point in an expression enclosed between floor brackets.

2. Choose Evaluate.

▶ Evaluate

$$\lfloor 5.6 \rfloor = 5 \qquad \lfloor -11.3 \rfloor = -12 \qquad \lfloor \tfrac{43}{5} \rfloor = 8 \qquad \lfloor \pi + e \rfloor = 5$$

You can find the smallest integer greater than or equal to a number by using the ceiling function, denoted $\lceil z \rceil$.

▶ **To put ceiling brackets around an expression**

1. Select the expression with click and drag.

2. Click ▨ or choose Insert + Brackets.

3. Click the left ceiling bracket ⌈ , and choose OK.

▶ **To find a smallest integer value**

1. Place the insertion point in a number enclosed between ceiling brackets.

2. Choose Evaluate.

▶ Evaluate

$$\lceil 5.6 \rceil = 6 \qquad \lceil -11.3 \rceil = -11 \qquad \lceil \tfrac{43}{5} \rceil = 9 \qquad \lceil \pi + e \rceil = 6$$

The floor and ceiling brackets are also available in the Special Delimiters panel under ▨ , although these are not expanding brackets.

Checking Equality and Inequality

You can verify equalities and inequalities with the command Check Equality or with the function istrue. There are three possible responses: true, false, and undecidable. The latter means that the test is inconclusive and the equality may be either true or false. The computational engines may use probabilistic methods to check equality, and there is a very small probability that an equation judged as true is actually false. Some expressions cannot be compared by this method—hence the inconclusive response.

Checking Equalities and Inequalities with Check Equality

▶ **To check whether an equality is true or false**

1. Leave the insertion point in the equation.

2. Choose Check Equality.

▶ Check Equality

$e^{i\pi} = -1$ is true $\pi = 3.14$ is false $\arcsin \sin x = x$ is undecidable

You can also use Check Equality to check an *inequality* between two numbers. Set the difference of the two numbers equal to the absolute value of the difference, place the insertion point in the equation, and choose Check Equality.

▶ Check Equality

$\frac{9}{8} - \frac{8}{9} = \left| \frac{9}{8} - \frac{8}{9} \right|$ is true $\pi^e - e^\pi = |\pi^e - e^\pi|$ is false

These results verify that $\frac{9}{8} - \frac{8}{9} \geq 0$, or $\frac{9}{8} \geq \frac{8}{9}$; and that $\pi^e - e^\pi < 0$, or $\pi^e < e^\pi$.

Checking Equalities and Inequalities Using istrue

Type *istrue* in mathematics mode to get the function name istrue, or create it as a Math Name in the Insert + Math Name dialog. Evaluate this function at an equation or inequality to test it.

▶ Evaluate

$\text{istrue} \left(\frac{9}{8} < \frac{8}{9} \right) = false$ $\text{istrue} \left(\pi^e < e^\pi \right) = true$

$\text{istrue} \left(2 + 2 = 4 \right) = true$ $\text{istrue} \left(\left(\sqrt{2} \right)^2 = 2 \right) = true$

Checking Equalities and Inequalities Using Logical Operators

The operators $\wedge$ (and) and $\vee$ (or) can be used as logical operators. The statement $\alpha \wedge \beta$ is true if and only if both α and β are true. The statement $\alpha \vee \beta$ is true if and only if at least one of α and β is true. Using a tautology such as $0 = 0$ or $1 = 1$ as one of the statements, you can test the truth or falsity of another equation or inequality.

▶ Evaluate

$$(5^6 < 6^5) \wedge (1 = 1) = false \qquad (5^6 > 6^5) \wedge (1 = 1) = true$$
$$(5^6 > 6^5) \vee (1 = 1) = true \qquad (5^6 < 6^5) \vee (1 = 1) = true$$
$$(1 = 1) \vee (1 = 0) = true \qquad (e^\pi = \pi^e) \wedge (0 = 0) = false$$

Checking Inequalities with Evaluate Numerically

In some cases, you can recognize an inequality by inspection after applying **Evaluate Numerically** to each of the numbers.

▶ Evaluate Numerically

$$\frac{9}{8} = 1.125 \qquad \frac{8}{9} = 0.88889 \qquad \text{so } \frac{9}{8} > \frac{8}{9}$$
$$\pi^e = 22.459 \qquad e^\pi = 23.141 \qquad \text{so } \pi^e < e^\pi$$

Union, Intersection, and Difference

You can find the union of two or more finite sets with **Evaluate**, by using the symbol $\cup$ between the sets.

▶ Evaluate

$$\{1, 2, 3\} \cup \{a, b, c\} = \{1, 2, 3, a, b, c\} \qquad \{1, 2, 3\} \cup \{3, 5\} \cup \{7\} = \{1, 2, 3, 5, 7\}$$
$$\{\sqrt{2}, \pi, 3.9, r\} \cup \{a, b, c\} = \{\pi, r, a, b, c, 3.9, \sqrt{2}\}$$

You can find the intersection of two or more finite sets with **Evaluate**, using the symbol $\cap$ between the sets.

▶ Evaluate

$$\{1, 2, 3\} \cap \{2, 4, 6\} = \{2\} \qquad \{a, b, c, d\} \cap \{d, e, f\} = \{d\}$$
$$\{1, 2, 3\} \cap \{a, b, c\} = \emptyset \qquad \{1, 2, 3\} \cap \{\} = \emptyset$$

If two sets have no elements in common, their intersection is the *empty set*, denoted by empty brackets $\{\}$ or the symbol $\emptyset$. To enter the symbol $\emptyset$ for the empty set, select it from the **Miscellaneous Symbols** panel under .

You can find the difference of two finite sets with **Evaluate**, by placing between the sets a BACKSLASH \ or the set minus symbol \ from the **Binary Operations** panel.

▶ Evaluate

$$\{1,2,3,4\} \setminus \{2,4\} = \{1,3\} \qquad \{a,b,c,d\} \setminus \{d,e,f\} = \{a,b,c\}$$

$$\{1,2,3\} \setminus \{a,b,c\} = \{1,2,3\} \qquad \{1,2,3\} \setminus \{1,2,3\} = \emptyset$$

You can evaluate combinations of union, intersection, and difference after grouping expressions appropriately with expanding parentheses.

▶ Evaluate

$$\{1,2,3,c\} \cap (\{2,4,6\} \cup \{a,b,c\}) = \{2,c\}$$

$$(\{1,2,3,c\} \cap \{2,4,6\}) \cup (\{1,2,3,c\} \cap \{a,b,c\}) = \{2,c\}$$

$$(\{2,4,6\} \cup \{a,b,c\}) \setminus \{2,a,b\} = \{c,4,6\}$$

Complex Numbers

Complex numbers are numbers of the form $a + b\sqrt{-1}$ where a and b are real numbers. The arithmetic of the complex numbers is obtained by observing all of the standard rules of arithmetic of real numbers together with the identity $\left(\sqrt{-1}\right)^2 = -1$. Complex numbers were developed, in part, because they complete, in a useful and elegant fashion, the study of the solutions of polynomial equations.

The default notation for $\sqrt{-1}$ is i, and we use the letter i throughout this document. If you prefer to use j you can change the default as follows. If you carry out this procedure, your system will recognize and return j rather than i for the square root of minus one.

▶ **To modify the system to recognize j for $\sqrt{-1}$ in all documents**

1. Choose Tools + Computation Setup and click the General tab.

2. Under Imaginary Unit, check Change from i to j, and choose OK.

You can override the default for an individual document without changing the global defaults.

▶ **To modify the system to recognize j for $\sqrt{-1}$ for an individual document**

1. Choose Compute + Settings and click the General tab.

2. Click Set Document Values.

3. Under Imaginary Unit, check Change from i to j, and choose OK.

Basic Operations

For addition, subtraction, multiplication, and division write the expressions in standard form and apply Evaluate. Your system will return the result in the form $a + bi$ or $a + ib$, with a and b real numbers.

▶ Evaluate

$$(23 - 5i) + (1 + 16i) = 24 + 11i \qquad \frac{i}{1+i} = \frac{1}{2} + \frac{1}{2}i$$

$$(1 + i)(3 - 2i) = 5 + i \qquad \frac{i}{1+i}\frac{2+i}{3-i} = \frac{1}{2}i$$

$$\frac{2.5 + 3i}{3.59 + 16i} = 0.21189 - 0.10871i \qquad (2 + 3i) \div (6i) = \frac{1}{2} - \frac{1}{3}i$$

For some complex expressions, the keyboard command CTRL + Y (Send String) produces different forms than CTRL + E or the menu item Evaluate. The choice CTRL + Y carries out a lower level of evaluation.

▶ Evaluate or CTRL + E

$$e^{2+3i+5-5i} = e^7 \cos 2 - ie^7 \sin 2$$

▶ CTRL + Y

$$e^{2+3i+5-5i} = e^{7-2i}$$

Real Powers and Roots of Complex Numbers

To compute real powers and roots of complex numbers, use common notation and apply Evaluate.

▶ Evaluate

$$i^2 = -1 \qquad (3 + 2i)^{-4} = -\frac{119}{28\,561} - \frac{120}{28\,561}i \qquad \sqrt{-5} = i\sqrt{5}$$

$$(3 + 2i)^4 = -119 + 120i \qquad \left((3 + 2i)^4\right)^{\frac{1}{4}} = 3 + 2i$$

$$\left(\tfrac{2}{5} - \tfrac{3}{4}i\right)^5 = \frac{113\,221}{400\,000} + \frac{43\,737}{128\,000}i \qquad (0.4 - 0.75i)^5 = 0.283\,05 + 0.341\,70i$$

Note that the fraction returned for $\left(\tfrac{2}{5} - \tfrac{3}{4}i\right)^5$ is the exact answer, and the floating number returned for $(0.4 - 0.75i)^5$ is the best 5-digit approximation to the exact answer.

You must use caution when working with roots of negative numbers, as the following example illustrates.

▶ Evaluate

$$\sqrt{(-2)(-3)} = \sqrt{6} \qquad \sqrt{(-2)}\sqrt{(-3)} = -\sqrt{6}$$

The simple rule $\sqrt{a \cdot b} = \sqrt{a} \cdot \sqrt{b}$ is no longer valid in this situation. To avoid errors in these cases, first express square roots of negative numbers as complex numbers and then operate with them.

For roots of symbolic numbers, Evaluate does not always provide the information you want. For example, applying Evaluate to $\sqrt[3]{i}$ produces $\sqrt[3]{i}$ back again. Use Rewrite + Rectangular to put a such a complex number in standard form.

▶ Rewrite + Rectangular

$$\sqrt[3]{i} = \tfrac{1}{2}\sqrt{3} + \tfrac{1}{2}i \qquad\qquad (8i)^{\frac{1}{3}} = \tfrac{1}{2}\sqrt{3}\sqrt[3]{8} + \left(\tfrac{1}{2}\sqrt[3]{8}\right)i$$

$$\sqrt{2+3i} = \sqrt{\tfrac{1}{2}\sqrt{13}+1} + i\sqrt{\tfrac{1}{2}\sqrt{13}-1}$$

$$\sqrt{a+ib} = \sqrt{\tfrac{1}{2}a + \tfrac{1}{2}\sqrt{a^2+b^2}} + i\sqrt{-\tfrac{1}{2}a + \tfrac{1}{2}\sqrt{a^2+b^2}}$$

$$(a+bi)^{-1} = \frac{a}{a^2+b^2} - \frac{b}{a^2+b^2}i$$

For a complex root of a negative real number, use floating point numbers or apply Evaluate Numerically.

▶ Evaluate

$$\sqrt[3]{-27} = \sqrt[3]{-27} \qquad \sqrt[3]{-0.008} = 0.1 + 0.173\,21i$$

▶ Evaluate Numerically

$$\sqrt[3]{-27} = 1.5 + 2.598\,1i \qquad \sqrt[3]{-0.008} = 0.1 + 0.173\,21i$$

$$(-8)^{\frac{1}{3}} = 1.0 + 1.732\,1i \qquad \sqrt[3]{-8} = 1.0 + 1.732\,1i$$

Complex Powers and Roots of Complex Numbers

A complex number $z = x + iy$ can be identified with a point (x, y) in the plane. Representing this point with polar coordinates gives $x = r\cos t$, $y = r\sin t$, where $r \geq 0$. Thus any complex number can be written in the form

$$z = x + iy = r\cos t + ir\sin t = r\left(\cos t + i\sin t\right)$$

where

$$r = |z| = \sqrt{x^2 + y^2} \quad \text{and} \quad \tan t = \frac{y}{x}$$

The angle t is called the *amplitude* or *argument* of z. Note that the argument is not unique; any two arguments of z differ by an integer multiple of 2π. The function that gives the argument between $-\pi$ and π is denoted $\arg(z)$. Enter arg in mathematics mode and it will automatically turn to a gray Math Name when you type the g.

▶ Evaluate, Simplify

$$\arg\left((2+3i)^{3i}\right) = \arctan\left(\frac{1}{\cos\left(\frac{3}{2}\ln 13\right)}\sin\left(\frac{3}{2}\ln 13\right)\right) - \pi = \frac{3}{2}\ln 13 - 2\pi$$

$$\arg\left(3^{\frac{1}{i+1}}\right) = -\arctan\left(\frac{1}{\cos\left(\frac{1}{2}\ln 3\right)}\sin\left(\frac{1}{2}\ln 3\right)\right) = -\frac{1}{2}\ln 3$$

▶ Evaluate Numerically

$$\arg\left((5i)^{2+i}\right) = -1.5322 \qquad \arg\left(3^{1/(i+1)}\right) = -0.54931$$

Euler's Formula

Rewrite + Sin and Cos yields *Euler's formula*

$$e^{iy} = \cos y + i\sin y$$

To obtain the general formula for a^{x+iy} for positive real numbers a, first evaluate assume $(a, \text{positive})$ so that a will be interpreted as a positive real number (see page 260 for information on making assumptions about variables). Then Rewrite + Rectangular followed by Factor yields the general formula

$$a^{x+iy} = e^{x\ln a}\left(\cos(y\ln a) + i\sin(y\ln a)\right)$$

▶ Rewrite + Rectangular

$$3^{\frac{1}{i+1}} = \sqrt{3}\cos\left(\frac{1}{2}\ln 3\right) + \left(-\sqrt{3}\sin\left(\frac{1}{2}\ln 3\right)\right)i$$

$$(2+3i)^{3i} = e^{-3\arctan\frac{3}{2}}\cos\left(\frac{3}{2}\ln 13\right) + ie^{-3\arctan\frac{3}{2}}\sin\left(\frac{3}{2}\ln 13\right)$$

▶ Evaluate Numerically

$$(5i)^{2+i} = 0.20077 - 5.1931i \qquad (2+3i)^{2+3i} = 0.60757 - 0.30876i$$

Real and Imaginary Parts of a Complex Number

You can find the real and imaginary parts of a complex number with the functions Re and Im. When you enter these functions in mathematics mode, they will turn gray.

▶ Evaluate

$$\text{Re}\left(\frac{2+3i}{3-5i}\right) = -\frac{9}{34} \qquad \text{Im}\left(\frac{2+3i}{3-5i}\right) = \frac{19}{34}$$

$$\text{Re}\left(\frac{3.6+6i}{5-3.25i}\right) = -4.2179 \times 10^{-2} \qquad \text{Re}\left(\frac{a+bi}{c+di}\right) = \frac{ac}{c^2+d^2} + \frac{bd}{c^2+d^2}$$

▶ Evaluate

$$\frac{2+3i}{3-5i} = -\frac{9}{34} + \frac{19}{34}i \qquad \frac{3.6+6i}{5-3.25i} = -4.2179 \times 10^{-2} + 1.1726i$$

Use **Expand** to separate the real and complex parts of a complex number in exponential form.

▶ Expand

$$e^{3(x-5i)+2x} = e^{5x}e^{-15i}$$

Put the complex factor in standard form with **Rewrite**.

▶ Rewrite + Sin and Cos **or** Rewrite + Rectangular

$$e^{-15i} = \cos 15 - i\sin 15$$

Absolute Value

The *absolute value* of a complex number z, the distance of z from zero, is denoted $|z|$. That is, if $z = a + ib$ with a and b real numbers, then $|z| = \sqrt{(a^2 + b^2)}$.

▶ **To put vertical bars around an expression**

1. Select the expression with the mouse.

2. Click , or choose **Insert + Brackets**, or type CTRL + VERTICAL LINE.

3. Select the vertical bar and choose **OK**.

▶ **To take the absolute value of a complex number**

1. Place the insertion point in an expression enclosed between vertical bars.

2. Choose **Evaluate**.

▶ Evaluate

$$|2 + 3i| = \sqrt{13} \qquad \left|\sqrt{1 + 2i}\right| = \sqrt[4]{5}$$

$$|2.5 - 16.3i| = 16.491 \qquad \left|e^{i\pi}\right| = 1$$

$$|a + bi| = \sqrt{a^2 + b^2}$$

▶ Evaluate Numerically

$$|2 + 3i| = 3.6056 \qquad \left|\sqrt{1 + 2i}\right| = 1.4953$$

Complex Conjugate

The complex conjugate of a complex number $a + ib$ (where a and b are real numbers) is the complex number $a - ib$. To find the complex conjugate of $a + ib$, evaluate the expression $(a + ib)^{*}$.

▶ Evaluate

$$(5 - 14i)^{*} = 5 + 14i \qquad \left(\frac{3.6 + 6i}{5 - 3.25i}\right)^{*} = -4.2179 \times 10^{-2} - 1.1726i$$

$$(a + ib)^{*} = a - ib \qquad (a + ib)(a + ib)^{*} = (a - ib)(a + ib)$$

▶ Expand

$$(a + ib)(a + ib)^{*} = a^2 + b^2$$

Numerical Approximations of Complex Numbers

Your system returns an exact, or symbolic, answer whenever appropriate and otherwise returns a numerical approximation. You can force a numerical result to any evaluation by choosing Evaluate Numerically from the Compute menu, or by choosing Float from the Rewrite submenu, or by starting with numbers in decimal notation. You obtain numerical approximations in response to any operation when you enter numbers with decimals.

You can change the number of digits displayed in these approximations and the threshold for scientific notation by changing settings in the Computation Setup dialog on the Tools menu (see page 29). You can also change the number of digits used in computations in the Engine Setup dialog on the Tools menu. Numerical approximations are discussed in greater detail in Section 3.4, starting on page 28.

Units and Measurements

The available units include units from the System of International Units (SI units), an internationally agreed upon system of coherent units that is now in use for all scientific and most technological purposes in many countries. SI units are of three kinds: the base, supplementary, and derived units. There are seven base units for the seven dimensionally independent physical quantities: *length, mass, time, electric current, thermodynamic temperature, amount of substance,* and *luminous intensity*. Units from some other commonly used systems are also implemented. You can define other units in terms of the ones available in the Unit Name list.

Units

Units appear on your screen as dark green characters (unless you have changed this default to another color). Units are in mathematics mode and are active mathematical objects.

▶ **To enter a unit in a document**

1. Place the insertion point at the position where you want the unit name.

2. Choose Insert + Unit Name, or click ⬛ on the Math Templates bar.

3. Select a category from the Physical Quantity list.

4. Select a name from the Unit Name list.

5. Choose Insert.

The unit name will appear at the position of the insertion point. The Unit Name dialog will remain on your screen for further use. To close it, click the × in the upper right corner of the dialog.

▶ **To change a unit in a document**

1. Select the unit name you want to replace, either with click and drag or by placing the insertion point to the right of the unit name.

2. Choose Insert + Unit Name.

3. Select a category from the Physical Quantity list.

4. Select a name from the Unit Name list.

5. Choose Replace.

The new unit name will replace the previous unit name. The Unit Name dialog will remain on your screen for further use. To close it, click the × in the upper right corner of the dialog.

Units are automatically recognized and can be entered from the keyboard. See keyboard shortcuts below.

▶ **To enter a unit from the keyboard**

1. Place the insertion point at the position where you want the unit name.

2. If the insertion point is not in mathematics mode, click $\boxed{T}$, choose **Insert + Mathematics,** press INSERT, or press CTRL + M to place the insertion point in mathematics mode.

3. Type 'u' followed by the unit symbol, with the following exceptions.
   - Type 'mc' for 'micro' in place of μ which will appear in the unit symbol.
   - Type 'uhr' for the hour symbol h.
   - Type 'uda' for the day symbol d.
   - Type 'use' for the second symbol s.
   - Type 'ume' for the meter symbol m.
   - Type 'uan' for the angstrom symbol Å.
   - Type 'uCo' for the Coulomb symbol C.
   - Type 'uTe' for the Tesla symbol T.
   - Type 'uli' for the Liter symbol l.
   - Type 'ohm' (after the prefix) for the symbols for ohm (and its derivatives) Ω.
   - Type '$ucel$' and '$ufahr$' for degrees Celsius $^\circ$C and degrees Fahrenheit $^\circ$F, respectively.
   - Type '$udeg$' for the degree symbol (plane angle) $^\circ$. (Also see page 92.)
   - Type '$udmn$' and 'uds' for (degree) minute $\prime$ and (degree) second $\prime\prime$, respectively.

Autorecognition is case sensitive, so type upper case where indicated. The unit symbol should turn green when you type the last character.

Unit prefixes are as follows:

Prefix	Factor	Symbol	Prefix	Factor	Symbol
kilo	10^3	k	milli	10^{-3}	m
mega	10^6	M	micro	10^{-6}	$\mu\ (mc)$
giga	10^9	G	nano	10^{-9}	n
tera	10^{12}	T	pico	10^{-12}	p
peta	10^{15}	P	femto	10^{-15}	f
exa	10^{18}	E	atto	10^{-18}	a

Physical Quantities, Symbols and Keyboard Shortcuts

Units available in the Unit Name dialog and their keyboard shortcuts are shown in the following tables. Below the name of the physical quantity, the three columns list the Unit Name, unit symbol, and keyboard shortcut:

Physical Quantity		
Unit Name	Unit Symbol	*Keyboard Shortcut*

Activity

Becquerel	Bq	*uBq*
Curie	Ci	*uCi*

Amount of substance

Attomole	amol	*uamol*
Examole	Emol	*uEmol*
Femtomole	fmol	*ufmol*
Gigamole	Gmol	*uGmol*
Kilomole	kmol	*ukmol*
Megamole	Mmol	*uMmol*
Micromole	μmol	*umcmol*
Millimole	mmol	*ummol*
Mole	mol	*umol*
Nanomole	nmol	*unmol*
Petamole	Pmol	*uPmol*
Picomole	pmol	*upmol*
Teramole	Tmol	*uTmol*

Area

Acre	acre	*uacre*
Hectare	hectare	*uhectare*
Square foot	ft^2	*uft* (insert superscript)
Square inch	in^2	*uin* (insert superscript)
Square meter	m^2	*ume* (insert superscript)

Current

Ampere	A	*uA*
Kiloampere	kA	*ukA*
Microampere	μA	*umcA*
Milliampere	mA	*umA*
Nanoampere	nA	*unA*

Electric capacitance

Farad	F	*uF*
Microfarad	μF	*umcF*
Millifarad	mF	*umF*
Nanofarad	nF	*unF*
Picofarad	pF	*upF*

Electric charge

Coulomb	C	*uCo*

Electric conductance

Kilosiemens	kS	*ukS*
Microsiemens	μS	*umcS*
Millisiemens	mS	*umS*
Siemens	S	*uS*

Electrical potential difference

Kilovolt	kV	*ukV*
Megavolt	MV	*uMV*
Microvolt	μV	*umcV*
Millivolt	mV	*umV*
Nanovolt	nV	*unV*
Picovolt	pV	*upV*
Volt	V	*uV*

Electric resistance

Gigaohm	GΩ	*uGohm*
Kiloohm	kΩ	*ukohm*
Megaohm	MΩ	*uMohm*
Milliohm	mΩ	*umohm*
Ohm	Ω	*uohm*

Energy

British thermal unit	Btu	*uBtu*
Calorie	cal	*ucal*
Electron volt	eV	*ueV*
Erg	erg	*uerg*
Gigaelectronvolt	GeV	*uGeV*
Gigajoule	GJ	*uGJ*
Joule	J	*uJ*
Kilocalorie	kcal	*ukcal*
Kilojoule	kJ	*ukJ*
Megaelectronvolt	MeV	*uMeV*
Megajoule	MJ	*uMJ*
Microjoule	μJ	*umcJ*
Millijoule	mJ	*umJ*
Nanojoule	nJ	*unJ*

Force

Dyne	dyn	*udyn*
Kilonewton	kN	*ukN*
Meganewton	MN	*uMN*
Micronewton	μN	*umcN*
Millinewton	mN	*umN*
Newton	N	*uN*
Ounce-force	ozf	*uozf*
Pound-force	lbf	*ulbf*

Frequency

Exahertz	EHz	*uEHz*
Gigahertz	GHz	*uGHz*
Hertz	Hz	*uHz*
Kilohertz	kHz	*ukHz*
Megahertz	MHz	*uMHz*
Petahertz	PHz	*uPHz*
Terahertz	THz	*uTHz*

Length

Angstrom	Å	*uan*
Attometer	am	*uame*
Centimeter	cm	*ucm*
Decimeter	dm	*udme*
Femtometer	fm	*ufme*
Foot	ft	*uft*
Inch	in	*uin*
Kilometer	km	*ukme*
Meter	m	*ume*
Micrometer	μm	*umcme*
Mile	mi	*umi*
Millimeter	mm	*umme*
Nanometer	nm	*unme*
Picometer	pm	*upme*

Illuminance

Footcandle	fc	*ufc*
Lux	lx	*ulx*
Phot	phot	*uphot*

Luminance

Stilb	sb	*usb*

Luminous flux

Lumen	lm	*ulm*

Luminous intensity

Candela	cd	*ucd*

Magnetic flux

Maxwell	Mx	*uMx*
Microweber	μWb	*umcWb*
Milliweber	mWb	*umWb*
Nanoweber	nWb	*unWb*
Weber	Wb	*uWb*

Magnetic inductance

Henry	H	*uHe*
Microhenry	μH	*umcH*
Millihenry	mH	*umH*

Magnetic flux density

Gauss	G	*uGa*
Microtesla	μT	*umcT*
Millitesla	mT	*umT*
Nanotesla	nT	*unT*
Picotesla	pT	*upT*
Tesla	T	*uTe*

Mass

Atomic mass unit	u	*uu*
Centigram	cg	*ucg*
Decigram	dg	*udg*
Gram	g	*ugr*
Kilogram	kg	*ukg*
Microgram	μg	*umcg*
Milligram	mg	*umg*
Pound-mass	lb	*ulbm*
Slug	slug	*uslug*

Plane angle

Degree	°	*udeg*
Microradian	μrad	*umcrad*
Milliradian	mrad	*umrad*
Minute	′	*udmn*
Radian	rad	*urad*
Second	″	*uds*

Power

Gigawatt	GW	*uGWa*
Horsepower	hp	*uhp*
Kilowatt	kW	*ukWa*
Megawatt	MW	*uMWa*
Microwatt	μW	*umcWa*
Milliwatt	mW	*umWa*
Nanowatt	nW	*unWa*
Watt	W	*uWa*

Pressure

Atmosphere	atm	*uatm*
Bar	bar	*ubar*
Kilobar	kbar	*ukbar*
Kilopascal	kPa	*ukPa*
Megapascal	MPa	*uMPa*
Micropascal	μPa	*umcPa*
Millibar	mbar	*umbar*
Millimeters of Mercury (0 °C)	mmHg	*ummHg*
Pascal	Pa	*uPa*
Torr	torr	*utorr*

Solid angle

Steradian	sr	*usr*

Temperature

Celsius	°C	*ucel*
Fahrenheit	°F	*ufahr*
Kelvin	K	*uK*

Time

Attosecond	as	*uas*
Day	d	*uda*
Femtosecond	fs	*ufs*
Hour	h	*uhr*
Microsecond	μs	*umcs*
Millisecond	ms	*ums*
Minute	min	*umn*
Nanosecond	ns	*uns*
Picosecond	ps	*ups*
Second	s	*use*
Year	y	*uy*

Volume

Cubic foot	ft^3	*uft*	(insert superscript)
Cubic inch	in^3	*uin*	(insert superscript)
Cubic meter	m^3	*ume*	(insert superscript)
Gallon (US)	gal	*ugal*	
Liter	l	*uli*	
Milliliter	ml	*uml*	
Pint	pint	*upint*	
Quart	qt	*uqt*	

Compound Units

▶ **To enter compound units**

- Compound names are written as fractions or products, such as $\dfrac{\text{ft}}{\text{s}}$, ft lbf, and acre ft.

  You can use the **Math Name** dialog to define additional unit names. The standard default color for these names is gray, the same as a **Math Name** used for a function name.

▶ **To define additional unit names**

1. Place the insertion point at the position where you want the unit name.

2. Choose **Insert + Math Name**

3. Type the name you want in the **Name** box and choose **OK**.

4. Enter a defining equation, using one of the available units.

5. With the insertion point in the equation, choose **Definitions + New Definition**.

▶ **Definitions + New Definition**

$$\text{century} = 100\,\text{y} \qquad\qquad \text{decade} = 10\,\text{y}$$

▶ **Evaluate, Simplify**

$$5\,\text{century} = 500\,\text{y} = 15\,778\,463\,000\,\text{s} \qquad \tfrac{1}{2}\,\text{decade} = 5\,\text{y} = 157784630\,\text{s}$$
$$2\,\text{century} + 1\,\text{decade} = 210\,\text{y} = 6626\,954\,460\,\text{s}$$

Arithmetic Operations With Units

You can carry out normal arithmetic operations with units using **Evaluate**. If the units differ, the results will be returned in terms of the basic unit in the category. Measurements will be returned in the metric system.

▶ **Evaluate**

$$6\,\text{ft} + 8\,\text{ft} = 14\,\text{ft} \qquad\qquad 6\,\text{ft} \times 8\,\text{ft} = 48\,\text{ft}^2$$
$$4\,\text{ft} + 16\,\text{in} = 1.625\,6\,\text{m} \qquad 4\,\text{d} + 3\,\text{mi} = 345\,600\,\text{s} + 4827.0\,\text{m}$$
$$10\,\text{A} \times 5\,\text{T} = 50\,\text{A}\,\text{T} \qquad\qquad \dfrac{10\,\text{mi}}{15\,\text{s}} = \tfrac{2}{3}\dfrac{\text{mi}}{\text{s}}$$

Converting Units

You can convert from one unit to another using **Solve + Exact**.

▶ **To convert units**

- Place the insertion point in an equation $47\,\text{ft} = x\,\text{m}$, choose **Solve + Exact**.
  or
- Place the insertion point in an equation $47\frac{\text{ft}}{\text{m}} = x$, choose **Solve + Exact**.

▶ **Solve + Exact**

> $7\,\text{ft} = x\,\text{in}$, Solution is: 84.0 $7\,\text{ft} = x\,\text{m}$, Solution is: 2.134
>
> $458.4^\circ = x\,\text{rad}$, Solution is: 8.0006 $8\,\text{rad} = x^\circ$, Solution is: 458.4
>
> $1\,\text{acre\,ft} = x\,\text{gal}$, Solution is: 3.259×10^5

The difference between the notions of pound-mass (lb) and pound-force (lbf) is illustrated in the following examples.

▶ **Solve + Exact**

> $1\,\text{lbf} = x\,\text{lb}$, Solution is: $9.8066\frac{\text{m}}{\text{s}^2}$ $47\frac{\text{lb}}{\text{kg}} = x$, Solution is: 21.319
>
> $47\,\text{lb} = x\,\text{kg}$, Solution is: 21.319 $47\,\text{lbf} = x\,\text{kg}$, Solution is: $209.07\frac{\text{m}}{\text{s}^2}$

Exercises

1. Find all the primes between 100 and 120.

2. Find two positive integers between 1000 and 1100 whose greatest common divisor is 23.

3. Compare the prime factorizations of 19!, 20!, 21!, 22!, 23!, 24!, and 25! with the number of zeros at the end of the evaluated factorials. Use these results to predict how many zeros will appear at the end of 100!. Check your prediction by direct evaluation. Revise your method of prediction, if necessary, and predict the number of zeros for each of 125!, 200!, 500!, 625!, and 1000!.

4. Evaluate numerically the power $\left(1 + \frac{1}{n}\right)^n$ for $n = 2, 4, 8, 16, 32, 64, 128$, and 256. What well-known number is starting to emerge?

5. Experiment with numbers to test the potential identities
$$a \wedge (b \vee c) = (a \wedge b) \vee (a \wedge c)$$
$$a \vee (b \wedge c) = (a \vee b) \wedge (a \vee c)$$

6. Test the potential identity
$$A \cap (B \cup C) = (A \cap B) \cup (A \cap C)$$
using the sets $A = \{1, 3, 5, 7, 9\}$, $B = \{1, 4, 9, 16\}$, and $C = \{2, 3, 5, 7, 11\}$.

7. The weight of a block of aluminum is 403.2 lbf and the density is $168\frac{\text{lbf}}{\text{ft}^3}$. What is its volume?

8. If a toy rocket shoots vertically upward with an initial velocity of 80 m/s, at t seconds after the rocket takes off, until it returns to the ground, it is at the height $80t - 16t^2$ m. Find the time it takes for the rocket to return to the ground. When does it reach its highest point?

Solutions

1. Test the odd integers between 100 and 120 by factoring:

$101 = 101$ $103 = 103$ $105 = 3 \times 5 \times 7$ $107 = 107$ $109 = 109$
$111 = 3 \times 37$ $113 = 113$ $115 = 5 \times 23$ $117 = 3^2 13$ $119 = 7 \times 17$

It follows that the primes in this range are 101, 103, 107, 109, and 113.

2. **Expand** gives $\frac{1000}{23} = 43\frac{11}{23}$. Note that $44 \cdot 23 = 1012$ and $45 \cdot 23 = 1035$. Checking, we see that $\gcd(1012, 1035) = 23$. Find more pairs.

3. Note that **Factor** and **Evaluate** yield

$$19! = 2^{16}3^8 5^3 7^2 11 \times 13 \times 17 \times 19 = 121\,645\,100\,408\,832\,000$$
$$20! = 2^{18}3^8 5^4 7^2 11 \times 13 \times 17 \times 19 = 2432\,902\,008\,176\,640\,000$$
$$21! = 2^{18}3^9 5^4 7^3 11 \times 13 \times 17 \times 19 = 51\,090\,942\,171\,709\,440\,000$$
$$22! = 2^{19}3^9 5^4 7^3 11^2 13 \times 17 \times 19 = 1124\,000\,727\,777\,607\,680\,000$$
$$23! = 2^{19}3^9 5^4 7^3 11^2 13 \times 17 \times 19 \times 23 = 25\,852\,016\,738\,884\,976\,640\,000$$
$$24! = 2^{22}3^{10}5^4 7^3 11^2 13 \times 17 \times 19 \times 23 = 620\,448\,401\,733\,239\,439\,360\,000$$
$$25! = 2^{22}3^{10}5^6 7^3 11^2 13 \times 17 \times 19 \times 23 = 15\,511\,210\,043\,330\,985\,984\,000\,000$$

and hence 19! ends in 3 zeros, 20! ends in 4, and 25! ends in 6 zeros. It appears that the exponent of 5 counts the number of trailing zeros. The number of fives in 100! is given by $\frac{100}{5} + \frac{100}{25} = 24$, and direct evaluation shows that 100! ends in 24 zeros. In general, the number of trailing zeros in $n!$ is given by the sum

$$\left\lfloor \frac{n}{5} \right\rfloor + \left\lfloor \frac{n}{5^2} \right\rfloor + \left\lfloor \frac{n}{5^3} \right\rfloor + \left\lfloor \frac{n}{5^4} \right\rfloor + \cdots$$

where $\lfloor x \rfloor$ denotes the greatest integer $\le x$. In particular,

$$\left\lfloor \frac{125}{5} \right\rfloor + \left\lfloor \frac{125}{5^2} \right\rfloor + \left\lfloor \frac{125}{5^3} \right\rfloor = 25 + 5 + 1 = 31$$
$$\left\lfloor \frac{200}{5} \right\rfloor + \left\lfloor \frac{200}{5^2} \right\rfloor + \left\lfloor \frac{200}{5^3} \right\rfloor = 40 + 8 + 1 = 49$$
$$\left\lfloor \frac{500}{5} \right\rfloor + \left\lfloor \frac{500}{5^2} \right\rfloor + \left\lfloor \frac{500}{5^3} \right\rfloor = 100 + 20 + 4 = 124$$
$$\left\lfloor \frac{625}{5} \right\rfloor + \left\lfloor \frac{625}{5^2} \right\rfloor + \left\lfloor \frac{625}{5^3} \right\rfloor + \left\lfloor \frac{625}{5^4} \right\rfloor = 125 + 25 + 5 + 1 = 156$$
$$\left\lfloor \frac{1000}{5} \right\rfloor + \left\lfloor \frac{1000}{5^2} \right\rfloor + \left\lfloor \frac{1000}{5^3} \right\rfloor + \left\lfloor \frac{1000}{5^4} \right\rfloor = 200 + 40 + 8 + 1 = 249$$

Verify that these count the number of trailing zeros in 125!, 200!, 500!, 625!, and 1000!, respectively. It takes roughly three-quarters of a 17-inch computer screen to display 1000!, and the results are not displayed here.

4. Note that $\left(1 + \frac{1}{2}\right)^2 = 2.25$ $\left(1 + \frac{1}{4}\right)^4 = 2.4414$

$\left(1 + \frac{1}{8}\right)^8 = 2.5658$ $\left(1 + \frac{1}{16}\right)^{16} = 2.6379$

$\left(1 + \frac{1}{32}\right)^{32} = 2.677$ $\left(1 + \frac{1}{64}\right)^{64} = 2.6973$

$\left(1 + \frac{1}{128}\right)^{128} = 2.7077$ $\left(1 + \frac{1}{256}\right)^{256} = 2.713$

The number $e = 2.7182818284590452354$ is beginning to emerge.

5. With the numbers 1, 2, and 3 we have

$1 \wedge (2 \vee 3) = 1$ and $(1 \wedge 2) \vee (1 \wedge 3) = 1$

$2 \wedge (3 \vee 1) = 2$ and $(2 \wedge 3) \vee (2 \wedge 1) = 2$

$3 \wedge (1 \vee 2) = 2$ and $(3 \wedge 1) \vee (3 \wedge 2) = 2$

Similarly,

$1 \vee (2 \wedge 3) = 2$ and $(1 \vee 2) \wedge (1 \vee 3) = 2$

$2 \vee (1 \wedge 3) = 2$ and $(2 \vee 1) \wedge (2 \vee 3) = 2$

$3 \vee (1 \wedge 2) = 3$ and $(3 \vee 1) \wedge (3 \vee 2) = 3$

These provide experimental evidence that the following are identities:

$$a \wedge (b \vee c) = (a \wedge b) \vee (a \wedge c)$$
$$a \vee (b \wedge c) = (a \vee b) \wedge (a \vee c)$$

6. Note that

$$\{1, 3, 5, 7, 9\} \cap (\{1, 4, 9, 16\} \cup \{2, 3, 5, 7, 11\}) = \{1, 3, 5, 7, 9\}$$

and

$$(\{1, 3, 5, 7, 9\} \cap \{1, 4, 9, 16\}) \cup (\{1, 3, 5, 7, 9\} \cap \{2, 3, 5, 7, 11\}) = \{1, 3, 5, 7, 9\}$$

7. The volume of the block of aluminum is

$$\frac{403.2 \text{ lbf}}{168 \frac{\text{lbf}}{\text{ft}^3}} = 0.06796 \text{ m}^3$$

The volume in cubic feet is the solution to the equation

$$0.06796 \text{ m}^3 = x \text{ ft}^3$$

The solution is

$$x = 2.4$$

8. The rocket returns to the ground when its height is $0\,\text{m}$. Solving

$$\left(80t - 16t^2\right) \text{ m} = 0\,\text{m}$$

gives the two solutions $t = 0$ and $t = 5$. The rocket thus returns to the ground in $5\,\text{s}$. The rocket reaches its highest point in half this time, that is, in $\frac{5}{2}\,\text{s} = 2.5\,\text{s}$. The maximum height of the rocket is $80\,(2.5) - 16\,(2.5)^2 = 100.0\,\text{m}$.

3 Algebra

Algebraic operations are generalizations of arithmetic operations. Algebraic expressions are obtained by starting with variables and constants and combining them using addition, subtraction, multiplication, division, exponentiation, and roots. The simplest types of algebraic expressions use only addition, subtraction, and multiplication; these are called polynomials. The general form of a polynomial of degree n in the variable x is

$$a_n x^n + a_{n-1} x^{n-1} + \cdots + a_1 x + a_0$$

where $a_0, a_1, \ldots, a_n$ are constants and $a_n \neq 0$.

Polynomials and Rational Expressions

You can perform the usual operations on polynomials in a variety of ways. The general procedure is as follows.

▶ **To work with a polynomial expression**

1. Enter the expression in mathematics mode.

2. Leave the insertion point in the expression.

3. Apply one of the commands from the Compute menu.

The commands that operate on polynomials include Evaluate, Simplify, Factor, Expand, Combine + Powers, and, from the Polynomials submenu, Collect, Divide, Partial Fractions, Roots, Sort, and Companion Matrix.

Sums, Differences, Products, and Quotients of Polynomials

▶ **To perform basic operations on polynomials**

1. Enter the expression in mathematics mode.

2. Leave the insertion point in the expression.

1. Click the Evaluate button ⬛ on the Compute toolbar; or choose Evaluate; or press CTRL + E.

▶ Evaluate

$$\left(3x^2 + 3x\right) + \left(8x^2 + 7\right) = 11x^2 + 3x + 7$$

$$\left(3x^2 + 3x\right) / \left(8x^2 + 7\right) = \frac{1}{8x^2 + 7}\left(3x^2 + 3x\right) \qquad x \div y = \frac{x}{y}$$

Several of the other commands listed earlier have the same effect on these particular expressions.

To expand a product of polynomials, leave the insertion point in the expression and click the Expand button ![button] on the Compute toolbar, or choose Expand.

▶ Expand

$$\left(3x^2 + 3x - 1\right)\left(8x^2 + 7\right) = 21x + 13x^2 + 24x^3 + 24x^4 - 7$$

$$\left(x + 1\right)^{-1}\left(x - 1\right)^{-1} = \frac{1}{\left(x + 1\right)\left(x - 1\right)}$$

You can also evaluate the function expand.

▶ Evaluate

$$\text{expand}\left(\left(3x^2 + 3x - 1\right)\left(8x^2 + 7\right)\right) = 21x + 13x^2 + 24x^3 + 24x^4 - 7$$

To enter this function, while in mathematics mode, type **xpnd**. Assuming Automatic Substitution is enabled, this will automatically transform to the gray function name expand.

Choose Tools + Automatic Substitution for a list of the function names on the Automatic Substitution list.

Summation Notation

A polynomial in general form can be written in summation notation

$$\sum_{k=0}^{n} a_k x^k = a_n x^n + a_{n-1}x^{n-1} + \cdots + a_1 x + a_0$$

▶ **To enter a polynomial as a summation**

1. Click $\boxed{\Sigma}$; or choose Insert + Operators and, from the panel, choose $\boxed{\Sigma}$.

2. With the insertion point immediately to the right of $\sum$, choose $\boxed{N_x}$, or choose Insert + Subscript. Type $k = 0$ in the input box.

3. Press TAB; or press SPACEBAR and choose $\boxed{\mathsf{N^x}}$; or press SPACEBAR and choose Insert + Superscript. Type n in the input box.

4. Press SPACEBAR or an arrow key to return to the line, and type $a_k x^k$.

▶ Evaluate

$$\sum_{k=0}^{5} a_k x^k = a_0 + x a_1 + x^2 a_2 + x^3 a_3 + x^4 a_4 + x^5 a_5$$

Sums and Differences of Rational Expressions

▶ **To combine a sum or difference of expressions over a common denominator**

1. Enter the expression in mathematics mode.

2. Leave the insertion point in the expression.

3. Apply **Factor**, and then apply **Rewrite + Rational**.

▶ Factor, Rewrite + Rational

$$\frac{3x^2 + 3x}{8x^2 + 7} + \frac{5x^2 + 3}{2x^2 + x + 7}$$

$$= \left(8x^2 + 7\right)^{-1} \left(x + 2x^2 + 7\right)^{-1} \left(21x + 83x^2 + 9x^3 + 46x^4 + 21\right)$$

$$= \frac{1}{\left(8x^2 + 7\right)\left(x + 2x^2 + 7\right)} \left(21x + 83x^2 + 9x^3 + 46x^4 + 21\right)$$

▶ **To put a product of rational expressions over a common denominator**

1. Enter the expression in mathematics mode.

2. Leave the insertion point in the expression.

3. Apply **Rewrite + Rational**

or

Click the **Simplify** button $\boxed{\substack{\mathsf{x+x} \\ \mathsf{=2x}}}$ on the **Compute** toolbar, or choose **Simplify**.

▶ **Rewrite + Rational**

$$\left(8x^2 + 7\right)^{-1}\left(x + 2x^2 + 7\right)^{-1}\left(4x^3 - 5\right) = \frac{4x^3 - 5}{\left(8x^2 + 7\right)\left(x + 2x^2 + 7\right)}$$

▶ **Simplify**

$$\left(8x^2 + 7\right)^{-1}\left(x + 2x^2 + 7\right)^{-1}\left(4x^3 - 5\right) = \frac{4x^3 - 5}{7x + 70x^2 + 8x^3 + 16x^4 + 49}$$

You can find the standard form for the equation of a circle by "completing the square." You can take advantage of the feature of computing in place (see page 12) for this computation.

Example To find the center and radius of the circle $x^2 - 6x + 18 + y^2 + 10y = 0$, first subtract the constant term 18 from both sides of the equation to get

$$x^2 - 6x + 18 + y^2 + 10y - 18 = 0 - 18$$

Select the left side of this equation and, while holding down the CTRL key, apply **Simplify**. Then do the same to the right side. This gives the equation

$$x^2 - 6x + y^2 + 10y = -18$$

Select the terms containing x with the mouse and click ⬚ . Then, do the same to the terms containing y. This puts the equation in the form

$$\left(x^2 - 6x\right) + \left(y^2 + 10y\right) = -18.$$

To complete the squares, add the square of one-half the coefficient of x to both sides and, do the same for the coefficient of y.

$$\left(x^2 - 6x + \left(\frac{-6}{2}\right)^2\right) + \left(y^2 + 10y + \left(\frac{10}{2}\right)^2\right) = -18 + \left(\frac{-6}{2}\right)^2 + \left(\frac{10}{2}\right)^2$$

Select the term $\left(x^2 - 6x + \left(\frac{-6}{2}\right)^2\right)$ and, while holding down the CTRL key, apply **Factor**. Do the same for the term with y.

$$\left(x - 3\right)^2 + \left(y + 5\right)^2 = -18 + \left(\frac{-6}{2}\right)^2 + \left(\frac{10}{2}\right)^2$$

Select the right side of the equation, and while holding down the CTRL key, apply **Simplify**.

$$\left(x - 3\right)^2 + \left(y + 5\right)^2 = 16$$

You can read the solution to this problem from this form of the equation. The center of the circle is $(3, -5)$ and the radius is $\sqrt{16} = 4$.

Partial Fractions

The command **Partial Fractions** appears on both the **Polynomials** and **Calculus** sub-menus. With this command, you can write a rational expression as a sum of simpler fractions—essentially the reverse of the operation demonstrated in the previous section.

The **Partial Fractions** command expands a rational expression into a sum of rational expressions having denominators that are multiples of powers of linear and irreducible quadratic factors of the denominator. In this case *irreducible* means the roots are neither rational nor rational combinations of the coefficients of the polynomials.

The numerators of the partial fractions are constants or, in the case the denominator is a power of an irreducible quadratic, linear. Thus each partial fraction is of the form

$$\frac{A}{(ax+b)^n} \quad \text{or} \quad \frac{Ax+B}{(ax^2+bx+c)^m}$$

If more than one variable occurs in the expression, specify your choice of variable in the dialog box that appears. The other variables will be treated as arbitrary constants.

▶ **To write a rational expression as a sum of rational expressions**

1. Enter the rational expression in mathematics mode.

2. Leave the insertion point in the expression.

3. Choose **Polynomials + Partial Fractions** or **Calculus + Partial Fractions**.

▶ **Polynomials + Partial Fractions**

$$\frac{36}{(x-2)(x-1)^2(x+1)^2} = \frac{4}{x-2} - \frac{9}{(x-1)^2} - \frac{3}{(x+1)^2} - \frac{4}{x+1}$$

$$\frac{x^3+x^2+1}{x(x-1)(x^2+x+1)(x^2+1)^3} = \frac{1}{8(x-1)} - \frac{1}{x} + \frac{-x-1}{x+x^2+1}$$

$$+ \frac{1}{(x^2+1)^3}\left(\tfrac{1}{2} - \tfrac{1}{2}x\right) + \frac{1}{(x^2+1)^2}\left(\tfrac{3}{4}x + \tfrac{3}{4}\right) + \frac{1}{x^2+1}\left(\tfrac{15}{8}x - \tfrac{1}{8}\right)$$

(Variable: y) $\dfrac{y}{(x-y)^2(x+1)} = \dfrac{1}{(x+1)(y-x)} + \dfrac{x}{(x+1)(y-x)^2}$

This operation does not accept decimal or floating-point numbers, so write the coefficients as integers or quotients of integers. Use **Rewrite + Rational** if you have expressions with decimal or floating-point numbers (see page 25).

Products and Powers of Polynomials

The command **Expand** can be used to expand products or powers of polynomials.

▶ **Expand (or click** **)**

$$\left(3x^2 + 3x\right)\left(8x^2 + 7\right) = 21x + 21x^2 + 24x^3 + 24x^4$$

Alternately, while in mathematics mode, type **xpnd** (when you type the final letter, the function expand will appear), enter the polynomial inside parentheses, and choose **Evaluate**.

▶ **Evaluate**

$$\text{expand}\left(\left(3x^2 + 3x\right)\left(8x^2 + 7\right)\right) = 21x + 21x^2 + 24x^3 + 24x^4$$

Division by Polynomials

You can convert a quotient of polynomials $\dfrac{f(x)}{g(x)}$ with rational coefficients to the form $q(x) + \dfrac{r(x)}{g(x)}$, where $r(x)$ and $q(x)$ are polynomials and $\deg r(x) < \deg g(x)$.

▶ **To divide polynomials**

1. Enter a quotient of polynomials.

2. Leave the insertion point in the expression.

3. From the **Polynomials** submenu, choose **Divide**.

▶ **Polynomials + Divide**

$$\frac{3x^5 + 3x^3 - 4x^2 + 5}{8x^2 + 7} = \frac{3}{64}x + \frac{3}{8}x^3 + \frac{1}{8x^2 + 7}\left(\frac{17}{2} - \frac{21}{64}x\right) - \frac{1}{2}$$

Note   This algorithm is the familiar long-division algorithm for polynomials.

Collecting and Ordering Terms

The **Sort** command on the **Polynomials** submenu collects numeric coefficients of terms of a polynomial expression and returns the terms in order of decreasing degree. The **Collect** command on the **Polynomials** submenu collects all coefficients of terms of a polynomial expression, but does not necessarily sort the terms by degree. Specify your choice of polynomial variable in the dialog box that appears.

▶ **Polynomials + Sort**

$$x^2 + 3x + 5 - 3x^3 + 5x^2 + 4x^3 + 13 + 2x^4 = 2x^4 + x^3 + 6x^2 + 3x + 18$$
$$5t^2 + 3xt^2 - 16t^5 + y^3 - 2xt^2 + 9 = t^2x + 5t^2 - 16t^5 + y^3 + 9 \text{ (Variable: } x)$$
$$5t^2 + 3xt^2 - 16t^5 + y^3 - 2xt^2 + 9 = -16t^5 + xt^2 + 5t^2 + y^3 + 9 \text{ (Variable: } t.)$$

▶ **Polynomials + Collect**

$$5t^2 + 3xt^2 - 16t^5 + y^3 - 2xt^2 + 9 = 5t^2 - 16t^5 + y^3 + t^2x + 9 \text{ (Variable: } x.)$$
$$5t^2 + 3xt^2 - 16t^5 + y^3 - 2xt^2 + 9 = y^3 - 16t^5 + t^2(x + 5) + 9 \text{ (Variable: } t.)$$

For some expressions, you may want to apply both commands.

▶ **Polynomials + Collect, Polynomials + Sort**

$$x^3b + e + x^2c + x^2 + x^4k + x - x^3d + a = a + x + e + kx^4 + x^2(c + 1) + x^3(b - d)$$
$$= kx^4 + x^3(b - d) + x^2(c + 1) + x + a + e$$

(Variable: x)

▶ **Polynomials + Sort, Polynomials + Collect**

$$x^3b + e + x^2c + x^2 + x^4k + x - x^3d + a = kx^4 + x^3(b - d) + x^2(c + 1) + x + a + e$$
$$= a + x + e + kx^4 + x^2(c + 1) + x^3(b - d) \qquad \text{(Variable: } x)$$

Factoring Polynomials

The ability to factor polynomials is an important algebraic tool. You will find that the factoring capabilities of your computer algebra system are powerful and useful. You can factor polynomials with integer or rational roots and with other roots directly related to the coefficients of the expanded polynomial.

To factor a polynomial, you must type it without using decimal notation. Numbers such as 1.5 are interpreted as floating-point numbers, and Factor does not handle polynomials with floating-point coefficients. Replace decimal numbers with fractions (such as $1.5 = \frac{15}{10}$) using **Rewrite + Rational**, and then apply **Factor** to the resulting polynomial.

▶ Factor

$$5x^5 + 5x^4 - 10x^3 - 10x^2 + 5x + 5 = 5(x-1)^2(x+1)^3$$

$$\tfrac{1}{16}x^2 - \tfrac{7}{5}x + \tfrac{1}{6}ix - \tfrac{56}{15}i = \tfrac{1}{240}(5x - 112)(3x + 8i)$$

$$120x^3 + 20\left(-3 + 2\sqrt{3}\right)x^2 - \tfrac{5}{2}\left(8\sqrt{3} - 3\right)x + \tfrac{5}{2}\sqrt{3} = \tfrac{5}{2}\left(3x + \sqrt{3}\right)(4x - 1)^2$$

Factor is effective primarily for polynomials with integer or rational coefficients, although it also factors polynomials whose roots are closely related to the coefficients, as demonstrated in two of the preceding examples. Technically, the polynomial is factored over the field generated by its coefficients. If all the coefficients are rational, then the polynomial is factored over the rationals.

If you know the form of the root, you can multiply by an appropriate expression to obtain a factorization.

▶ Factor

$$5x^2 + x + 3 = (x + 5x^2 + 3)$$

$$i\sqrt{59}\left(5x^2 + x + 3\right) = \left(5i\sqrt{59}\right)\left(x + \tfrac{1}{10}i\sqrt{59} + \tfrac{1}{10}\right)\left(x - \tfrac{1}{10}i\sqrt{59} + \tfrac{1}{10}\right)$$

Alternatively, while in mathematics mode, type **factor**, enter the polynomial inside parentheses, and choose Evaluate. For the command expand, type **xpnd** in mathematics mode. If your system is not set for automatic recognition, you can enter **factor** or **expand** as a Math Name.

▶ Evaluate

$$\text{factor}\left(5x^5 + 5x^4 - 10x^3 - 10x^2 + 5x + 5\right) = 5(x-1)^2(x+1)^3$$

$$\text{expand}\left(5(x-1)^2(x+1)^3\right) = 5x^5 + 5x^4 - 10x^3 - 10x^2 + 5x + 5$$

You can factor not only the difference of two squares and the sum and difference of two cubes, but also the difference of any two equal powers.

▶ Factor

$$x^2 - y^2 = (x - y)(x + y) \qquad x^3 - y^3 = (x - y)\left(x^2 + xy + y^2\right)$$

$$x^4 - y^4 = (x - y)(x + y)\left(x^2 + y^2\right)$$

You can also factor the sum of any two equal odd powers.

▶ Factor

$$x^3 + y^3 = \left(x^2 - xy + y^2\right)(x + y)$$
$$x^5 + y^5 = \left(x^4 - x^3 y + x^2 y^2 - xy^3 + y^4\right)(x + y)$$
$$x^7 + y^7 = \left(y^6 - xy^5 + x^2 y^4 - x^3 y^3 + x^4 y^2 - x^5 y + x^6\right)(x + y)$$

Greatest Common Divisor of Two Polynomials

You find the greatest common divisor of two or more polynomials in the same way as you found the greatest common divisor (see page 22) of two or more integers.

▶ **To find the greatest common divisor of two or more polynomials**

   1. Type **gcd** in mathematics mode, or choose it from the functions menu.

   2. Click (□) , type the polynomials separated by commas, and choose Evaluate.

▶ Evaluate

$$\gcd(5x^2 - 5x, 10x - 10) = 5x - 5$$
$$\gcd\left(x^2 + 3x + yx + 3y, x^2 - 4yx - 5y^2, 3x^2 + 2yx - y^2\right) = x + y$$

You can check these results by factoring the polynomials and comparing the factors.

▶ Factor

$$x^2 - 4yx - 5y^2 = (x + y)(x - 5y) \qquad 3x^2 + 2yx - y^2 = (x + y)(3x - y)$$

The least common multiple function (see page 22) is also available for polynomials.

▶ **To find the least common multiple of two or more polynomials**

   1. Type **lcm** in mathematics mode. (It will turn gray.)

   2. Click (□) ; or choose Insert + Brackets and select parentheses; or press CTRL + 9.

   3. In the input box, type the polynomials separated by commas.

   4. Choose Evaluate.

▶ Evaluate

$$\text{lcm}(yx + 3x - 5y - 15, xz - 53x - 5z + 265) = 265y - 159x - 15z - 53xy + 3xz - 5yz + xyz + 795$$

Apply **Factor** to the result to reveal the relationship among the polynomials.

▶ Factor

$$265y - 159x - 15z - 53xy + 3xz - 5yz + xyz + 795 = (x - 5)(y + 3)(z - 53)$$

$$yx + 3x - 5y - 15 = (x - 5)(y + 3)$$

$$xz - 53x - 5z + 265 = (x - 5)(z - 53)$$

Roots of Polynomials

If zero is obtained when a number is substituted for the variable in a polynomial, then that number is a *root* of the polynomial. In other words, the roots of a polynomial $p(x)$ are the solutions to the equation $p(x) = 0$. For example, 1 is a root of $x^2 - 1$. It is important to remember that a number r is a root of a polynomial if and only if $x - r$ is a factor of that polynomial.

The factorization

$$x^3 - \frac{8}{3}x^2 - \frac{5}{3}x + 2 = (x - 3)\left(x - \frac{2}{3}\right)(x + 1)$$

displays the three roots $3, \frac{2}{3}, -1$. These roots are precisely the values of the x-coordinate where the graph of $y = x^3 - \frac{8}{3}x^2 - \frac{5}{3}x + 2$ crosses the x-axis. The following graph depicts this polynomial expression. Techniques for such displays are discussed in Chapter 6.

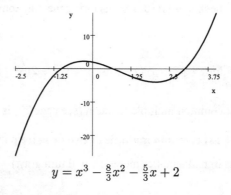

$$y = x^3 - \tfrac{8}{3}x^2 - \tfrac{5}{3}x + 2$$

The factorization of the complex polynomial

$$x^3 - \frac{13}{5}ix^2 - 8x^2 + \frac{29}{5}ix + \frac{81}{5}x + 6i - \frac{18}{5}$$
$$= (x - 3)\left(x + \frac{2}{5}i\right)(x - (5 + 3i))$$

displays the three roots $3, -\frac{2}{5}i, 5 + 3i$.

You can find all real and complex roots of a real or complex polynomial with rational coefficients by applying Roots from the Polynomials menu.

▶ **To find the roots of a polynomial**

1. Type the polynomial and leave the insertion point in the expression.

2. From the Polynomials submenu, choose Roots.

▶ Polynomials + Roots

$$5x^2 + 2x - 3, \text{roots:} \begin{bmatrix} -1 \\ 3 \\ \frac{3}{5} \end{bmatrix} \qquad x^2 + 1, \text{roots:} \begin{array}{c} i \\ -i \end{array}$$

$$x^3 - \tfrac{13}{5}ix^2 - 8x^2 + \tfrac{29}{5}ix + \tfrac{81}{5}x + 6i - \tfrac{18}{5}, \text{roots:} \begin{array}{c} \left(\frac{5}{2} + \frac{13}{10}i\right) - \frac{1}{10}\sqrt{(336+850i)} \\ \frac{1}{10}\sqrt{(336+850i)} + \left(\frac{5}{2} + \frac{13}{10}i\right) \end{array}$$

You can simplify these complex radical expressions with Rewrite + Rectangular.

▶ Rewrite + Rectangular

$$\left(\frac{5}{2} + \frac{13}{10}i\right) - \frac{1}{10}\sqrt{(336+850i)} = \frac{2}{5}i(-1)$$

$$\frac{1}{10}\sqrt{(336+850i)} + \left(\frac{5}{2} + \frac{13}{10}i\right) = 5 + 3i$$

You can change settings so that only real roots will be computed.

▶ **To find real roots of a polynomial**

1. While in mathematics mode, type **assume(real)**.

2. With the insertion point in assume (real), choose Evaluate.

3. Place the insertion point in the polynomial.

4. From the Polynomials submenu, choose Roots.

▶ Evaluate

assume (real) = real

▶ Polynomials + Roots

$$x^3 - \tfrac{13}{5}ix^2 - 8x^2 + \tfrac{29}{5}ix + \tfrac{81}{5}x + 6i - \tfrac{18}{5}, \text{ roots: } 3$$

$$5x^2 + x + 3, \text{ roots: } \emptyset$$

The symbol $\emptyset$ denotes the empty set; that is, the case when there are no real solutions.

▶ **To return to the default mode**

- Apply Evaluate to unassume or to assume (complex).

It follows from the Fundamental Theorem of Algebra that the number of roots (including complex roots and multiplicities) is the same as the degree of the polynomial.

For polynomials with rational (real or complex) coefficients, the computer algebra system uses the usual formulas for finding roots symbolically for polynomials of degree 4 or less, and it finds the roots numerically for polynomials of higher degree. This implementation was dictated by the mathematical phenomenon that there is no general formula in terms of radical expressions for the roots of polynomials of degree 5 and higher. For polynomials of any degree with floating point or decimal coefficients, the computer algebra system finds the roots numerically.

Second-Degree Polynomials

You can obtain the familiar quadratic formula for roots of $ax^2 + bx + c$. The solution includes all cases. The logical symbol $\wedge$ is used for AND, so $a = 0 \wedge b = 0 \wedge c = 0$ is the case that all three coefficients, a, b, c, are zero. The symbol $\mathbb{C}$ denotes the set of all complex numbers. The symbol $\emptyset$ denotes the empty set; that is, the case when there are no solutions.

▶ Polynomials + Roots

$ax^2 + bx + c,$ (Variable: x)

$$\text{roots: } \begin{cases} \mathbb{C} & \text{if } a = 0 \wedge b = 0 \wedge c = 0 \\ \emptyset & \text{if } a = 0 \wedge b = 0 \wedge c \neq 0 \\ \left\{ -\dfrac{1}{b}c \right\} & \text{if } a = 0 \wedge b \neq 0 \\ \left\{ \dfrac{1}{a}\left(-\dfrac{1}{2}b - \dfrac{1}{2}\sqrt{-4ac + b^2} \right), \dfrac{1}{a}\left(-\dfrac{1}{2}b + \dfrac{1}{2}\sqrt{-4ac + b^2} \right) \right\} & \text{if } a \neq 0 \end{cases}$$

Third- and Fourth-Degree Polynomials

The roots of third- and fourth-degree polynomials can be complicated, with multiple embedded radicals in the expressions. To put those roots in simpler form, you may want numerical approximations. You get numerical results if you enter at least one coefficient in decimal notation. You can also get a numerical form directly from the symbolic one by applying Evaluate Numerically to the matrix of roots. The following example shows both a symbolic solution and a numerical solution (with Digits Shown in Results set to 6).

▶ Polynomials + Roots

$x^3 + 3x + 1$, roots:

$$\frac{1}{2\sqrt[3]{\frac{1}{2}\sqrt{5}-\frac{1}{2}}} - \frac{1}{2}\sqrt[3]{\frac{1}{2}\sqrt{5}-\frac{1}{2}} + \frac{1}{2}i\sqrt{3}\left(\frac{1}{\sqrt[3]{\frac{1}{2}\sqrt{5}-\frac{1}{2}}} + \sqrt[3]{\frac{1}{2}\sqrt{5}-\frac{1}{2}}\right)$$

$$\frac{1}{2\sqrt[3]{\frac{1}{2}\sqrt{5}-\frac{1}{2}}} - \frac{1}{2}\sqrt[3]{\frac{1}{2}\sqrt{5}-\frac{1}{2}} - \frac{1}{2}i\sqrt{3}\left(\frac{1}{\sqrt[3]{\frac{1}{2}\sqrt{5}-\frac{1}{2}}} + \sqrt[3]{\frac{1}{2}\sqrt{5}-\frac{1}{2}}\right)$$

$$\sqrt[3]{\frac{1}{2}\sqrt{5}-\frac{1}{2}} - \frac{1}{\sqrt[3]{\frac{1}{2}\sqrt{5}-\frac{1}{2}}}$$

▶ Polynomials + Roots

$x^3 + 3x + 1.0$, roots:
-0.322185
$0.161093 - 1.75438i$
$0.161093 + 1.75438i$

Substituting the exact roots for x in the polynomial $x^3 + 3x + 1$ gives zero, as it should. Applying Evaluate has little effect, but Simplify gives the following result.

▶ Simplify

$$\left(\frac{1}{2\sqrt[3]{\frac{1}{2}\sqrt{5}-\frac{1}{2}}} - \frac{1}{2}\sqrt[3]{\frac{1}{2}\sqrt{5}-\frac{1}{2}} + \frac{1}{2}i\sqrt{3}\left(\frac{1}{\sqrt[3]{\frac{1}{2}\sqrt{5}-\frac{1}{2}}} + \sqrt[3]{\frac{1}{2}\sqrt{5}-\frac{1}{2}}\right)\right)^3$$

$$+3\left(\frac{1}{2\sqrt[3]{\frac{1}{2}\sqrt{5}-\frac{1}{2}}} - \frac{1}{2}\sqrt[3]{\frac{1}{2}\sqrt{5}-\frac{1}{2}} + \frac{1}{2}i\sqrt{3}\left(\frac{1}{\sqrt[3]{\frac{1}{2}\sqrt{5}-\frac{1}{2}}} + \sqrt[3]{\frac{1}{2}\sqrt{5}-\frac{1}{2}}\right)\right)$$

$+1 = 0$

Using the numerical approximations to the roots, you get a very small, but nonzero, value. You can get closer approximations to the roots by increasing the number of displayed digits before finding the roots. (See page 30.)

▶ Evaluate

$$(-0.322185)^3 + 3(-0.322185) + 1.0 = 1.174312318 \times 10^{-6}$$

$$(-0.32218535462608559291)^3 + 3(-0.32218535462608559291) + 1$$
$$= 4.870126439 \times 10^{-21}$$

▶ Polynomials + Roots

$$x^4 + 3x^3 - 2x^2 + x + 1.0, \text{ roots: } \begin{bmatrix} 0.519\,28 - 0.613\,32i \\ 0.519\,28 + 0.613\,32i \\ -3.609\,6 \\ -0.428\,98 \end{bmatrix}$$

$$x^4 - 7x^3 + 2x^2 + 64x - 96, \text{ roots: } \begin{bmatrix} 2 \\ -3 \\ 4 \\ 4 \end{bmatrix}$$

Polynomials of Degree 5 and Higher

Numerical approximations are always returned for roots of polynomials of degree 5 and higher. You can change the number of digits shown in the display of these roots by making changes in the Computation Setup or Compute Settings. (See page 30.)

▶ Polynomials + Roots

$$5x^5 + 5x^4 - 10x^3 - 10x^2 + 5x + 5, \text{ roots: } \begin{matrix} 1.0 \\ 1.0 \\ -1.0 \\ -1.0 \\ -1.0 \end{matrix}$$

$$x^8 + x^7 + x^6 + x^5 + x^4 + x^3 + x^2 + x + 1, \text{ roots: } \begin{matrix} 0.766\,04 + 0.642\,79i \\ 0.766\,04 - 0.642\,79i \\ 0.173\,65 - 0.984\,81i \\ 0.173\,65 + 0.984\,81i \\ -0.5 + 0.866\,03i \\ -0.5 - 0.866\,03i \\ -0.939\,69 + 0.342\,02i \\ -0.939\,69 - 0.342\,02i \end{matrix}$$

Defining Variables and Functions

The Definitions commands enable you to define a symbol to be a mathematical object and to define a function using an expression or a collection of expressions. The first four operations on the Define submenu—New Definition, Undefine, Show Definitions, Clear Definitions—are explained briefly in this section for the types of expressions and functions that occur in precalculus. See Chapter 5, starting on page 109, for greater detail on these operations and other aspects of definitions. For examples of these operations pertinent to topics such as calculus, vector calculus, and matrix algebra, see the chapter covering the topic.

Assigning Values to Variables

You can assign a value to a variable using Definitions.

▶ **To assign the value** 5 **to** z

1. Write $z = 5$ in mathematics mode and, leaving the insertion point in the equation,

2. Click the New Definition button ![button] on the Compute toolbar; or choose Definitions + New Definition; or press CTRL + =.

   Thereafter, until you undefine the variable (see page 70) the system recognizes z as 5, evaluating the expression $3 + z$ as 8. See page 121 for choices of behavior when you close and reopen the document.

   Variables normally have single-character names. (See page 109 for other possibilities.) The value assigned can, however, be any mathematical expression. For example, you could define a variable to be any of the following:

- Number: $a = 245$
- Polynomial: $p = x^3 + 3x^2 - 5x + 1$
- Quotient of polynomials: $b = \dfrac{x^2 - 1}{x^2 + 1}$
- Matrix: $z = \begin{bmatrix} a & b \\ c & d \end{bmatrix}$

   The symbol p defined here represents the expression $x^3 + 3x^2 - 5x + 1$. It is not a function, and in particular, $p(2)$ is not the expression evaluated at $x = 2$. In fact, $p(2)$ is interpreted simply as the product $2p = 2x^3 + 6x^2 - 10x + 2$.

Defining Functions Of One Variable

You follow a similar procedure to define a function. Write a function name followed by parentheses containing the variable, and set this equal to an expression.

▶ **To define the function f whose value at x is $ax^2 + bx + c$**

1. Write $f(x) = ax^2 + bx + c$.

2. Place the insertion point in the equation.

3. Click the **New Definition** button $\boxed{f^{(x)}}$ on the **Compute** toolbar; or, from the **Definitions** submenu, choose **New Definition**; or press CTRL + =.

Thereafter, until the function is undefined, the symbol f represents the defined function and behaves like a function. For example, apply **Evaluate** to $f(t)$ to obtain $f(t) = at^2 + bt + c$.

Note Making the definition $f(y) = ay^2 + by + c$ defines the *same* function as the definition $f(x) = ax^2 + bx + c$. *The symbol used for the function argument in making the definition does not matter.* This is the crux of the subtle but essential difference between expressions and functions.

The two expressions $y = x^2 + \sqrt{x}$ and $y = t^2 + \sqrt{t}$ are different because y is replaced by an expression in x under the first definition and y is replaced by an expression in t under the second definition. However, the functions defined by $f(x) = x^2 + \sqrt{x}$ and $f(t) = t^2 + \sqrt{t}$ are identical.

If g and h are previously defined functions, then the following equations are examples of legitimate definitions:

$$
\begin{aligned}
f(x) &= 2g(x) \\
f(x) &= g(x) + h(x) \\
f(x) &= g(x)h(x) \\
f(x) &= g(h(x))
\end{aligned}
$$

Make a definition for g and h, and then apply **Evaluate** to $f(t)$ for each definition of f. Each time you redefine f, the new definition replaces the old one. Also, once you have defined both $g(x)$ and $f(x) = 2g(x)$, then changing the definition of $g(x)$ redefines $f(x)$.

Note The algebra of functions includes objects such as $f \pm g$, $f \circ g$, fg, and f^{-1}. For the value of $f + g$ at x, write $f(x) + g(x)$; for the value of the composition of two defined functions f and g, write $f(g(x))$ or $(f \circ g)(x)$; and for the value of the product of two defined functions, write $f(x)g(x)$. You can obtain the inverse for some functions $f(x)$ by applying **Solve + Exact** to the equation $f(y) = x$ and specifying y as the **Variable to Solve for**.

Example   Define $f(x) = x^2 + 3x + 5$ and $g(x) = x^3 - 1$. Then, Evaluate produces

$$
\begin{aligned}
f(3) &= 23 \\
g(3) &= 26 \\
f(g(3)) &= 759 \\
g(f(3)) &= 12166 \\
f(4 + 5i) &= 8 + 55i \\
f(f(f(4 + 5i))) &= 7495808 - 6124745i
\end{aligned}
$$

You can sometimes find the inverse of a function $y = f(x)$, if it exists, by interchanging x and y and solving for y.

▶ **To find the inverse of $f(x) = 5x - 3$**

1. Evaluate $f(y)$ to get $f(y) = 5y - 3$.

2. Solve the equation $5y - 3 = x$ for y using Solve + Exact.

This computation yields the solution $y = \frac{1}{5}x + \frac{3}{5}$. Thus

$$f^{-1}(x) = \frac{1}{5}x + \frac{3}{5}$$

To check this result, define $f(x) = 5x - 3$ and $g(x) = \frac{1}{5}x + \frac{3}{5}$. (The symbol f^{-1} will not work as a function name.) Evaluating the expressions $f(g(x))$ and $g(f(x))$ gives $f(g(x)) = x$ and $g(f(x)) = x$, demonstrating that the function g is the inverse of the function f.

Defining Functions Of Several Variables

▶ **To define a function of several variables**

1. Write an equation such as $f(x, y, z) = ax + y^2 + 2z$ or $g(x, y) = 2x + \sin 3xy$.

2. Leave the insertion point in the equation.

3. Click the New Definition button on the Compute toolbar, or from the Definitions submenu, choose New Definition.

As in the case of functions of one variable, the computer algebra system operates on expressions obtained by evaluating the function.

Showing and Removing Definitions

After making definitions of functions or expressions, you need to know techniques for keeping track of them, saving them, and deleting them.

▶ **To view the list of currently defined variables and functions**

- Click the Show Definitions button [f(x)] on the Compute toolbar or, from the Definitions submenu, choose Show Definitions.

  A window opens showing the active definitions. In general, the defined variables and functions are listed in the order in which the definitions were made.

▶ **To remove a definition from a document**

1. Place the insertion point in the equation you wish to undefine or, select the name of the function or expression (anywhere that it appears).

2. From the Definitions submenu, choose Undefine.

  From the Definitions submenu, you can choose Show Definitions to check that the definition has been removed from the list of defined functions and expressions.

▶ **To remove all definitions in a document**

- From the Definitions submenu, choose Clear Definitions.

  Definitions that you do not remove remain active as long as a document is open. As a default, definitions are saved and then restored when you reopen a document. You can change this setting so that exiting a document will remove your definitions. See page 121 for a discussion of Save Definitions, Restore Definitions, and Clear Definitions.

Errors Resulting from (Overlooked) Definitions

It is easy to forget that a symbol has been defined to be some expression. If you use that symbol later, you can get surprising results. For example, if you define $a = x^2$, forget about it, and later compute $f(a)$ for some function f that you have just defined, you are in for a surprise. In complicated computations the error may not be apparent.

Tip   Check the Show Definitions list from time to time. If your mathematics is behaving strangely, this list is a place to look for a possible explanation.

Solving Polynomial Equations

There are four options on the Solve submenu: Exact, Integer, Numeric, and Recursion. The option Exact is general in nature and is used in most situations. It returns symbolic solutions when it can and numerical solutions otherwise. If any of the components of the problem use floating-point (decimal) notation, the response is a numerical solution.

The three options Integer, Numeric, and Recursion are used in more specialized situations. These will be discussed later.

Equations with One Variable

All solutions are given for polynomial equations with symbolic coefficients, including complex solutions.

▶ **To solve an equation with one variable**

1. Place the insertion point in the equation.

2. Click the **Solve Exact** button $\boxed{\text{x?}}$ on the **Compute** toolbar or, from the **Solve** submenu, choose **Exact**.

Your system returns a solution.

Note that in the following examples, integer or rational coefficients yield algebraic solutions and real (floating-point) coefficients yield decimal approximations.

▶ Solve + Exact

$5x^2 + 3x = 1$, Solution is: $-\frac{1}{10}\sqrt{29} - \frac{3}{10}, \frac{1}{10}\sqrt{29} - \frac{3}{10}$

$5x^2 + 3x = 1.0$, Solution is: $-0.838\,52, 0.238\,52$

$s^2 + 10s + \frac{1681}{64} = 0$, Solution is: $-5 - \frac{9}{8}i, -5 + \frac{9}{8}i$

$s^2 + (10.0)\,s + \frac{1681}{64} = 0$, Solution is: $-5.0 - 1.\,125i, -5.0 + 1.\,125i$

$x^3 - 3x^2 + x - 3 = 0$, Solution is: $3, -i, i$

$\pi x^2 - i\pi x - 3i\sqrt{2} + 3x\sqrt{2} = 0$, Solution is: $i, -\frac{3}{\pi}\sqrt{2}$

When there are multiple roots, only distinct roots are displayed.

▶ Solve + Exact

$(x - 5)^3 (x + 1) = 0$, Solution is: $-1, 5$

You can solve equations with rational expressions, and equations involving absolute values.

▶ Solve + Exact

$\dfrac{14}{a + 2} - \dfrac{1}{a - 4} = 1$, Solution is: $5, 10$

$|3x - 2| = 5$, Solution is: $\left\{ \dfrac{5}{3}e^{2iX_1\pi} + \dfrac{2}{3} \mid X_1 \in [0, 1] \right\}$

If you want only real roots, first evaluate assume (real). When you enter these words in mathematics mode, they automatically gray. You can also use Insert + Math Name to enter assume (real). (See page 260 for more information on the assume function.)

▶ Evaluate

assume (real) = real

▶ Solve + Exact

$x = i$, No solution found.

$x^3 - 3x^2 + x - 3 = 0$, Solution is: 3

$|3x - 2| = 5$, Solution is: $\frac{7}{3}, -1$

Evaluate unassume (real) to return to the default mode.

▶ Evaluate

unassume (real)

In general, explicit solutions in terms of radicals for polynomial equations of degree greater than 4 do not exist. In these cases, implicit solutions are given in terms of "RootsOf." When the equation is a polynomial equation with degree 3 or 4, the explicit solution can be very complicated—and too large to preview, print, or save. To avoid this problem, you can set the engine to return large complicated solutions in implicit form for smaller degree polynomials as well.

▶ **To raise or lower polynomial degree for implicit solutions**

1. Choose Tools + Engine Setup

2. On the General page, under Solve Options, change Maximum Degree to 1, 2, 3, or 4.

With a setting of 1, only rational or other relatively simple solutions are computed for all polynomials. With a setting of 2 or 3, this behavior occurs for polynomials of degree greater than 2.

▶ Solve + Exact (Maximum Degree set to 1)

$5x^2 + 3x = 1$, Solution is: ρ_1 where ρ_1 is a root of $3X_8 + 5X_8^2 - 1, X_8$

$x^4 + x$, Solution is: $\{-1, 0\} \cup \rho_1$ where ρ_1 is a root of $X_9^2 - X_9 + 1, X_9$

▶ Solve + Exact (Maximum Degree set to 2 or 3)

$5x^2 + 3x = 1$, Solution is: $-\dfrac{1}{10}\sqrt{29} - \dfrac{3}{10}, \dfrac{1}{10}\sqrt{29} - \dfrac{3}{10}$

$x^4 + x$, Solution is: $-1, 0, \dfrac{1}{2} - \dfrac{1}{2}i\sqrt{3}, \dfrac{1}{2}i\sqrt{3} + \dfrac{1}{2}$

$x^4 + x - 1 = 0$, Solution is: ρ_1 where ρ_1 is a root of $X_{10} + X_{10}^4 - 1, X_{10}$

▶ Solve + Exact (Maximum Degree set to 4)

$x^4 + x$, Solution is: $-1, 0, \dfrac{1}{2} - \dfrac{1}{2}i\sqrt{3}, \dfrac{1}{2}i\sqrt{3} + \dfrac{1}{2}$

$x^4 + x - 1 = 0$, Solution is: [This solution displays on the screen, but is too large to preview, print, or save.]

The function solve takes an equation as input. Evaluate solve at an equation and the output is a list of solutions. To make the function name, type **solve** while in mathematics mode and it will automatically gray, or create the name with Insert + Math Name.

▶ Evaluate

$\text{solve}\left(5x^2 + 3x = 1\right) = \left\{\left[x = -\dfrac{1}{10}\sqrt{29} - \dfrac{3}{10}\right], \left[x = \dfrac{1}{10}\sqrt{29} - \dfrac{3}{10}\right]\right\}$

$\text{solve}\left(s^2 + (10.0)\,s + \dfrac{1681}{64} = 0\right) = \left\{[s = -5.0 - 1.\,125i], [s = -5.0 + 1.\,125i]\right\}$

Checking the Answer
The **Definitions** command discussed previously (see page 67) provides a convenient way of testing solutions. After working through this example, choose **Definitions + Clear Definitions**.

Example   Check the solutions to several of the preceding equations.
- Define $a = 5$. Evaluate the expression $\dfrac{14}{a+2} - \dfrac{1}{a-4}$ to get $\dfrac{14}{a+2} - \dfrac{1}{a-4} = 1$.
- Define $x = 0.238\,52$. Evaluation gives $5x^2 + 3x - 1 = 1.\,895\,2 \times 10^{-5}$. (Note that this solution is not exact. The value of the expression for this x is "close to zero.")
- Define $x = -\dfrac{3}{10} + \dfrac{1}{10}\sqrt{29}$. **Evaluate** followed by **Simplify** gives

$$5x^2 + 3x - 1 = \dfrac{3}{10}\sqrt{29} + 5\left(-\dfrac{3}{10} + \dfrac{1}{10}\sqrt{29}\right)^2 - \dfrac{19}{10} = 0$$

Equations with Several Variables

If there is more than one variable, enter the Variable(s) to Solve for in the dialog box that opens when you click the Solve Exact button $\boxed{x?}$ on the Compute toolbar or choose Exact from the Solve submenu.

▶ Solve + Exact

$$\frac{1}{x} + \frac{1}{y} = 1 \text{ (Enter } x\text{), Solution is :} \begin{cases} undecidable & \text{if } y = 1 \\ \dfrac{1}{-\dfrac{1}{y} + 1} & \text{if } y \neq 1 \end{cases}$$

$$\frac{1}{y} + \frac{1}{z} + \frac{1}{x} = 1 \text{ (Enter } z\text{), Solution is :} \begin{cases} undecidable & \text{if } -\dfrac{1}{x} - \dfrac{1}{y} + 1 = 0 \\ \dfrac{1}{-\dfrac{1}{x} - \dfrac{1}{y} + 1} & \text{if } -\dfrac{1}{x} - \dfrac{1}{y} + 1 \neq 0 \end{cases}$$

$$\frac{1}{r_1} + \frac{1}{r_2} = \frac{1}{R} \text{ (Enter } R\text{), Solution is :} \begin{cases} undecidable & \text{if } \dfrac{1}{r_1} + \dfrac{1}{r_2} = 0 \\ \dfrac{1}{\dfrac{1}{r_1} + \dfrac{1}{r_2}} & \text{if } \dfrac{1}{r_1} + \dfrac{1}{r_2} \neq 0 \end{cases}$$

Systems of Equations

You can create a system of equations either by entering equations in a one-column matrix or by entering equations in a multirow display.

▶ **To create a system of equations using a matrix**

1. Click the Matrix button $\boxed{\vdots}$ on the Math Objects toolbar or, from the Insert menu, choose Matrix.

2. Set the number of rows equal to the number of equations.

3. Set the number of columns to 1.

4. Choose OK.

5. Type the equations in the matrix, one equation to a row.

▶ **To create a system of equations using a display**

1. Click the Display button ▦ on the Math Objects toolbar, or choose Insert + Display.

2. Type the equations in the display, one equation to a row, adding rows as needed with the ENTER key.

Tip   From the View menu, choose Helper Lines or Input Boxes to help place equations in a matrix or display.

▶ **To solve a system of equations**

1. Create a system of equations.

2. Leave the insertion point in the system.

3. Click the Solve Exact button ⬛ on the Compute toolbar or, from the Solve submenu, choose Exact.

4. If a dialog box opens asking Variable(s) to Solve for, type the variable name(s) in the box, separated by commas.

Following are examples for systems of two equations.

▶ **Solve + Exact**

$$\begin{array}{l} 2x - y = 5 \\ x + 3y = 4 \end{array}, \text{ Solution is : } \left[x = \frac{19}{7}, y = \frac{3}{7} \right]$$

$$\begin{array}{l} x^2 - y^2 = 5 \\ x + y = 1 \end{array}, \text{ Solution is : } [x = 3, y = -2]$$

$$\begin{array}{l} x^2 - 3y = 7 \\ 6x + 4y = 9 \end{array}, \text{ Solution is: } \begin{array}{l} \left[x = \frac{1}{4}\sqrt{301} - \frac{9}{4}, y = \frac{45}{8} - \frac{3}{8}\sqrt{301} \right], \\ \left[x = -\frac{1}{4}\sqrt{301} - \frac{9}{4}, y = \frac{3}{8}\sqrt{301} + \frac{45}{8} \right] \end{array}$$

When the number of unknowns is larger than the number of equations, you are asked to specify variables in a dialog box.

▶ **Solve + Exact**

$$\begin{array}{l} 2x - y = 1 \\ x + 3z = 4 \\ w + x = -3 \end{array}, \text{ Variable(s) to Solve for: } x, y, z$$

$$\text{Solution is : } \left[x = -w - 3, y = -2w - 7, z = \frac{1}{3}w + \frac{7}{3} \right]$$

Numerical Solutions

You can find numerical solutions in two ways. You can choose **Exact** from the **Solve** submenu after entering at least one coefficient in floating-point form—that is, with a decimal.

▶ **Solve + Exact**

$x^2 + 7x - 5.2 = 0$, Solution is: $-7.6773, 0.67732$

$x^3 - 3.8x - 15.6 = 0$, Solution is: $-1.5 - 1.7176i, -1.5 + 1.7176i, 3.0$

You can choose **Numeric** on the **Solve** submenu. This gives all solutions, both real and complex, to a polynomial equation or system of polynomial equations.

▶ **Solve + Numeric**

$x^2 + 7x - 5.2 = 0$, Solution is : $\{x = -7.6773\}, \{x = 0.67732\}$

$x^3 - 3.8x - 15.6 = 0$,
   Solution is: $\{[x = 3.0], [x = -1.5 - 1.7176i], [x = -1.5000 + 1.7176i]$

$x^8 + 3x^2 - 1 = 0$,
   Solution is: $\{[x = -1.0023 + 0.63210i], [x = -1.0023 - 0.63210i],$
   $[x = 1.0023 + 0.63210i], [x = 1.0023 - 0.63210i], [x = -0.57394],$
   $[x = 0.57394], [x = -1.2408i], [x = 1.2408i]\}$

▶ **Solve + Numeric**

$\begin{bmatrix} x^2 + y^2 = 5 \\ x^2 - y^2 = 1 \end{bmatrix}$,
   Solution is: $\{[x = -1.7321, y = -1.4142], [x = -1.7321, y = 1.4142],$
   $[x = 1.7321, y = -1.4142], [x = 1.7321, y = 1.4142]\}$

The command **Solve + Numeric** is particularly useful when solving transcendental equations or systems of transcendental equations, or when you want to specify a search interval for the solution.

▶ **To find a numerical solution within a specified range of the variable**

1. Add a row to the bottom of the matrix or, press ENTER to generate a new input box in a display.

2. Write the intervals of your choice, and use the membership symbol ∈ to indicate that the variable lies in that interval.

▶ Solve + Numeric

$$x^2 + y^2 = 5$$
$$x^2 - y^2 = 1$$
$$x \in (-2, 0)$$, Solution is: $[x = -1.7321, y = 1.4142]$
$$y \in (0, 2)$$

▶ **To find all numerical solutions to a system of polynomial equations**

1. Change at least one of the coefficients to floating-point form.

2. From the Solve submenu, choose Exact.

▶ Solve + Exact

$$x^2 + y^2 = 5.0$$
$$x^2 - y^2 = 1.0$$, Solution is :

$\{y = -1.4142, x = 1.7321\}$
$\{y = -1.4142, x = -1.7321\}$
$\{y = 1.4142, x = 1.7321\}$
$\{y = 1.4142, x = -1.7321\}$

These four solutions are illustrated in the following graph as the four points of intersection of two curves. See Implicit Plots on page 180 for guidelines on making such graphs.

▶ Plot 2D + Implicit

$$x^2 + y^2 = 5$$
$$x^2 - y^2 = 1$$

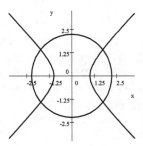

See page 29 for details on changing the appearance of these numerical solutions by resetting the number of digits shown in results and the threshold for scientific notation.

Inequalities

You can find exact solutions for many inequalities.

▶ **To solve an inequality**

1. Leave the insertion point in the inequality.

2. From the Solve submenu, choose Exact.

▶ Solve + Exact

$16 - 7y \geq 10y - 4$, Solution is: $\left(-\infty, \dfrac{20}{17}\right]$

$x^3 + 1 > x^2 + x$, Solution is: $(-1, 1) \cup (1, \infty)$

$|2x + 3| \leq 1$, Solution is: $[-2, -1]$

$\dfrac{7 - 2x}{x - 2} \geq 0$, Solution is: $\left(2, \frac{7}{2}\right]$

$x^2 + 2x - 3 > 0$, Solution is: $(1, \infty) \cup (-\infty, -3)$

These solutions are intervals—open, closed, or half-open and half-closed:
$$(a, b) = \{x : a < x < b\} \qquad [a, b] = \{x : a \leq x \leq b\}$$
$$(a, b] = \{x : a < x \leq b\} \qquad [a, b) = \{x : a \leq x < b\}$$
For two sets (intervals) A and B,
$$A \cup B = \{x : x \in A \text{ or } x \in B\}$$
$$A \cap B = \{x : x \in A \text{ and } x \in B\}$$

The solution to the last inequality, $x^2 + 2x - 3 > 0$, can also be read from the graph of the polynomial $y = x^2 + 2x - 3$. In the next plot, you see that the graph passes through the x-axis at $x = -3$ and $x = 1$, and the solution includes every point to the left of -3 or to the right of 1.

▶ Plot 2D + Rectangular

$x^2 + 2x - 3$

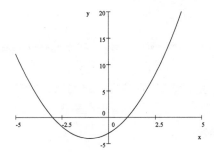

Substitution

Use common notation for variable substitution:
$$[F(x)]_{x=a} = F(a) \quad \text{and} \quad [F(x)]_{x=a}^{x=b} = F(b) - F(a)$$

Substituting for a Variable

To substitute a number or new expression for a variable, enclose an expression in square expanding brackets, enter an assignment for the variable in a subscript, and choose Evaluate.

▶ Evaluate

$$\left[x^2 + 2x - 3\right]_{x=a} = 2a + a^2 - 3 \qquad \left[x^2 + 2x - 3\right]_{x=5} = 32$$

$$\left[\tfrac{x^2-3x}{5}\right]_{x=y-z} = \frac{3}{5}z - \frac{3}{5}y + \frac{1}{5}(y-z)^2$$

The expression in the subscript is an assignment for the variable on the left of the equals sign. Notice that, in particular, $x = y + z$, $y = x - z$ and $z = x - y$ are not equivalent assignments.

▶ Evaluate

$$[x+y]_{x=y+z} = 2y + z \qquad [x+y]_{y=x-z} = 2x - z \qquad [x+y]_{z=x-y} = x + y$$

Evaluating at Endpoints

To substitute two numbers or new expressions for a variable and find the difference, first enclose an expression in square expanding brackets
$$[x]_{x=a}^{x=b}$$
Then, enter the numbers or new expressions in a subscript and superscript, and choose Evaluate.

▶ Evaluate

$$[x]_{x=a}^{x=b} = b - a \qquad\qquad [x^2 + 2x - 3]_{x=a}^{x=b} = 2b - 2a - a^2 + b^2$$

$$[x^2 + 2x - 3]_{x=3}^{x=5} = 20$$

▶ Evaluate, Factor

$$\left[\tfrac{x^2 - 3x}{5}\right]_{x=y-z}^{x=y+z} = \tfrac{1}{5}(y+z)^2 - \tfrac{6}{5}z - \tfrac{1}{5}(y-z)^2 = \tfrac{2}{5}z(2y-3)$$

From the expanding brackets panel, you can choose a nonprinting dashed vertical line for the left bracket and a square bracket or vertical line for the right bracket. You can also enter from the keyboard a right vertical line only, with subscript, or with both subscript and superscript.

▶ Evaluate

$$[x^2 - 3]_{x=a}^{x=b} = b^2 - a^2 \qquad x^2 - 3]_{x=a}^{x=b} = b^2 - a^2$$

$$x^2 - 3\Big|_{x=a}^{x=b} = b^2 - a^2 \qquad x + 3|_{x=y+z} = y + z + 3$$

Note   All of the expressions above are enclosed in expanding brackets. The left brackets that do not appear are the nonprinting brackets that appear as a vertical dashed line in the Brackets panel.

Exponents and Logarithms

You can work with exponential and logarithmic functions in their common notation: e^x, $\exp x$, $\log_5 x$, $\ln x$, and so forth. The exponential and logarithmic functions are inverses of one another, as exemplified by the identities

$$e^{\ln x} = x$$

and

$$\ln e^x = x$$

Exponents and Exponential Functions

Exponential functions are used in modeling many real-life situations. The laws of exponents are an important feature of these functions.

Combining Exponentials

Expressions involving exponential functions with base e are combined by applying the following commands:

▶ Combine + Exponential

$$(e^x)^y = e^{yx} \qquad e^x e^y = e^{x+y}$$

▶ Expand

$$e^{x+3\ln y} = y^3 e^x \qquad e^{x+y} = e^x e^y$$

Combining Powers

Some expressions involving powers with arbitrary base a are combined applying the following command:

▶ Combine + Powers

$$a^x a^y = a^{x+y}$$

Laws of Exponents

You can demonstrate some of the laws of exponents with either Simplify or Combine + Powers. You can also use Combine + Exponential for the first of these equations. These laws work for real or complex exponents and for other expressions as well.

▶ Simplify

$$2^x 2^y = 2^{x+y} \qquad 3^{x^2-3xy} 3^{2x+5} = 3^{2x-3xy+x^2+5}$$
$$\frac{a^x}{a^y} = a^{x-y} \qquad 10^{x+i} 10^{x-i} = 10^{2x}$$

The function exp is defined by $\exp x = e^x$. Typing **exp** in mathematics mode automatically returns the grayed Math Name exp. Exponential expressions are normally returned to your document in the form $e^{f(x)}$ rather than $\exp f(x)$. However, when the exponent $f(x)$ is sufficiently complicated, the linear form $\exp f(x)$ is returned. The following example illustrates these two behaviors.

▶ Combine + Exponentials

$$\left(e^{a+b}\right)^3 = e^{3a+3b} \qquad e^{x^2-3xy} e^{2x+5} = \exp\left(2x - 3xy + x^2 + 5\right)$$

Evaluating Exponential Functions

Use Evaluate or Evaluate Numerically for numerical approximations. Note that the use of floating-point notation triggers a numeral evaluation. Change the Computation Setup dialog on the Tools menu for the number of digits you want. In these examples, Digits Shown in Results is set to 8 (see page 30).

Expression	Evaluate	Evaluate Numerically
e^2	e^2	7.3890561
$e^{0.0025}$	1.0025031	1.0025031
5^4	625	625.0
$2^{\sqrt{5}}$	$2^{\sqrt{5}}$	4.7111131

Logarithms and Logarithmic Functions

The function $\ln x$ is interpreted as the natural logarithm base e. Logarithms to other bases are entered with a subscript on the function log. Thus, evaluation gives $\log_5 25 = 2$ and $\log_{10} 10^3 = 3$. The symbol $\log x$ is interpreted as the natural logarithm (base e) unless you make a change in the Computation Setup dialog. You can change the system default, and you can also override this default for individual documents.

1. From the Tools menu, choose Computation Setup and click the General tab.

2. Under Base for Log Function, check Change from e to 10 and choose OK.

▶ **To change the base for the function name** log **for an individual document**

1. From the Compute menu, choose Settings and click the General tab.

2. Click Set Document Values.

3. Under Base for Log Function, check Change from e to 10 and choose OK.

All logarithmic functions are converted symbolically to natural logarithms by evaluation as follows.

▶ Evaluate

$$\log_{10} x = \frac{\ln x}{\ln 10} \qquad \log_2 x = \frac{\ln x}{\ln 2} \qquad \log_{\frac{1}{2}} x = -\frac{\ln x}{\ln 2}$$

If you have changed the default for log from base e to base 10, Evaluate still converts an expression to natural logarithms.

▶ Evaluate

$$\log x = \frac{\ln x}{\ln 10}$$

Thus, the behavior of the system for natural logarithms is of particular interest.

Properties of Logarithms
You can demonstrate properties of logarithms with Simplify and Combine.

▶ **Simplify**

$$\ln x^y = y \ln x$$

▶ **Combine + Logs**

$$\ln x + \ln y = \ln xy$$

$$\ln a - \ln b = \ln \frac{a}{b}$$

Evaluating Logarithms

Use Evaluate or Evaluate Numerically for numerical approximations. Note that Evaluate converts to natural logarithms symbolically, and use of floating-point notation triggers a numerical evaluation. Change the Computation Setup and Engine Setup dialogs on the Tools menu (or the Document Computation Settings dialog under Compute + Settings) for the degree of accuracy you need. In these examples, Digits Shown in Results is set to 5.

Expression	Evaluate	Evaluate Numerically
$\ln 2$	$\ln 2$	0.69315
$\log_{10} 5$	$\frac{\ln 5}{\ln 10}$	0.69897
$\ln 0.0025$	-5.9915	-5.9915
$\ln 1.0025$	2.4969×10^{-3}	2.4969×10^{-3}

Solving Exponential and Logarithmic Equations

For symbolic solutions to exponential or logarithmic equations, choose Solve + Exact. Enter Variable(s) to Solve for if requested.

For numerical solutions, you can either enter a coefficient in decimal notation and choose Solve + Exact or, apply Evaluate Numerically to the symbolic solutions. In the case of a single variable, you can choose Solve + Numeric.

▶ **Solve + Exact**

$$3^x = 8, \text{ Solution is: } \left\{ \frac{1}{\ln 3} \left(2i X_4 \pi + \ln 8 \right) \mid X_4 \in \mathbb{Z} \right\}$$

$$e^x = \frac{y+1}{y-1}, \text{ (Solve for } x)$$

$$\text{Solution is: } \begin{cases} \emptyset & \text{if } y = -1 \\ \left\{ 2i X_9 \pi + \left(\ln \frac{1}{y-1} (y+1) \right) \mid X_9 \in \mathbb{Z} \right\} & \text{if } y \neq -1 \end{cases}$$

For simplified results, you may want to use the options **Principal Value Only** and **Ignore Special Cases**.

▶ **To obtain solutions in the following form,**

1. Choose **Tools + Engine Setup** and select the **General** page.

2. Check both **Principal Value Only** and **Ignore Special Cases**.

▶ **Solve + Exact**

$3^x = 8$, Solution is: $\dfrac{1}{\ln 3} \ln 8$

$e^x = \dfrac{y+1}{y-1}$, (Solve for x), Solution is: $\ln \dfrac{1}{y-1}(y+1)$

$P = Qe^{kt}$, (Solve for k), Solution is: $\dfrac{1}{t} \ln \dfrac{P}{Q}$

For numerical solutions, you can either enter a coefficient in decimal notation and choose **Solve + Exact** or, apply **Evaluate Numerically** to the symbolic solutions. For a particular solution, you can choose **Solve + Numeric**.

▶ **Solve + Exact**

$3^x = 8.0$, Solution is: 1.8928

$\log(3x + y) = 8.0$, Solution is: $993.65 - 0.33333y$

▶ **Solve + Numeric**

$3^x = 8$, Solution is: $\{[x = 1.8928]\}$

$\log(3x + y) = 8$, Solution is: $\{[x = 891.78, y = 305.6]\}$

Exercises

1. Given that when $x^2 - 3x + 5k$ is divided by $x + 4$ the remainder is 9, find the value of k using **Divide** on the **Polynomials** submenu and **Solve + Exact**.

2. Define functions $f(x) = x^3 + x \ln x$ and $g(x) = x + e^x$. Evaluate $f(g(x))$, $g(f(x))$, $f(x)g(x)$, and $f(x) + g(x)$.

3. Find the equation of the line passing through the two points (x_1, y_1), (x_2, y_2).

4. Find the equation of the line passing through the two points $(2, 5)$, $(3, -7)$.

5. Find the equation of the line passing through the two points $(1, 2)$, $(2, 4)$.

6. Find the slope of the line determined by the equation $sx + ty = c$.

7. Find the center and semi-axes of the ellipse $16\, x^2 + 4y^2 + 96x - 8y + 84 = 0$.

8. Factor the difference of powers $x^n - y^n$ for several values of n, and deduce a general formula.

9. Applying **Factor** to $x^2 + \left(\sqrt{5} - 3\right) x - 3\sqrt{5}$ gives the factorization
$$x^2 + \left(\sqrt{5} - 3\right) x - 3\sqrt{5} = \left(x + \sqrt{5}\right)(x - 3)$$
showing that the system can factor some polynomials with irrational roots. However, applying **Factor** to $x^2 - 3$ and $x^3 + 3x^2 - 5x + 1$ does not do anything. Find a way to factor these polynomials.

10. If principal P is invested at an interest rate r compounded annually, in t years it grows to an amount A given by $A = P\left(1 + r\right)^t$. Find the number of years required for the amount to double at interest rates of 3%, 5%, 8%, and 10%.

   - Analytic solution: Solving the appropriate equation for t, define r to be each of the different interest rates, and evaluate the expression you obtained for t.
   - Experimental solution: Treat A as a function of r and t, define different values for r and t, and make an appropriate table of values for each interest rate r.

11. Find all real and complex solutions to the system of equations
$$\begin{aligned} 2x^2 - y &= 1 \\ x + 3y^3 &= 4 \end{aligned}$$

12. The curves determined by quadratic equations of the form $Ax^2 + Bxy + Cy^2 + Dx + Ey + F = 0$ are often referred to as *conic sections*, as each can be described as the intersection of a plane with a cone. The *discriminant* of the expression $Ax^2 + Bxy + Cy^2 + Dx + Ey + F$ is the number
$$\triangle = B^2 - 4AC$$
The type of curve determined by the quadratic equation depends on the *sign* of the discriminant. In nondegenerate cases, the quadratic equation gives an ellipse if $\triangle < 0$, a parabola if $\triangle = 0$, and a hyperbola if $\triangle > 0$, where a circle is considered as a special case of an ellipse. The degenerate cases give a pair of lines, a line, or a point, or they have imaginary solutions. A rotation of one or both axes in the plane leads to a change of variables such as the change $u = ax + by$, $v = cx + dy$ from (x, y) to (u, v). Show that the sign of the discriminant (and thus the basic shape of the curve) is invariant under such changes for a quadratic equation $Ax^2 + Bxy + Cy^2 + Dx + Ey + F = 0$ if $ad \neq bc$.

Solutions

1. Using **Polynomials + Divide**,

$$\frac{x^2 - 3x + 5k}{x + 4} = x + \frac{1}{x + 4}(5k + 28) - 7$$

Thus, the remainder is $5k + 28$. Applying **Solve + Exact** to $5k + 28 = 9$ gives the solution $k = -\frac{19}{5}$.

2. Defining functions $f(x) = x^3 + x \ln x$ and $g(x) = x + e^x$ and evaluating gives

$$
\begin{aligned}
f(g(x)) &= (x + e^x) \ln (x + e^x) + (x + e^x)^3 \\
g(f(x)) &= x \ln x + x^3 + e^{x \ln x + x^3} \\
f(x)g(x) &= (x + e^x)(x \ln x + x^3) \\
f(x) + g(x) &= x + e^x + x \ln x + x^3
\end{aligned}
$$

3. For any two distinct points (x_1, y_1) and (x_2, y_2) in the plane, there is a unique line $ax + by + c = 0$ through these two points. Substituting these points in the equation for the line gives the two equations $ax_1 + by_1 + c = 0$ and $ax_2 + by_2 + c = 0$. Apply **Solve + Exact** to this system

$$
\begin{aligned}
ax_1 + by_1 + c &= 0 \\
ax_2 + by_2 + c &= 0
\end{aligned}
$$

of linear equations, solving for the variables a, b.

$$\text{Solution is: } \left[a = \frac{cy_1 - cy_2}{x_1 y_2 - x_2 y_1}, b = \frac{cx_1 - cx_2}{x_2 y_1 - x_1 y_2} \right]$$

Consequently, the equation for the line is

$$c\left(\frac{y_1 - y_2}{x_1 y_2 - x_2 y_1}\right) x - c\left(\frac{x_1 - x_2}{x_2 y_1 - x_1 y_2}\right) y + c = 0$$

or, clearing fractions and collecting coefficients by factoring in place,

$$(y_1 - y_2) x - (x_1 - x_2) y + (x_1 y_2 - y_1 x_2) = 0$$

4. For the points $(2, 5), (3, -7)$, the system of equations is

$$
\begin{aligned}
2a + 5b + c &= 0 \\
3a - 7b + c &= 0
\end{aligned}
$$

Apply **Solve + Exact** to obtain

$$\text{Solution is: } \left[a = -\frac{12}{29}c, b = -\frac{1}{29}c \right]$$

Consequently, the equation for the line is

$$-\frac{12}{29}cx - \left(-\frac{1}{29}\right) cy + c = 0$$

or, clearing fractions and simplifying,

$$-12x + y + 29 = 0$$

5. Since the point $(0, 0)$ lies on the line, you do not get a unique solution to the system of equations for the pair a, b. Thus, choosing **Solve + Exact** and specifying a, b for the variables gives no response. However, specifying a, c for **Variable(s) to Solve**

for gives the solution
$$[a = -2b, c = 0]$$
Thus, the equation for the line is
$$-2bx + by = 0$$
or, dividing by b and applying **Simplify**,
$$(-2bx + by)\frac{1}{b} = -2x + y = 0$$
Note: An interesting method for finding the equation of a line through two specified points using determinants is described in a *Matrix Algebra* exercise on page 350.

6. The *slope-intercept* form of the equation for a line is $y = mx + b$, where m is the slope and b the y-intercept. If a line is given as a linear equation in the form $sx + ty = c$, you can find the slope by solving the equation for y. Apply **Expand** to the solution $y = -\frac{sx-c}{t}$ to get $y = \frac{c}{t} - \frac{s}{t}x$, revealing the slope to be $-\frac{s}{t}$.

7. To find the center and semi-axes of the ellipse $16x^2 + 4y^2 + 96x - 8y + 84 = 0$,

   a. Subtract 84 from both sides of the equation to get
   $$\left(16\,x^2 + 4y^2 + 96x - 8y + 84\right) - 84 = 0 - 84$$
   b. Select the left side and while holding down the CTRL key, apply **Simplify**; then, do the same to the right side to obtain
   $$96x - 8y + 16x^2 + 4y^2 = -84$$
   c. Drag the terms containing x together, select them and click $\boxed{(\square)}$; then, do the same to the terms containing y, to get
   $$\left(16x^2 + 96x\right) + \left(4y^2 - 8y\right) = -84$$
   d. Factor out the leading coefficients by dragging them outside the parentheses and dividing other coefficients by their value, to get
   $$16\left(x^2 + \frac{96}{16}x\right) + 4\left(y^2 - \frac{8}{4}y\right) = -84$$
   e. Add the product of the coefficient of x^2 with the square of one-half the coefficient of x to both sides; then, do the same for y, to get
   $$16\left(x^2 + \tfrac{96}{16}x + \left(\tfrac{1}{2}\tfrac{96}{16}\right)^2\right) + 4\left(y^2 - \tfrac{8}{4}y + \left(\tfrac{1}{2}\tfrac{8}{4}\right)^2\right)$$
   $$= -84 + 16\left(\tfrac{1}{2}\tfrac{96}{16}\right)^2 + 4\left(\tfrac{1}{2}\tfrac{8}{4}\right)^2$$
   f. Select the term $16\left(x^2 + \tfrac{96}{16}x + \left(\tfrac{1}{2}\tfrac{96}{16}\right)^2\right)$ and while holding down the CTRL key, apply **Factor**; then, do the same for the term with y, to get
   $$16\left(x + 3\right)^2 + 4\left(y - 1\right)^2 = -84 + 16\left(\frac{1}{2}\frac{96}{16}\right)^2 + 4\left(\frac{1}{2}\frac{8}{4}\right)$$
   g. Select the right side and while holding down the CTRL key, apply **Simplify**, to get
   $$16\left(x + 3\right)^2 + 4\left(y - 1\right)^2 = 64$$

h. Divide each term by the right-hand side, to get

$$\frac{16\,(x+3)^2}{64} + \frac{4\,(y-1)^2}{64} = \frac{64}{64}$$

i. Select each term and while holding down the CTRL key, apply **Factor**, to get

$$\frac{1}{4}\,(x+3)^2 + \frac{1}{16}\,(y-1)^2 = 1$$

You can read the answer from this form of the equation: The center of the ellipse is $(-3,1)$, and the semi-axes are $\sqrt{4} = 2$ and $\sqrt{16} = 4$.

8. Apply **Factor** to several differences.

$$x^2 - y^2 = (x-y)(x+y)$$
$$x^3 - y^3 = (x-y)\left(x^2 + xy + y^2\right)$$
$$x^4 - y^4 = (x-y)(x+y)\left(x^2 + y^2\right)$$
$$x^5 - y^5 = (x-y)\left(x^4 + x^3 y + x^2 y^2 + xy^3 + y^4\right)$$
$$x^6 - y^6 = (x-y)(x+y)\left(x^2 + xy + y^2\right)\left(x^2 - xy + y^2\right)$$
$$x^7 - y^7 = (x-y)\left(x^6 + x^5 y + x^4 y^2 + x^3 y^3 + x^2 y^4 + xy^5 + y^6\right)$$

After looking at only these few examples, you might find it reasonable to conjecture that, for n odd,

$$x^n - y^n = (x-y)\sum_{k=0}^{n-1} x^{n-k-1} y^k$$

We leave the general conjecture for you. Experiment.

9. Using the clue from the example that the system will factor over roots that appear as coefficients, apply **Factor** to the product $\sqrt{3}\left(x^2 - 3\right)$ to get $\sqrt{3}\left(x^2 - 3\right) = \sqrt{3}\left(x - \sqrt{3}\right)\left(x + \sqrt{3}\right)$. Now you can divide out the extraneous $\sqrt{3}$ to get

$$x^2 - 3 = \left(x - \sqrt{3}\right)\left(x + \sqrt{3}\right)$$

For the polynomial $x^3 + 3x^2 - 5x + 1$, apply **Polynomials + Roots** to find the roots: $\left[1, -2 + \sqrt{5}, -2 - \sqrt{5}\right]$. You can multiply by $\sqrt{5}$ to factor this polynomial: $\sqrt{5}\left(x^3 + 3x^2 - 5x + 1\right) = \left(\sqrt{5}\right)(x-1)\left(x - \sqrt{5} + 2\right)\left(x + \sqrt{5} + 2\right)$. Then, dividing out the extraneous factor of $\sqrt{5}$, you have

$$x^3 + 3x^2 - 5x + 1 = (x-1)\left(x + 2 + \sqrt{5}\right)\left(x + 2 - \sqrt{5}\right)$$

10. **Analytic solution:** The question asks when $A = 2P$. Solving the equation $2P = P\,(1+r)^t$ for t gives the solution $t = \frac{\ln 2}{\ln(1+r)}$. **Evaluate Numerically** gives the number of years for each of the four different interest rates as follows:

For 3%, $\dfrac{\ln 2}{\ln\,(1 + 0.03)} = 23.4$ years;	For 5%, $\dfrac{\ln 2}{\ln\,(1 + 0.05)} = 14.2$ years
For 8%, $\dfrac{\ln 2}{\ln\,(1 + 0.08)} = 9.01$ years;	For 10%, $\dfrac{\ln 2}{\ln\,(1 + 0.10)} = 7.27$ years

Experimental solution: For the interest rate 3%, you might start with the following data:

$$10 \text{ years:} \quad A = P\left(1 + 0.03\right)^{10} = 1.34P$$
$$15 \text{ years:} \quad A = P\left(1 + 0.03\right)^{15} = 1.56P$$
$$20 \text{ years:} \quad A = P\left(1 + 0.03\right)^{20} = 1.81P$$
$$25 \text{ years:} \quad A = P\left(1 + 0.03\right)^{25} = 2.09P$$

Then, look at the years between 20 and 25:

$$21 \text{ years:} \quad A = P\left(1 + 0.03\right)^{21} = 1.86P$$
$$22 \text{ years:} \quad A = P\left(1 + 0.03\right)^{22} = 1.92P$$
$$23 \text{ years:} \quad A = P\left(1 + 0.03\right)^{23} = 1.97P$$
$$24 \text{ years:} \quad A = P\left(1 + 0.03\right)^{24} = 2.03P$$

These results indicate that it takes between 23 and 24 years to double an investment at 3% annual interest. Take similar steps for the other interest rates.

11. With the insertion point in the array
$$2x^2 - y = 1$$
$$x + 3y^3 = 4$$
choose **Solve + Exact**. You receive the response

Solution is : $\left[y = 1, x = 1\right], \left[x = 4 - 3y^3, y = \rho_1\right]$
where ρ_1 is a root of $18X_{39}^3 - 30X_{39}^2 - 30X_{39} + 18X_{39}^4 + 18X_{39}^5 - 31, X_{39}$
Leave your insertion point in the polynomial
$$18X_{39}^3 - 30X_{39}^2 - 30X_{39} + 18X_{39}^4 + 18X_{39}^5 - 31$$
and choose **Polynomials + Roots**. You receive the following solution.

$$\text{roots:} \quad \begin{array}{c} -0.60663 - 0.98268i \\ -0.60663 + 0.98268i \\ -0.48801 - 0.92069i \\ -0.48801 + 0.92069i \\ 1.1893 \end{array}$$

Define the function $x(t) = -3t^3 + 4$ with **Definitions + New Definition**. Select the vector of roots, and click the parentheses button. Type an x at the left of the vector, leave the insertion point in the expression, and apply **Evaluate** for the following:

$$x \begin{pmatrix} -0.60663 - 0.98268i \\ -0.60663 + 0.98268i \\ -0.48801 - 0.92069i \\ -0.48801 + 0.92069i \\ 1.1893 \end{pmatrix} = \begin{pmatrix} -3\left(-0.60663 - 0.98268i\right)^3 + 4 \\ -3\left(-0.60663 + 0.98268i\right)^3 + 4 \\ -3\left(-0.48801 - 0.92069i\right)^3 + 4 \\ -3\left(-0.48801 + 0.92069i\right)^3 + 4 \\ -1.0466 \end{pmatrix}$$

Leave the insertion point in the matrix on the right and click **Expand**.

$$\begin{pmatrix} -3\left(-0.60663 - 0.98268i\right)^3 + 4 \\ -3\left(-0.60663 + 0.98268i\right)^3 + 4 \\ -3\left(-0.48801 - 0.92069i\right)^3 + 4 \\ -3\left(-0.48801 + 0.92069i\right)^3 + 4 \\ -1.0466 \end{pmatrix} = \begin{pmatrix} -0.60247 + 0.40783i \\ -0.60247 - 0.40783i \\ 0.62562 - 0.36793i \\ 0.62562 + 0.36793i \\ -1.0466 \end{pmatrix}$$

To display this result, you can concatenate the two vectors: Place them side by side and from the **Matrices** submenu, choose **Concatenate**. Then, select the (two-column) matrix, choose **Edit + Insert Row(s)** to add two new rows at the top. Label the columns with x and y and, in the other new row, add the solution $x = 1, y = 1$.

$\mathbf{x = -3y^3+4}$	$\mathbf{y}$
1	1
$-0.60247 + 0.40783i$	$-0.60663 - 0.98268i$
$-0.60247 - 0.40783i$	$-0.60663 + 0.98268i$
$0.62562 - 0.36793i$	$-0.48801 - 0.92069i$
$0.62562 + 0.36793i$	$-0.48801 + 0.92069i$
-1.0466	1.1893

12. First, solve the system of equations for x and y with **Solve + Exact** to get the solution
$$x = -\frac{bv - du}{ad - bc} \qquad y = \frac{av - cu}{ad - bc}$$
(Note that the condition $ad \neq bc$ is necessary for this system of equations to have a solution.) Now find the quadratic equation in u and v by replacing x and y with their equivalent expressions in terms of u and v. These replacements produce the following polynomial:
$$A\left(-\frac{vb - du}{-cb + ad}\right)^2 + B\left(-\frac{vb - du}{-cb + ad}\right)\left(\frac{-cu + av}{-cb + ad}\right) + C\left(\frac{-cu + av}{-cb + ad}\right)^2$$
$$+ D\left(-\frac{vb - du}{-cb + ad}\right) + E\left(\frac{-cu + av}{-cb + ad}\right) + F = 0$$
Simplify this equation and put it into standard quadratic form in terms of the variables u and v:
$$\left(d^2 A - Bcd + c^2 C\right) u^2 + (-2bdA + aBd + bBc - 2acC)\,uv$$
$$+ \left(-abB + Ab^2 + a^2 C\right) v^2 + (ad - cb)(Dd - cE)\,u$$
$$+ (ad - cb)(Ea - Db)\,v + F(-cb + ad)^2 = 0$$
To find the discriminant, replace $\tilde{A}, \tilde{B}, \tilde{C}$ in the expression $\tilde{\triangle} = \tilde{B}^2 - 4\tilde{A}\tilde{C}$ with the coefficients:
$$\tilde{A} = d^2 A - Bcd + c^2 C$$
$$\tilde{B} = -2bdA + aBd + bBc - 2acC$$
$$\tilde{C} = -abB + Ab^2 + a^2 C$$
This can be simplified by the following steps:
$$\tilde{\triangle} = \tilde{B}^2 - 4\tilde{A}\tilde{C} = (-2bdA + aBd + bBc - 2acC)^2$$
$$- 4\left(d^2 A - Bcd + c^2 C\right)\left(-abB + Ab^2 + a^2 C\right)$$
$$= a^2 B^2 d^2 - 2aB^2 dbc + b^2 B^2 c^2 + 8bdAacC - 4d^2 Aa^2 C - 4c^2 CAb^2$$
$$= B^2 (ad - cb)^2 - 4AC (ad - cb)^2$$
$$= \left(B^2 - 4AC\right)(ad - cb)^2$$
$$= \triangle (ad - cb)^2$$
It is easy to see from the equation
$$\tilde{\triangle} = \triangle (ad - cb)^2$$
that $\triangle$ and $\tilde{\triangle}$ have the same sign when $ad \neq bc$.

4 Trigonometry

Trigonometry developed from the study of triangles, particularly right triangles, and the relations between the lengths of their sides and the sizes of their angles. The trigonometric functions that measure the relationships between the sides of similar triangles have far-reaching applications that extend beyond their use in the study of triangles.

Trigonometric Functions

Most of the trigonometric computations in this chapter use six basic trigonometric functions. The two fundamental trigonometric functions, sine and cosine, can be defined in terms of the unit circle—the set of points in the Euclidean plane of distance 1 from the origin. A point on this circle has coordinates $(\cos t, \sin t)$, where t is a measure (in radians) of the angle at the origin between the positive x-axis and the ray from the origin through the point measured in the counterclockwise direction. The other four basic trigonometric functions can be defined in terms of these two—namely,

$$\tan x = \frac{\sin x}{\cos x} \qquad \sec x = \frac{1}{\cos x}$$

$$\cot x = \frac{\cos x}{\sin x} \qquad \csc x = \frac{1}{\sin x}$$

For $0 < t < \frac{\pi}{2}$, these functions can be found as a ratio of certain sides of a right triangle that has one angle of radian measure t.

The symbols used for the six basic trigonometric functions—sin, cos, tan, cot, sec, csc—are abbreviations for the words sine, cosine, tangent, cotangent, secant, and cosecant, respectively. You can enter these trigonometric functions and many other functions either from the keyboard in mathematics mode or from the dialog box that opens when you choose Insert + Math Name or click [sin cos]. When you enter one of these functions from the keyboard in mathematics mode, the function name automatically turns gray when you type the final letter of the name.

Note   Ordinary functions require parentheses around the function argument, while trigonometric functions commonly do not. The default behavior of your system allows trigonometric functions without parentheses. For further information on understanding or changing this behavior, see page 141.

To find values of the trigonometric functions of real variables, use Evaluate or Evaluate Numerically.

▶ Evaluate

$$\sin \tfrac{3\pi}{4} = \tfrac{1}{2}\sqrt{2} \qquad \sin 60° = \tfrac{1}{2}\sqrt{3}$$

▶ Evaluate Numerically

$$\sin \tfrac{3\pi}{4} = 0.70711 \qquad \sin 60° = 0.86603$$

Your choice for **Digits Shown in Results** in the **General** page of the **Tools + Computation Setup** (or **Compute + Settings**) dialog determines the number of places displayed in the response to **Evaluate Numerically** (see page 30).

Radians and Degrees

The notation you use determines whether the argument of a trigonometric function is interpreted as radians or degrees.

▶ Evaluate Numerically

$$\sin 30 = -0.98803 \qquad \sin 30° = 0.5$$

The degree symbol is available in two forms—a green **Unit Name** or a small red circle entered as a superscript. With no degree symbol, the function argument is interpreted as radians, and with either a green or red degree symbol, the function argument is interpreted as degrees.

- To enter the small green circle and other units of angle measure, click the **Unit Name** button ⏢ on the **Math Templates** toolbar or choose **Insert + Unit Name**, and choose **Plane Angle**. Select the desired **Unit Name** and choose **Insert** or **Replace**. See page 48 for a list of unit names related to plane angles and keyboard shortcuts for entering them.

- The small red circle appears on the **Symbol Cache** toolbar and on the **Binary Operations** symbol panel, and must be entered as a superscript. The red symbol for minutes is the apostrophe or "prime" entered in mathematics from the keyboard. For seconds, enter this symbol twice.

Evaluating an expression entered with red degree, minute, and second symbols produces symbolic results and all operations convert angle measure to radians:

$$33°16' = \frac{499}{2700}\pi \qquad \sin 45° = \frac{1}{2}\sqrt{2}$$

Evaluating an expression entered with the green **Unit Names** gives numerical results:

$$33°16' = 0.580\,61 \text{ rad}$$

See page 93 for a discussion of conversion between radians and degrees.

Solving Trigonometric Equations

You can use both Exact and Numeric from the Solve submenu to find solutions to trigonometric equations. These operations also convert degrees to radians. Use of decimal notation in the equation gives you a numerical solution, even with Exact.

- With radians or with red degree symbols, Solve + Exact gives symbolic solutions and Solve + Numeric gives numerical solutions:

Equation	Solve + Exact	Solve + Numeric
$x = \sin \frac{\pi}{4}$	$x = \frac{1}{2}\sqrt{2}$	$x = 0.70711$
$x = \sin 4.7$	$x = -0.999\,92$	$x = -0.999\,92$
$x = 3°54'$	$x = \frac{13}{600}\pi$	$x = 6.8068 \times 10^{-2}$

- With green Unit Name symbols, Solve + Exact gives numerical solutions.

Equation	Solve + Exact
$\sin 22° = \frac{14}{c}$	$c = 37.373$
$x = 3°54'$	$x = 6.8068 \times 10^{-2}\,\text{rad}$

The two methods may give different answers when there are multiple solutions. The Numeric command from the Solve submenu offers the advantage that you can specify the range in which you wish the solution to lie. Enter the equation and the range in different rows of a display or a one-column matrix.

▶ Solve + Numeric

$$x = 10 \sin x$$
$$x \in (5, 7.5)$$, Solution is: $\{x = 7.0682\}$

The interval $(5, 7.5)$ was specified for the solution in the preceding example. By specifying other intervals, you can find all seven solutions: $\{x = 0\}$, $\{x = \pm 2.8523\}$, $\{x = \pm 7.0682\}$, $\{x = \pm 8.4232\}$, as depicted in the following graph. The Solve + Exact command for solving equations gives only the solution $x = 0$ for this equation.

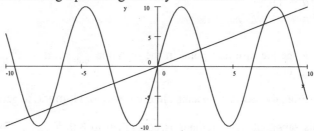

When any operation is applied to an angle represented in degrees by a mathematics superscript, such as $48°$, degrees are automatically converted to radians. To go in the other direction, replace 2π radians with $360°$ and convert other angles proportionately. You can also solve directly for the number of degrees. Both methods follow.

▶ **To convert radians to degrees (using ratios)**

1. Write the equation $\dfrac{\theta}{360} = \dfrac{x}{2\pi}$, where x represents radians.

2. Leave the insertion point in this equation.

3. From the Solve submenu, choose Exact or Numeric.

4. Name θ as the Variable to Solve for.

5. Choose OK to get $\theta = \dfrac{180}{\pi}x$.

▶ **To convert radians to degrees (directly)**

1. Write an equation such as $2 = \theta°$ or, using Insert + Unit Name, $2\,\mathrm{rad} = \theta°$.

2. Leave the insertion point in this equation.

3. From the Solve submenu, choose Exact or Numeric, to get $\theta = \frac{360}{\pi}$ or $\theta = 114.59$.

Example   Convert $x = \frac{13}{600}\pi$ to degrees as follows.

1. Place the insertion point in the equation $\dfrac{\theta}{360} = \dfrac{\left(\frac{13}{600}\pi\right)}{2\pi}$.

2. From the Solve submenu, choose Exact to get $\theta = \frac{39}{10}$ degrees, or choose Numeric to get $\theta = 3.9$ degrees.

   or

1. Place the insertion point in the equation $\frac{13}{600}\pi = \theta°$. (The degree symbol is red.)

2. From the Solve submenu, choose Exact to get $\theta = \frac{39}{10}$, or choose Numeric to get $\theta = 3.9$.

   or

1. Place the insertion point in the equation $\frac{13}{600}\pi\,\mathrm{rad} = \theta°$. (The degree symbol is green.)

2. From the Solve submenu, choose Exact or Numeric to get $\theta = 3.9$.

This says that $\frac{13}{600}\pi$ radians is equivalent to 3.9 degrees. To change 0.9 degrees to minutes,

- Apply Evaluate to 0.9×60 to get 54.0.

   or

- Apply Solve to $0.9° = x'$ to get $x = 54.0$.

This gets $\frac{13}{600}\pi$ rad $= 3°54'$. To check this result, apply **Evaluate** to $3°54'$ to get

$$3°54' = \frac{13}{600}\pi \qquad \text{or} \qquad 3°54' = 6.8068 \times 10^{-2}\,\text{rad}$$

See page 50 for additional examples of converting units.

Trigonometric Identities

This section illustrates the effects of some operations on trigonometric functions. First, simplifications and expansions of various trigonometric expressions illustrate many of the familiar trigonometric identities.

Definitions in Terms of Basic Trigonometric Functions

Apply **Simplify** to the secant and cosecant to find their definition in terms of the sine and cosine functions.

▶ Simplify

$$\sec x = \frac{1}{\cos x} \qquad \csc x = \frac{1}{\sin x}$$

Apply **Rewrite + Sin and Cos** to a trigonometric expression to rewrite it in terms of the sine and cosine functions.

▶ Rewrite + Sin and Cos

$$\tan x = \frac{1}{\cos x}\sin x \qquad \cot x = \frac{\cos x}{\sin x}$$

$$\sec x = \frac{1}{\cos x} \qquad \csc x = \frac{1}{\sin x}$$

Apply **Rewrite + Tan** to a trigonometric expression to rewrite it in terms of the tangent function.

▶ Rewrite + Tan

$$\cos x \sin x - 2\sec x \csc x = 2\frac{\tan\frac{1}{2}x}{\left(\tan^2\frac{1}{2}x+1\right)^2}\left(1 - \tan^2\frac{1}{2}x\right) - \frac{1}{\tan\frac{1}{2}x}\frac{\left(\tan^2\frac{1}{2}x+1\right)^2}{1-\tan^2\frac{1}{2}x}$$

Pythagorean Identities

▶ Simplify

$$\sin^2 x + \cos^2 x = 1$$

You can find identities by using **Rewrite + Sin and Cos** followed by basic techniques for simplifying such expressions.

▶ Rewrite + Sin and Cos

$$\tan^2 x - \sec^2 x = \frac{1}{\cos^2 x} \sin^2 x - \frac{1}{\cos^2 x}$$

$$\cot^2 x - \csc^2 x = \frac{\cos^2 x}{\sin^2 x} - \frac{1}{\sin^2 x}$$

$$\frac{\tan x + \tan y}{1 - \tan x \tan y} = \frac{\frac{1}{\cos x} \sin x + \frac{1}{\cos y} \sin y}{1 - \frac{1}{\cos x \cos y} \sin x \sin y}$$

Addition Formulas

▶ Expand

$$\sin(x + y) = \cos x \sin y + \cos y \sin x \qquad \cos(x + y) = \cos x \cos y - \sin x \sin y$$

Combine + Trig Functions reverses this operation in some cases.

▶ Combine + Trig Functions

$$\sin x \cos y + \cos x \sin y = \sin(x + y) \qquad \cos x \cos y - \sin x \sin y = \cos(x + y)$$

Multiple-Angle Formulas

There are a multitude of possible forms for these expressions. To get expressions in the form you want, you can experiment with different combinations of commands.

▶ Expand

$$\sin 2\theta = 2 \sin \theta \cos \theta \qquad \tan 2\theta = 2 (\sin \theta) \frac{\cos \theta}{2 \cos^2 \theta - 1}$$

$$\cos 2\theta = \cos^2 \theta - \sin^2 \theta$$

You can uncover other multiple-angle formulas with Expand. Following are some examples.

▶ Expand

$$\sin 6\theta = 6 \cos \theta \sin^5 \theta + 6 \cos^5 \theta \sin \theta - 20 \cos^3 \theta \sin^3 \theta$$

$$\sin(2a + 3b) = 2 \cos a \cos^3 b \sin a - 6 \cos a \cos b \sin a \sin^2 b - \cos^2 a \sin^3 b$$
$$+ \sin^2 a \sin^3 b + 3 \cos^2 a \cos^2 b \sin b - 3 \cos^2 b \sin^2 a \sin b$$

Combining and Simplifying Trigonometric Expressions

Products and powers of trigonometric functions and hyperbolic functions are combined into a sum of trigonometric functions or hyperbolic functions whose arguments are integral linear combinations of the original arguments.

▶ Combine + Trig Functions

$$\sin x \sin y = \tfrac{1}{2} \cos (x - y) - \tfrac{1}{2} \cos (x + y) \qquad \sin^2 x = \tfrac{1}{2} - \tfrac{1}{2} \cos 2x$$

$$\sin x \cos y = \tfrac{1}{2} \sin (x + y) + \tfrac{1}{2} \sin (x - y) \qquad \cos^2 x = \tfrac{1}{2} \cos 2x + \tfrac{1}{2}$$

$$\sin^5 x \cos^5 x = \tfrac{1}{512} \sin 10x + \tfrac{5}{256} \sin 2x - \tfrac{5}{512} \sin 6x$$

The command Simplify combines and simplifies trigonometric expressions, as in the following examples.

▶ Simplify

$$\cos^2 x + \tfrac{1}{4} \sin^2 2x - \sin^2 x \cos^2 x + 2 \sin^2 x = \tfrac{3}{2} - \tfrac{1}{2} \cos 2x$$

$$\sin 3a + 4 \sin^3 a = 3 \sin a$$

You may need to apply repeated operations to get the result you want. The order in which you apply the operations is not necessarily critical.

▶ Simplify, Expand, Factor

$$(\sec t)(1 + \cos 2t) = \tfrac{1}{\cos t}(\cos 2t + 1) = \cos t + \tfrac{1}{\cos t} - \tfrac{1}{\cos t} \sin^2 t$$
$$= (\cos t)^{-1}(\cos^2 t - \sin^2 t + 1)$$

Now recognize that $1 - \sin^2 t = \cos^2 t$, make the replacement, and simplify again:

▶ Simplify

$$= (\cos t)^{-1}(\cos^2 t + \cos^2 t) = 2 \cos t$$

Solution of Triangles

To *solve* a triangle means to determine the lengths of the three sides and the measures (in degrees or radians) of the three angles.

Solving a Right Triangle

You can solve a right triangle with sides a, b, c and opposite angles α, β, γ, respectively, if you know the value of one side and one acute angle, or the value of any two sides.

Example   To solve the right triangle with one side of length $c = 2$ and one angle $\alpha = \frac{\pi}{9}$,

1. Choose New Definition from the Definitions menu for each of the given values $\alpha = \frac{\pi}{9}$ and $c = 2$.

2. Apply Evaluate to $\beta = \frac{\pi}{2} - \alpha$ to get $\beta = \frac{7}{18}\pi$.

3. Apply Evaluate (or Evaluate Numerically) to $a = c \sin \alpha$ to get $a = 2 \sin \frac{1}{9}\pi$ ($= 0.68404$).

4. Apply Evaluate to $b = c \cos \alpha$ to get $b = 2 \cos \frac{1}{9}\pi$ ($= 1.8794$).

Note   To obtain the solutions in the simple form shown in the following examples, choose Tools + Engine Setup. On the General page, under Solve Options, check Principal Value Only.

Example   To solve a right triangle given two sides, say $a = 19$ and $c = 23$,

1. Apply Definitions + New Definition to each of the given values, $a = 19$ and $c = 23$.

2. Place the insertion point in the equation $a^2 + b^2 = c^2$ and, from the Solve submenu, choose Exact (Numeric) to get $b = 2\sqrt{42}$ ($= 12.96$).

3. Place the insertion point in each of the equations $\sin \alpha = \dfrac{a}{c}$, $\cos \beta = \dfrac{a}{c}$ in turn, and choose Solve + Exact to get

$$\alpha = \arcsin \frac{19}{23}, \beta = \arccos \frac{19}{23}$$

or

Place the insertion point in each of the one-column matrices $\begin{bmatrix} \sin \alpha = a/c \\ \alpha \in (0, \pi/2) \end{bmatrix}$ and $\begin{bmatrix} \cos \beta = a/c \\ \beta \in (0, \pi/2) \end{bmatrix}$ in turn, and choose Solve + Numeric to get

$$\alpha = 0.9721, \ \beta = 0.5987$$

Solving General Triangles

The *law of sines*

$$\frac{a}{\sin \alpha} = \frac{b}{\sin \beta} = \frac{c}{\sin \gamma}$$

enables you to solve a triangle if you are given one side and two angles, or if you are given two sides and an angle opposite one of these sides.

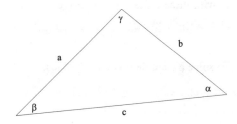

Example To solve a triangle given one side and two angles,

1. Use **New Definition** on the **Definitions** submenu to define $\alpha = \frac{\pi}{9}$, $\beta = \frac{2\pi}{9}$, and $c = 2$

2. Evaluate $\gamma = \pi - \alpha - \beta$ to get $\gamma = \frac{2}{3}\pi$.

3. Use **New Definition** on the **Definitions** submenu to define $\gamma = \frac{2}{3}\pi$.

4. Apply **Solve + Exact** to $\dfrac{a}{\sin \alpha} = \dfrac{c}{\sin \gamma}$ and to $\dfrac{b}{\sin \beta} = \dfrac{c}{\sin \gamma}$ to get

$$a = \frac{4}{3}\sqrt{3}\sin \frac{1}{9}\pi \text{ and } b = \frac{4}{3}\sqrt{3}\sin \frac{2}{9}\pi$$

You can apply **Solve + Numeric** to get numerical solutions, or you can evaluate the preceding solutions numerically.

Using both the law of sines and the *law of cosines,*
$$a^2 + b^2 - 2ab \cos \gamma = c^2$$
you can solve a triangle given two sides and the included angle, or given three sides.

Example To solve a triangle given two sides and the included angle,

1. Define each of $a = 2.34$, $b = 3.57$, and $\gamma = \frac{29}{216}\pi$.

2. Apply **Solve + Exact** to $a^2 + b^2 - 2ab \cos \gamma = c^2$ to get $c = 1.7255$.

3. Define $c = 1.7255$.

4. Apply **Solve + Exact** to both $\dfrac{a}{\sin \alpha} = \dfrac{c}{\sin \gamma}$ and $\dfrac{b}{\sin \beta} = \dfrac{c}{\sin \gamma}$, or apply **Solve +**

   Numeric to each of the matrices $\begin{pmatrix} \dfrac{a}{\sin \alpha} = \dfrac{c}{\sin \gamma} \\ \alpha \in (0, \pi/2) \end{pmatrix}$ and $\begin{pmatrix} \dfrac{b}{\sin \beta} = \dfrac{c}{\sin \gamma} \\ \beta \in (0, \pi/2) \end{pmatrix}$ to

   get $\alpha = 0.58859$ and $\beta = 1.0104$.

A triangle with three sides given is solved similarly: interchange the actions on γ and c in the steps just described.

Example   To solve a triangle given three sides,

1. Define $a = 2.53$, $b = 4.15$, and $c = 6.19$.

2. Apply **Solve + Exact** to $a^2 + b^2 - 2ab\cos\gamma = c^2$ to get $\gamma = 2.3458$.

3. Define $\gamma = 2.3458$.

4. Apply **Solve + Exact** to both $\dfrac{a}{\sin\alpha} = \dfrac{c}{\sin\gamma}$ and $\dfrac{b}{\sin\beta} = \dfrac{c}{\sin\gamma}$, or apply **Solve +**

   Numeric to each of the matrices $\begin{pmatrix} \dfrac{a}{\sin\alpha} = \dfrac{c}{\sin\gamma} \\ \alpha \in (0, \pi/2) \end{pmatrix}$ and $\begin{pmatrix} \dfrac{b}{\sin\beta} = \dfrac{c}{\sin\gamma} \\ \beta \in (0, \pi/2) \end{pmatrix}$ to

   get $\alpha = 0.29632$, and $\beta = 0.49948$.

Inverse Trigonometric Functions and Trigonometric Equations

The following type of question arises frequently when working with the trigonometric functions: for which angles x is $\sin x = y$? There are many correct answers to these questions, since the trigonometric functions are periodic. The inverse trigonometric functions provide answers to such questions that lie within a restricted domain. The inverse sine function, for example, produces the angle x between $-\frac{\pi}{2}$ and $\frac{\pi}{2}$ that satisfies $\sin x = y$. This solution is denoted by $\arcsin x$ or $\sin^{-1} x$.

The inverse trigonometric functions and a number of other functions are available in the dialog box that opens when you click the **Math Name** button on the **Math Objects** toolbar. They can also be entered from the keyboard in mathematics mode.

Example   To find the angle x (between $-\frac{\pi}{2}$ and $\frac{\pi}{2}$) for which $\tan x = 100$,
- Leave the insertion point in the expression $\arctan 100$.
- Apply **Evaluate Numerically**

This gives $\arctan 100 = 1.56079666$

You can also find an angle satisfying $\tan x = 100$ by applying **Solve + Numeric** to the equation. This technique does not necessarily find the solution between $-\frac{\pi}{2}$ and $\frac{\pi}{2}$. It gives a solution $x = 1.56079666 + 2n\pi$ for some integer n. You can specify the interval for the solution in a one-column matrix, as follows.

▶ Solve + Numeric

$$\tan x = 100$$
$$x \in \left(-\frac{\pi}{2}, \frac{\pi}{2}\right) \text{ , Solution is : } \{x = 1.56079666\}$$

Using this technique, you can find numeric solutions to a variety of trigonometric equations in specified intervals.

Solve + Exact produces a general description of solutions to trigonometric equations.

▶ Solve + Exact

$\sin t = \sin 2t$, Solution is:

$$\{X_3\pi \mid X_3 \in \mathbb{Z}\} \cup \{\tfrac{1}{3}\pi + 2X_5\pi \mid X_5 \in \mathbb{Z}\} \cup \{2X_7\pi - \tfrac{1}{3}\pi \mid X_7 \in \mathbb{Z}\}$$

$\tan^2 x - \cot^2 x = 1$, Solution is:

$$\left\{2\pi X_{120} + 2\left(\arctan\sqrt{\sqrt{5}-2}\right) \mid X_{120} \in \mathbb{Z}\right\}$$
$$\cup\left\{2\pi X_{106} - 2\left(\arctan\sqrt{\sqrt{5}+2}\right) \mid X_{106} \in \mathbb{Z}\right\}$$
$$\cup\left\{2\pi X_{108} + 2\left(\arctan\sqrt{\sqrt{5}+2}\right) \mid X_{108} \in \mathbb{Z}\right\}$$
$$\cup\left\{2\pi X_{118} - 2\left(\arctan\sqrt{\sqrt{5}-2}\right) \mid X_{118} \in \mathbb{Z}\right\}$$
$$\cup\left\{2\pi X_{110} - 2\left(\arctan\sqrt{2-\sqrt{5}}\right) \mid X_{110} \in \mathbb{Z}\right\}$$
$$\cup\left\{2\pi X_{112} + 2\left(\arctan\sqrt{2-\sqrt{5}}\right) \mid X_{112} \in \mathbb{Z}\right\}$$
$$\cup\left\{2\pi X_{114} - 2\left(\arctan\sqrt{-\sqrt{5}-2}\right) \mid X_{114} \in \mathbb{Z}\right\}$$
$$\cup\left\{2\pi X_{116} + 2\left(\arctan\sqrt{-\sqrt{5}-2}\right) \mid X_{116} \in \mathbb{Z}\right\}$$

The intersection symbol ∩ is used for AND, the union symbol ∪ for OR. The letter $\mathbb{Z}$ denotes the set $\{\ldots, -2, -1, 0, 1, 2, \ldots\}$ of integers.

This is a good place to experiment with a plot to visualize the complete solution. You can see in the following plot, for example, the pattern of crossings of the graphs of $y = \tan^2 x - \cot^2 x$ and $y = 1$, depicting the solutions of the equation $\tan^2 x - \cot^2 x = 1$.

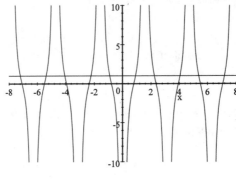

$$y = \tan^2 x - \cot^2 x, \quad y = 1$$

Solutions can be found in a specified range with **Solve + Numeric**, as demonstrated in the following example. Enter the equation and the range in different rows of a one-column matrix.

▶ **Solve + Numeric**

$$8\tan x - 13 + 5\tan^2 x = 3$$
$$x \in \left(-\frac{\pi}{2}, \frac{\pi}{2}\right)$$
, Solution is : $\{x = 0.85916\}$

Hyperbolic Functions

Certain functions, known as the hyperbolic sine, hyperbolic cosine, hyperbolic tangent, hyperbolic cotangent, hyperbolic secant, and hyperbolic cosecant occur as combinations of the exponential functions e^x and e^{-x} having the same relationship to the hyperbola that the trigonometric functions have to the circle. It is for this reason that they are called *hyperbolic* functions. The hyperbolic functions are "trigtype" functions, allowing you to enter arguments without parentheses. See page 141 for an explanation of this behavior.

$$\sinh x = \frac{e^x - e^{-x}}{2} \qquad\qquad \operatorname{csch} x = \frac{1}{\sinh x} = \frac{2}{e^x - e^{-x}}$$
$$\cosh x = \frac{e^x + e^{-x}}{2} \qquad\qquad \operatorname{sech} x = \frac{1}{\cosh x} = \frac{2}{e^x + e^{-x}}$$
$$\tanh x = \frac{\sinh x}{\cosh x} = \frac{e^{2x} - 1}{e^{2x} + 1} \qquad \coth x = \frac{\cosh x}{\sinh x} = \frac{e^{2x} + 1}{e^{2x} - 1}$$

The function names used for the basic hyperbolic functions are sinh, cosh, tanh, coth, sech, and csch. Most of these function names automatically gray when typed in mathematics mode. When they do not, choose **Insert + Math Name**, type the name in the **Name** box, and choose OK.

You can obtain exponential expressions for hyperbolic functions with the **Rewrite** command.

▶ **Rewrite + Exponential**

$$\sinh x = \tfrac{1}{2}e^x - \tfrac{1}{2}e^{-x}$$

$$\cosh x = \tfrac{1}{2}e^x + \tfrac{1}{2}e^{-x}$$

$$\tanh x = \frac{e^{2x}-1}{e^{2x}+1}$$

To find values of the hyperbolic functions, use **Evaluate Numerically.**

▶ **Evaluate Numerically**

$$\sinh \tfrac{3\pi}{4} = 5.2280 \qquad \tanh (1) = 0.761\,59$$

To solve equations involving hyperbolic functions, use **Solve + Exact** or **Solve + Numeric.**

▶ **Solve + Exact**

$$\sinh x + \cosh x = 3, \text{ Solution is: } \{2i\pi X_{123} + (\ln 3) \mid X_{123} \in \mathbb{Z}\}$$

▶ **Solve + Numeric**

$$\sinh x \cosh x = 3, \text{ Solution is: } \{[x = 1.2459]\}$$

You can obtain addition formulas.

▶ **Expand**

$$\sinh (x + y) = \sinh x \cosh y + \cosh x \sinh y$$
$$\cosh (x + y) = \cosh x \cosh y + \sinh x \sinh y$$

You can rewrite hyperbolic expressions in terms of $\sinh$ and $\cosh$.

▶ **Rewrite + Sinh and Cosh**

$$\operatorname{sech} x \tanh x - \operatorname{csch}^2 x = \frac{1}{\cosh^2 x} \sinh x - \frac{1}{\sinh^2 x}$$

The hyperbolic cosine function occurs naturally as a description of the curve formed by a hanging cable.

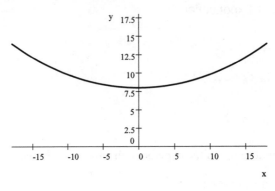

$$28\cosh\frac{x}{28} - 20$$

Inverse Hyperbolic Functions

Since the hyperbolic functions are defined in terms of exponential functions, the inverse hyperbolic functions can be expressed in terms of logarithmic functions.

$$\sinh^{-1} x \;=\; \operatorname{arcsinh} x = \ln\left(x + \sqrt{x^2 + 1}\right) \qquad x \in \mathbb{R}$$

$$\cosh^{-1} x \;=\; \operatorname{arccosh} x = \ln\left(x + \sqrt{x^2 - 1}\right) \qquad x \geq 1$$

$$\tanh^{-1} x \;=\; \operatorname{arctanh} x = \tfrac{1}{2}\ln\left(\frac{1+x}{1-x}\right) \qquad -1 < x < 1$$

▶ **To enter the inverse hyperbolic function names**

- Click the Math Name button ⬚ or choose Insert + Math Name, enter the function name in the Name box, and choose OK.

You can obtain the logarithmic expressions for these functions with the Rewrite command.

▶ Rewrite + Logarithm

$$\operatorname{arcsinh} x = \ln\left(x + \sqrt{x^2 + 1}\right)$$

$$\operatorname{arcsech} x = \ln\left(\frac{1}{x} + \sqrt{\frac{1}{x^2} - 1}\right)$$

To find values of the inverse hyperbolic functions, use Evaluate Numerically.

▶ Evaluate Numerically

$$\operatorname{arcsinh} 5 = 2.3124 \qquad \cosh^{-1} 10 = 2.9932$$

To solve equations involving inverse hyperbolic functions, use Solve + Exact or Solve + Numeric.

▶ Solve + Exact

$$\text{arcsinh}\, x - \text{arccosh}\, x = 0.3, \text{ Solution is: } \{x = 1.339\,5\}$$

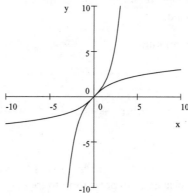

sinh x (solid), arcsinh x (dashed)

Complex Numbers

Complex numbers are numbers of the form $a + bi$ where a and b are real numbers and $i^2 = -1$. For general information on working with complex numbers, see page 38.

Polar Form

The *polar coordinate system* is a coordinate system that describes a point P in terms of its distance r from the origin and the angle θ between the polar axis (that is, the x-axis) and the line OP, measured in a clockwise direction from the polar axis. The point in the plane corresponding to a pair (a, b) of real numbers can be represented in polar coordinates $P(r, \theta)$ with

$$
\begin{aligned}
a &= r\cos\theta \\
b &= r\sin\theta
\end{aligned}
$$

where $r = \sqrt{a^2 + b^2}$ is the distance from the point (a, b) to the origin and θ is an angle satisfying $\tan\theta = \frac{a}{b}$.

In other words, any complex number in *rectangular form*

$$z = a + ib$$

can be written in *trigonometric form*

$$z = r\left(\cos\theta + i\sin\theta\right)$$

or *polar form*

$$z = re^{i\arctan a/b}$$

where $r = |z| = \sqrt{a^2 + b^2}$.

You can put a complex number $a + ib$ in polar form using the **Rewrite** command.

▶ **Rewrite + Polar**

$$3 + 5i = \sqrt{34}e^{i\arctan\frac{5}{3}} \qquad 16\pi - \sqrt{2}i = \sqrt{2}\sqrt{128\pi^2 + 1}\exp\left(-i\arctan\frac{1}{16\pi}\sqrt{2}\right)$$

You can change a complex number from polar form to rectangular form using the **Rewrite** command.

▶ **Rewrite + Rectangular**

$$\sqrt{34}e^{i\arctan\frac{5}{3}} = 3 + 5i$$

$$\sqrt{256\pi^2 + 2}\exp\left(-i\arctan\frac{\sqrt{2}}{16\pi}\right) = \sqrt{2}\frac{\sqrt{128\pi^2+1}}{\sqrt{\frac{1}{128\pi^2}+1}} - \frac{1}{8\pi}\frac{\sqrt{128\pi^2+1}}{\sqrt{\frac{1}{128\pi^2}+1}}i$$

For the identity
$$re^{it} = r\left(\cos t + i\sin t\right)$$
use **Rewrite** to change from polar form to trigonometric form.

▶ **Rewrite + Sin and Cos**

$$re^{it} = r\left(\cos t + i\sin t\right)$$

Example   To use the **Rewrite** command to get a numerical result in polar form,

1. Apply **Rewrite + Polar** to $5 + 6i$ to get $5 + 6i = \sqrt{61}e^{i\arctan\frac{6}{5}}$.

2. Select the exponent $\arctan\frac{6}{5}$ with click and drag and, while pressing down the CTRL key, choose **Evaluate Numerically** (or **Rewrite + Float**), to replace $\sqrt{61}e^{i\arctan\frac{6}{5}}$ by $\sqrt{61}e^{i0.876\,06}$.

DeMoivre's Theorem

DeMoivre's Theorem says that if $z = r\left(\cos\theta + i\sin\theta\right)$ and n is a positive integer, then
$$z^n = \left(r\left(\cos t + i\sin t\right)\right)^n = r^n\left(\cos nt + i\sin nt\right)$$
You can obtain this result for small values of n by the sequence of operations **Expand** followed by **Combine + Trig Functions** and then **Factor**.

▶ **Expand, Combine + Trig Functions, Factor**

$$\begin{aligned}\left(r\left(\cos t + i\sin t\right)\right)^3 &= r^3\cos^3 t + 3ir^3\cos^2 t\sin t - 3r^3\cos t\sin^2 t - ir^3\sin^3 t\\ &= r^3\cos 3t + ir^3\sin 3t\\ &= r^3\left(\cos 3t + i\sin 3t\right)\end{aligned}$$

You can get the same results in complete generality by working with re^{it}, since
$$\left(re^{it}\right)^n = r^n e^{itn}$$

Exercises

1. Define the functions $f(x) = x^3 + x\sin x$ and $g(x) = \sin x^2$. Evaluate $f(g(x))$, $g(f(x))$, $f(x)g(x)$, and $f(x) + g(x)$.

2. At Metropolis Airport, an airplane is required to be at an altitude of at least 800 ft above ground when it has attained a horizontal distance of 1 mi from takeoff. What must be the (minimum) average angle of ascent?

3. Experiment with expansions of $\sin nx$ in terms of $\sin x$ and $\cos x$ for $n = 1, 2, 3, 4, 5, 6$ and make a conjecture about the form of the general expansion of $\sin nx$.

4. Experiment with parametric plots of $(\cos t, \sin t)$ and $(t, \sin t)$. Attach the point $(\cos 1, \sin 1)$ to the first plot and $(1, \sin 1)$ to the second. Explain how the two graphs are related.

5. Experiment with parametric plots of $(\cos t, \sin t)$, $(\cos t, t)$, and $(t, \cos t)$, together with the point $(\cos 1, \sin 1)$ on the first plot, $(\cos 1, 1)$ on the second, and $(1, \cos 1)$ on the third. Explain how these plots are related.

Solutions

1. Defining functions $f(x) = x^3 + x\sin x$ and $g(x) = \sin x^2$ and evaluating gives
$$\begin{aligned}
f(g(x)) &= \sin^3 x^2 + \sin x^2 \sin(\sin x^2) \\
g(f(x)) &= \sin(x^3 + x\sin x)^2 \\
f(x)g(x) &= (x^3 + x\sin x)\sin x^2 \\
f(x) + g(x) &= x^3 + x\sin x + \sin x^2
\end{aligned}$$

2. You can find the minimum average angle of ascent by considering the right triangle with legs of length 800 ft and 5280 ft.

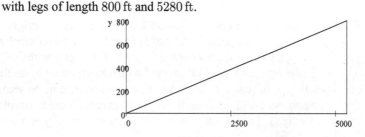

The angle in question is the acute angle with sine equal to $\dfrac{800}{\sqrt{800^2 + 5280^2}}$. Find the answer in radians with **Evaluate Numerically**:
$$\arcsin \frac{800}{\sqrt{800^2 + 5280^2}} = 0.15037$$

You can express this angle in degrees by using the following steps:

$$360 \times \frac{.15037}{2\pi} = 8.6157$$
$$0.6157 \times 60 = 36.942 \approx 37$$
$$\theta = 8°37'$$

or by solving the equations

$$0.15037 \, \text{rad} = \theta°, \text{Solution is: } \{\theta = 8.6156\}$$
$$0.6156° = x', \text{Solution is: } 36.936$$

3. Note that $\sin 2x = 2\sin x \cos x$
$$\sin 3x = 4\sin x \cos^2 x - \sin x$$
$$\sin 4x = 8\sin x \cos^3 x - 4\sin x \cos x$$
$$\sin 5x = 16\sin x \cos^4 x - 12\sin x \cos^2 x + \sin x$$
$$\sin 6x = 32\sin x \cos^5 x - 32\sin x \cos^3 x + 6\sin x \cos x$$

We leave the conjecture up to you.

4. The first figure shows a circle of radius 1 with center at the origin. The graph is drawn by starting at the point $(1, 0)$ and is traced in a counter-clockwise direction. The second figure shows the y-coordinates from the first figure as the angle varies from 0 to 2π. The point $(\cos 1, \sin 1)$ is marked with a small circle in the first figure. The corresponding point $(1, \sin 1)$ is marked with a small circle in the second figure.

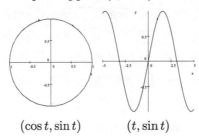

$$(\cos t, \sin t) \qquad (t, \sin t)$$

5. The first figure shows a circle of radius 1 with center at the origin. The graph is drawn by starting at the point $(1, 0)$ and is traced in a counter-clockwise direction. The second figure shows the x-coordinates of the first figure as the angle varies from 0 to 2π. The point $(\cos 1, \sin 1)$ is marked with a small circle in the first figure. The corresponding point $(\cos 1, 1)$ is marked with a small circle in the second figure. The third figure shows the graph from the second figure with the horizontal and vertical axes interchanged. The third figure shows the usual view of $y = \cos x$.

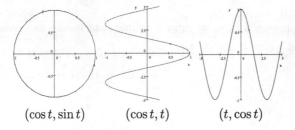

$$(\cos t, \sin t) \qquad (\cos t, t) \qquad (t, \cos t)$$

5 Function Definitions

The Definitions options provide a powerful tool, enabling you to define a symbol to be a mathematical object, and to define a function using an expression or a collection of expressions.

Function and Expression Names

A mathematical *expression* is a collection of valid expression names combined in a mathematically correct way. The notation for a *function* consists of a valid function name followed by a pair of parentheses containing a list of variables, called *arguments*. (Certain "trigtype" functions do not always require the parentheses around the argument. See page 141.) The argument of a function can also occur as a subscript (see page 110).

- Examples of mathematical expressions: x, $a^3 b^{-2} c$, $x \sin y + 3 \cos z$, $a_1 a_2 - 3 b_1 b_2$
- Examples of ordinary function notation: $a(x)$, $G(x, y, z)$, $f_5(a, b)$.

Valid Names for Functions and Expressions (Variables)

A variable or function name must be either

1. A single character (other than a standard constant), with or without a subscript

   or

2. A custom **Math Name** (see page 110), with or without a subscript.

   Expression names, but not function names, may include an arbitrary number of primes. Variables named with primed characters should be used with caution, as they are open to misinterpretation in certain contexts.

- Examples of valid expression names include a, α_X, f_{123}, g_θ, Ω_∞, e_2, r'', Waldo (custom name), John$_3$ (custom name with subscript).
- Examples of valid function names include a, α_X, f_{123}, g_θ, Ω_∞, e_2, sin, Alice (custom name), Lana$_2$ (custom name with subscript).
- Examples of invalid function names include ΔF (two characters), π, e (standard constants), f_{ab} (two-character subscript), r' (reserved for derivative).

   In the example of function names, the subscript on f_{123} is properly regarded as the number one hundred twenty-three, not "one, two, three."

Note   A subscript that is part of an expression or function name must be either a number or a single character.

Subscripts As Function Arguments

A subscript can be interpreted either as *part of the name* of a function or variable, or as a *function argument*. In the preceding examples, the subscripts that appear are part of the name. Observe the different behavior in the following examples.

1. Define $a_i = 3i$. In the Interpret Subscript dialog that appears, choose A function argument.

2. Define $b_i = 3i$. In the Interpret Subscript dialog that appears, choose Part of the name.

Evaluate then produces

$$a_i = 3i \qquad b_i = 3i \qquad a_2 = 6 \qquad b_2 = b_2$$

Choose Show Definitions and you will see that these definitions are listed as

$$a_i \;\; = \;\; 3i \text{ (variable subscript)}$$
$$b_i \;\; = \;\; 3i$$

Thus a_i denotes a function with argument i, and b_i is only a subscripted variable.

Note   A function cannot have both subscripted and in-line variables. For example, if you define $f_a(y) = 3ay$, then a is part of the function name and y is the function argument: $f_a(5) = 15a \qquad f_2(5) = 5f_2$.

Custom Names

In general, function or expression names must be single characters or subscripted characters. However, the system includes a number of predefined functions with names that appear to be multicharacter—such as gcd, inf, and lcm—but that behave like a single character. They will, for example, be deleted with a single backspace. You can create custom names with similar behavior that are legitimate function or expression names.

There are two types of custom names: Operator and Function or Variable. When you choose Name Type to be Operator, the custom name behaves like $\sum$ or $\int$ with regard to Operator Limit Placement. When you choose Name Type to be Function or Variable, it behaves like an ordinary character with regard to subscripts and superscripts. Observe the different behaviors of the two types for in-line and displayed situations.

- in-line operators: $\sum_{k=1}^{n}$, $\int_0^1$, $\operatorname{operator}_a^b$

- in-line function or variable: $\operatorname{function}_k^j$, $\operatorname{variable}_c^d$

- displayed operators, and displayed function or variable:

$$\sum_{k=1}^{n} \qquad \int_{0}^{1} \qquad \operatorname{operator}_{a}^{b} \qquad \operatorname{function}_{k}^{j} \qquad \operatorname{variable}_{c}^{d}$$

▶ **To create a custom name**

1. Click the **Math Name** button 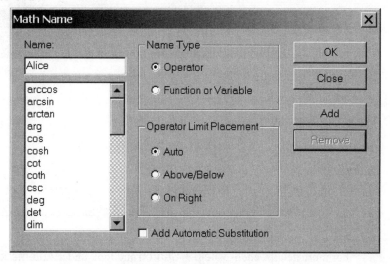, or choose **Insert + Math Name**.

2. Type a custom name in the text box under **Name**.

3. For **Name Type**, choose **Operator** or **Function or Variable**.

4. If you chose **Operator**, check your choice of **Operator Limit Placement**.

5. If you want this name to automatically gray when typed in mathematics mode, check **Add Automatic Substitution**.

6. Choose **OK**.

The gray custom name appears on the screen at the insertion point. You can use this name to define a function or expression. You can copy and paste or click and drag this grayed name on the screen, or you can recreate it with the **Math Name** dialog.

Automatic Substitution

Function names, such as sin, arcsin, gcd, etc., automatically turn gray when typed in mathematics. For an extensive list of these names, choose **Tools + Automatic Substitution**.

You can evoke the same behavior with new custom names using the **Automatic Substitution** dialog.

▶ **To make a custom name *automatically* gray**

1. From the **Tools** menu, choose **Automatic Substitution**.

2. Enter the keystrokes that you wish to use. (This may be an abbreviated form of the custom name.)

3. Click the **Substitution** box to place the cursor there and, leaving **Automatic Substitution** open, click the **Math Name** button 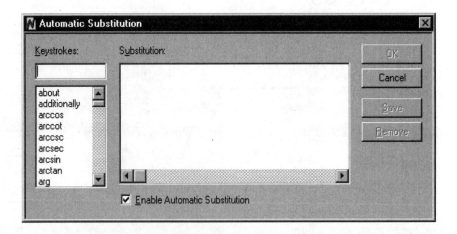.

4. Enter the custom name in the **Name** text box in the **Math Name** dialog or choose a custom name from the scroll-down list.

5. Choose **OK**. (The custom name appears in the **Substitution** box, in gray.)

6. Choose **Save** and choose **OK**.

For more details, choose **Help + Search** and look under **Automatic substitution**.

Defining Variables and Functions

When you choose **Definitions** on the **Compute** menu, the submenu that opens has seven items: **New Definition, Undefine, Show Definitions, Clear Definitions, Save Definitions, Restore Definitions,** and **Define MuPAD Name.** The choice **New Definition** can be applied both for defining functions and for naming expressions.

Assigning Values to Variables, or Naming Expressions

You can assign a value to a *variable* with **Definitions + New Definition**. There are two options for the behavior of the defined variable, depending on the symbol $=$ or $:=$ you use for assignment. The default behavior is *deferred evaluation*, meaning the definition is stored exactly as you make it. The alternate behavior is *full evaluation*, meaning the definition that is stored takes into account earlier definitions in force that might affect it. See page 114 for a discussion of the latter option.

Deferred Evaluation

▶ **To assign the value** 25 **to** z

1. Type $z = 25$ in mathematics.

2. Leave the insertion point in the equation.

3. Click the **New Definition** button [f(x)] on the **Compute** toolbar or, from the **Definitions** submenu, choose **New Definition**, or press CTRL + =.

Thereafter, until you exit the document or undefine the variable, the system recognizes z as 25. For example, evaluating the expression "$3 + z$" returns "$= 28$."

Another way to describe this operation is to say that an expression such as $x^2 + \sin x$ can be given a name. Enter $y = x^2 + \sin x$, leave the insertion point anywhere in the expression, and then, from the **Definitions** submenu, choose **New Definition**. Now, whenever you operate on an expression containing y, every occurrence of y is replaced by the expression $x^2 + \sin x$. For example, **Evaluate** applied to $y^2 + x^3$ produces

$$y^2 + x^3 = (x^2 + \sin x)^2 + x^3$$

Note that these *variables* or *names* are single characters. See page 110 for information on multicharacter names.

The *value assigned* can be any mathematical expression. For example, you could define a variable to be

- a number: $a = 245$

- a polynomial: $p = x^3 - 5x + 1$

- a quotient of polynomials: $b = \dfrac{x^2 - 1}{x^2 + 1}$

- a matrix: $z = \begin{bmatrix} a & b \\ c & d \end{bmatrix}$

- an integral: $d = \int x^2 \sin x\, dx$

You will find this feature useful for a variety of purposes.

Note   The symbol p defined previously represents the *expression* $x^3 - 5x + 1$. It is *not a function*, so, for example, $p(2)$ is not the polynomial evaluated at 2, but rather is twice p: $p(2) = 2p = 2x^3 - 10x + 2$.

Compound Definitions with Deferred Evaluation

It is legitimate to define expressions in terms of other expressions. For example, you can define $r = 3p - cq$ and then $s = nr + q$. Evaluating s will then give you $s = n(3p - cq) + q$. Redefining r will change the evaluation of s.

Full Evaluation and Assignment

With full evaluation, variables previously defined are evaluated before the definition is stored. Thus, definitions of expressions can depend on the order in which they are made.

- Use an equals sign preceded by a colon to make an assignment for full evaluation.

▶ **To assign the value** $25a$ **to** z

1. Type $z := 25a$ in mathematics.

2. Leave the insertion point in the equation.

3. Click the **New Definition** button ![f(x) button] on the **Compute** toolbar or, from the **Definitions** submenu, choose **New Definition**, or press CTRL + =.

Thereafter, until you exit the document or undefine the variable, if a has not been previously defined, the system recognizes z as $25a$. If a has previously been defined to be $x + y$, then the system recognizes z as $25(x + y)$. Try the following examples that contrast the two types of assignments, and look at the list displayed under **Definitions + Show Definitions** for each case.

Example   Make the definitions $a = 1$, $x := a$, $y = a$, and $a = 2$ (in that order), and evaluate x and y.
$$x = 1 \qquad y = 2$$
Make the definitions $a = b$, $x := a^2$, $y = a^2$, and $a = 6$ (in that order), and evaluate x and y.
$$x = b^2 \qquad y = 36$$

Example   Define $r = 3p - cq$ and then $s = nr + q$ and $t := nr + q$. Evaluating s and t will then give
$$s = n(3p - cq) + q \qquad t = n(3p - cq) + q$$
Now define $r = x + y$. Evaluating s and t will now give
$$s = n(x + y) + q \qquad t = n(3p - cq) + q$$

Functions of One Variable

By using function notation, you can use the same general procedure to define a *function* as was described for defining a variable.

▶ **To define the function f whose value at x is $ax^2 + bx + c$**

1. Enter the equation $f(x) = ax^2 + bx + c$.

2. Place the insertion point in the equation.

3. Click the **New Definition** button ![button] on the **Compute** toolbar or, from the **Definitions** submenu, choose **New Definition**.

Now the symbol f represents the defined function and it behaves like a function. For example, apply **Evaluate** to $f(t)$ to get $f(t) = at^2 + bt + c$, apply **Evaluate** to $f'(t)$ to get $f'(t) = 2at + b$, and apply **Evaluate** to $f(x + h)$ to get $f(x + h) = a(x + h)^2 + b(x + h) + c$.

Compound Definitions

If g and h are *previously defined* functions, then the following equations are examples of legitimate definitions:

$$f(x) = 2g(x) \qquad f(x) = g(x) + h(x) \qquad f(x) = g(x)h(x)$$
$$f(x) = g(h(x)) \qquad f(x) = (g \circ h)(x) \qquad f(x) = g(x)/h(x)$$

Once you have defined both $g(x)$ and $f(x) = 2g(x)$, then changing the definition of $g(x)$ will redefine $f(x)$.

Note   The algebra of functions includes objects such as $f + g$, $f - g$, $f \circ g$, fg, and f^{-1}. For the value of $f + g$ at x, write $f(x) + g(x)$; for the value of the composition of two defined functions f and g, write $f(g(x))$ or $(f \circ g)(x)$; and for the value of the product of two defined functions, write $f(x)g(x)$. You can obtain the inverse function (or inverse relation) for some functions $f(x)$ by applying **Solve + Exact** to the equation $f(y) = x$ and specifying y as the **Variable to Solve for**.

Example   You can make a table of values by applying a function to an $n \times 1$ matrix.

1. Choose **Insert + Matrix**, or click ![button] on the **Math Objects** toolbar.

2. Set the number of rows at 6 and the number of columns at 1, and choose **OK**.

3. To put parentheses around the matrix, select it and click ![button] or ![button] on the **Math Templates** toolbar or press CTRL + 9.

4. Enter x and five domain values in the matrix: $\begin{pmatrix} x \\ 0 \\ 1 \\ 2 \\ 3 \\ 4 \end{pmatrix}$.

5. Define $f(x) = x^2 + 3x + 5$ and apply f to the matrix with **Evaluate** for a matrix of values.

$$f \begin{pmatrix} x \\ 0 \\ 1 \\ 2 \\ 3 \\ 4 \end{pmatrix} = \begin{pmatrix} x^2 + 3x + 5 \\ 5 \\ 9 \\ 15 \\ 23 \\ 33 \end{pmatrix}$$

To put the x and y values in a two-column matrix, place copies of the initial and final matrices next to each another, put the insertion point in one of the matrices, and from the **Matrices** submenu, choose **Concatenate**.

$$\begin{pmatrix} x \\ 0 \\ 1 \\ 2 \\ 3 \\ 4 \end{pmatrix} \begin{pmatrix} x^2 + 3x + 5 \\ 5 \\ 9 \\ 15 \\ 23 \\ 33 \end{pmatrix}, \text{concatenate:} \begin{pmatrix} x & 3x + x^2 + 5 \\ 0 & 5 \\ 1 & 9 \\ 2 & 15 \\ 3 & 23 \\ 4 & 33 \end{pmatrix}$$

Tip   To make a table that will print with lines, create a table with **Insert + Table**   (or click 🔳) and copy the information into the table by selecting, clicking, and dragging each piece of data. Choose **Edit + Properties** and add lines according to instructions in the **Table Properties** dialog. This is only for purposes of editing—a table does not behave mathematically as a matrix.

In some cases you can use the function name directly to obtain a matrix of values.

▶ **Evaluate**

$$\arctan \begin{pmatrix} x \\ 0 \\ 1 \\ 2 \\ 3 \\ 4 \end{pmatrix} = \begin{pmatrix} \arctan x \\ 0 \\ \frac{1}{4}\pi \\ \arctan 2 \\ \arctan 3 \\ \arctan 4 \end{pmatrix}$$

When this does not produce a result, you can assign the function a name with **Definitions + New Definition** and use the defined name.

▶ Definitions + New Definition

$$\sigma(x) = \sin x$$

▶ Evaluate

$$\sigma(\begin{array}{cccccc} x & 0 & 1 & 2 & 3 & 4 \end{array}) = (\begin{array}{cccccc} \sin x & 0 & \sin 1 & \sin 2 & \sin 3 & \sin 4 \end{array})$$

Functions of Several Variables

Define functions of several variables by writing an equation such as $f(x, y, z) = ax + y^2 + 2z$ or $g(x, y) = 2x + \sin 3xy$, placing the insertion point in the equation, and choosing **New Definition** from the **Definitions** submenu or clicking the **New Definition** button on the **Compute** toolbar. Just as in the case of functions of one variable, the system always operates on expressions that it obtains from evaluating the function at a point.

Piecewise-Defined Functions

You can define functions of one variable that are described by different expressions on different parts of their domain. These functions are referred to as *piecewise-defined functions*, *case functions* or *multicase functions*. Most of the operations introduced in calculus are supported for piecewise-defined functions. You can evaluate, plot, differentiate, and integrate piecewise-defined functions.

Note   There are fairly strict conditions concerning the piecewise definition of functions.

- They must be specified in a two- or three-column matrix with at least two rows, with the *function values* in the first column, **if** or if in the second column of a three-column matrix (and **if**, or any text, or no text, in the second column of a two-column matrix), followed by the *range condition* in the last (second or third) column.

- The matrix must be fenced with a left brace and null right delimiter, as in the following examples.

The range for the function value in the bottom row is always interpreted as "otherwise," so it is not necessary to cover the entire number line in the ranges you specify.

▶ **To enter a piecewise-defined function**

1. In mathematics, type an expression of the form
$$f(x) =$$

2. From the **Brackets** ▨ list, choose ▯ for the left bracket and the null delimiter

(dashed vertical line) ⋮ for the right bracket. (The dashed vertical line does not normally appear in a printed document. It appears on screen as a dashed red line, but only when **View + Helper Lines** is turned on.)

3. With the insertion point inside the brackets, click ⌗ or choose **Insert + Matrix**.

4. Set the numbers for **Rows** (number of conditions) and **Columns** (3 or 2) and choose **OK**.

5. Type function values in the first column.

6. For 3 columns:

    a. Type **if** in the second column in *text or mathematics* mode.
    b. Type the range conditions in the third column.

or

For 2 columns:

    a. (Optional) Type **if** in text mode in the second column.
    b. Type the range conditions in the second column.

Functions should be entered as in any of the following examples. When entering such functions, check **Helper Lines** on the **View** submenu, to see important details. The first two functions are defined by three-column matrices:

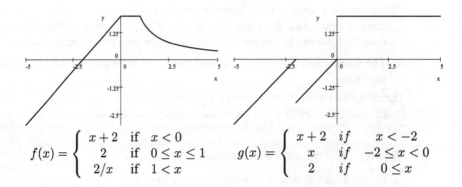

$$f(x) = \begin{cases} x+2 & \text{if} & x < 0 \\ 2 & \text{if} & 0 \le x \le 1 \\ 2/x & \text{if} & 1 < x \end{cases} \qquad g(x) = \begin{cases} x+2 & if & x < -2 \\ x & if & -2 \le x < 0 \\ 2 & if & 0 \le x \end{cases}$$

The following functions are defined by two-column matrices:

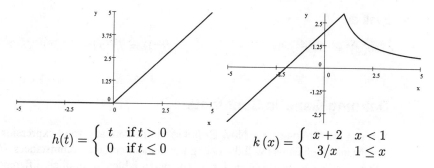

$$h(t) = \begin{cases} t & \text{if } t > 0 \\ 0 & \text{if } t \le 0 \end{cases} \qquad k(x) = \begin{cases} x+2 & x < 1 \\ 3/x & 1 \le x \end{cases}$$

See page 167 for guidelines to plotting piecewise-defined functions.

▶ **To define a piecewise-defined function**

1. Type the function values in a matrix enclosed in brackets as described.

2. Leave the insertion point in the function definition.

3. Click [f(x)] or, from the **Definitions** submenu, choose **New Definition**.

You can then choose **Evaluate** for results such as $f(-1) = 1$, $f(\frac{1}{2}) = 2$, $f(2) = 1$, $h(-1) = 0$ and

$$g(t+1) = \begin{cases} t+3 & \text{if} & t+1 < -2 \\ t+1 & \text{if} & -3-t \le 0 \,(\text{and}) \, t+1 < 0 \\ 2 & \text{if} & \text{otherwise} \end{cases}$$

Note To operate on piecewise-defined functions, such as to evaluate, plot, differentiate, or integrate such a function, you can make the definition and then work with the function name f or the expression $f(x)$. You can also place the insertion point in the defining matrix to carry out such operations.

Defining Generic Functions

You can use **Definitions + New Definition** to declare an expression of the form $f(x)$ to be a function without specifying any of the function values or behavior. Thus you can use the function name as input when defining other functions or performing various operations on the function.

▶ **Definitions + New Definition**

$f(x)$ $g(x) = x^2 - 3x$

▶ Evaluate

$$f(g(x)) = f\left(x^2 - 3x\right) \qquad\qquad g(f(x)) = f^2\left(x\right) - 3f\left(x\right)$$

Defining Generic Constants

You can use Definitions + New Definition to declare any valid expression name to be a constant. Such names will then be ignored under certain circumstances. For example, when identifying dependent and independent variables for implicit differentiation, a defined constant is not considered as a variable. Observe the difference below, where a is a defined variable and b is not.

▶ Definitions + New Definition

a

▶ Calculus + Implicit Differentiation

$axy = \sin y$ (Differentiation Variable x) Solution: $ay + axy' = (\cos(y))\,y'$

$bxy = \sin y$ (Differentiation Variable x) Solution: $b'xy + by + bxy' = (\cos(y))\,y'$

Handling Definitions

The choices on the Definitions submenu, in addition to New Definition, include Undefine, Show Definitions, Clear Definitions, Save Definitions, Restore Definitions, and Define MuPAD Name. The two choices New Definition and Show Definitions also appear on the Compute toolbar as ⊞ and ⊞ .

Showing Definitions

You view the complete list of currently defined variables and functions for the active document by clicking ⊞ on the Compute toolbar or by choosing Show Definitions from the Definitions submenu. A window opens showing the definitions in force in that document. The defined variables and functions are listed in the order in which the definitions were made.

Removing Definitions

You can remove from a document a definition that you created with Definitions + New Definition in any of the following ways:

- Select the defining equation, or select the name of the defined expression or function, and, from the Definitions submenu, choose Undefine.

- From the Definitions submenu, choose Clear Definitions (to cancel all definitions displayed under Show Definitions that were created with Definitions + New Definition).

- Make another definition with the same variable or function name as the definition you want to remove.

- On the Definition Options page of Computation Setup or Document Computation Settings, check Do Not Save. Close the document. (See the next section, *Saving and Restoring Definitions*, for more detail on definition options.)

For the first option, you can select the equation or name by placing the insertion point within or on the right side of the equation or name that you wish to remove, or you can select the entire equation, expression, or function name with the mouse. You can copy an equation or name from the list of definitions in the Show Definitions window if you do not have a copy readily at hand. (Items in this window are write protected. You cannot delete an item from this list to undefine it.)

Saving and Restoring Definitions

From the Computation Setup dialog you can change the global defaults for saving and restoring definitions. You can override the global defaults with a setting from the Compute + Settings dialog.

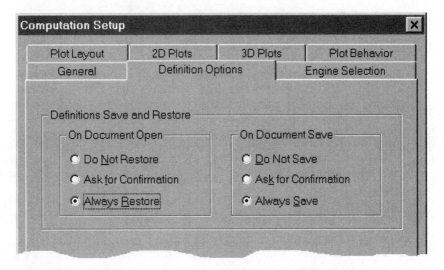

The choices in for local or global defaults are the following:

1. Do not save or restore definitions automatically (in which case you must actively choose to save or restore definitions when you wish to do so).

2. Display a prompt when you enter or exit the document asking whether you wish to save or restore definitions.

3. Save or restore definitions automatically.

▶ **To set global defaults for saving and restoring definitions**

- From the Tools menu, choose Computation Setup. Make your choices on the Definition Options page, and choose OK.

▶ **To set local defaults for saving and restoring definitions**

- From the Compute menu, choose Settings. On the Definition Options page, click Set Document Values, make your choices and choose OK.

You can also override default settings from the Definitions submenu, with Save Definitions and Restore Definitions. Choosing Save Definitions from the Definitions submenu has the effect of storing all the currently active definitions in the working copy of the current document. When the document is saved, the definitions are saved with it. Restore Definitions takes any definitions stored with the current document and makes them active.

If you choose Do Not Restore by any of the above methods, and you open, modify, and close the document without first choosing Define + Restore Definitions, then any definitions previously saved with the document will be lost and not recoverable.

Assumptions About Variables

The functions available for making assumptions are:

assume   additionally   about   unassume

These functions place restraints on specific variables or on all variables, or provide information on the restraints, or remove restraints. The function "assume" enables you to place a restraint on a variable. The function "additionally" allows you to place additional restraints. The function "about" returns information on the restraints. The function "unassume" removes restraints. Allowable assumptions include

real   complex   integer   positive   negative   nonzero

To enter the names of the functions and the names of the assumptions, put the insertion point in mathematics mode and type the name. It will turn gray when you type the last letter.

▶ **To change the global default domain for variables**

1. In mathematics, type assume. (It will automatically turn gray.)

2. Click the expanding parentheses button, or enter parentheses from the keyboard.

3. Enter an allowable assumption.

4. Choose Evaluate.

▶ Evaluate

$\text{assume}(\text{real}) = \text{real}$

▶ Solve + Exact

$x^2 = -1$, No solution found.

▶ **To return the global default domain to normal**

1. In mathematics, type assume.

2. Click the expanding parentheses button, or enter parentheses from the keyboard.

3. Choose Evaluate.

▶ Evaluate

$\text{unassume}()$

▶ Solve + Exact

$x^2 = -1$, Solution is: $i, -i$

The global default will also return to normal when you close and reopen *Scientific WorkPlace* or *Scientific Notebook*.

▶ **To check the status of the global default**

1. In mathematics, type about. (It will automatically turn gray.)

2. Click the expanding parentheses button, or enter parentheses from the keyboard.

3. Choose Evaluate.

▶ Evaluate

$\text{about}() = \text{Global}$

This response indicates that there are no special global assumptions in force; that is, the global default is normal.

▶ **To place a restraint on a variable**

1. In mathematics, type assume.

2. Click the expanding parentheses button, or enter parentheses from the keyboard.

3. Type the variable name, followed by a comma, followed by the desired assumption.

4. Choose **Evaluate**.

▶ **To restrain the variable n to be a positive integer**

- Place the insertion point in the expression assume $(n, \text{positive})$, and choose **Evaluate**.

Evaluation of the expressions assume$(n, \text{positive})$ and additionally$(n, \text{integer})$, followed by evaluation of the expression about(n), produces the following output.

▶ Evaluate

$\text{assume}(n, \text{positive}) = \text{positive}$

$\text{additionally}(n, \text{integer}) = \text{Type PosInt}$

$\text{about}(n) = \text{Type PosInt}$

To clear the assumptions about a variable,

- Select the variable and choose **Definitions + Undefine**

  or

- Evaluate unassume (name of variable).

▶ Definitions + Undefine

n

  or

▶ Evaluate

$\text{unassume}(n)$

After either procedure, you can check the status of the variable n with the function about.

▶ Evaluate

$\text{about}(n)$: No assumptions

Formula

The Formula dialog provides a way to enter an expression and a **Compute** operation. What appears on the screen is the result of the operation and depends upon active definitions of variables that appear in the formula. Formulas remain active in your document—that is, changing definitions of relevant variables will change the data on the screen.

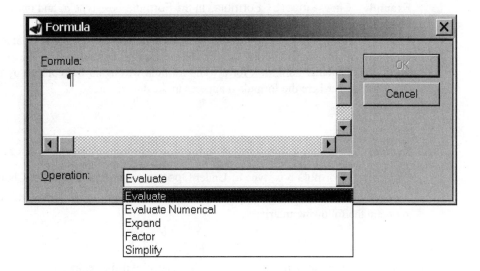

▶ **To insert a formula**

1. Click 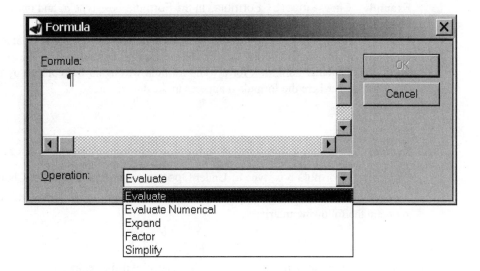 on the Field toolbar

   or

   From the Insert menu, choose Formula.

2. In the Formula area, enter a mathematics expression.

3. In the Operation area, enter the operation you want to perform on the expression. (Click the arrow at the right of the box for a list of available operations.)

4. Choose OK.

The results of the operation will be displayed on your screen.

   With Helper Lines on, a Formula can be identified by a colored background.

▶ **To change the formula background color**

1. Choose Tag + Appearance.

2. Under Tags, choose Style Defaults.

3. Under Tag Properties, choose Formula and choose Modify.

4. Select background color and choose OK.

5. Choose Save if you wish to make a permanent change in the screen style, and choose OK.

Example   Choose Insert + Formula. In the Formula box, type a, and under Operations choose Evaluate. Choose OK.

The a will appear on your screen at the position of the insertion point. Now, at any point
in your document, define $a = \sin x$. The formula a will be replaced by the expression
$\sin x$. Make another definition for a. The formula will again be replaced by the new
definition everywhere the formula a appears in the document.

Example   Insert a 2×2 matrix. With the insertion point in the first input box, click

d=rt . In the Formula box, type a. Under Operations, choose Evaluate. Choose OK.

Repeat for each matrix entry, typing, in turn, b, $a + 2b$, and $(a - b)^2$ in the formula box
to obtain the following matrix:

$$\begin{bmatrix} a & b \\ a + 2b & (a - b)^2 \end{bmatrix}$$

Now define $a = \sin x$ and $b = \cos x$. The matrix will be replaced by the following
matrix.

$$\begin{bmatrix} \sin x & \cos x \\ \sin x + 2\cos x & (\sin x - \cos x)^2 \end{bmatrix}$$

Define $a = \ln x$ and $b = e^x$. The matrix will be replaced by the following matrix.

$$\begin{bmatrix} \ln x & e^x \\ \ln x + 2e^x & (\ln x - e^x)^2 \end{bmatrix}$$

Example   Insert a table with 2 columns and 5 rows. Insert formulas x, y, z, and $x +
y + z$ in the column on the right, with Operation: Evaluate.

Date	Income
01/31/2002	x
02/28/2002	y
03/31/2002	z
Total	$x + y + z$

Define each of $x = 20.56$, $y = 18.92$, $z = 23.45$ to obtain the table

Date	Income
01/31/2002	20.56
02/28/2002	18.92
03/31/2002	23.45
Total	62.93

Multiple choice examinations with variations can be constructed using formulas.
A formula in a quiz question depends on definitions that are made globally for the
document—they are not local to each question or variant. For this reason, we recom-
mend using a Math Name (see page 110) instead of a single character name for each
variable.

The following example outlines a way for manually constructing a quiz with variants. For information on an automatic way to create such examinations with random variants, see **Help + Contents, Create Exams and Quizzes**.

Example The variables a1 and b1 shown in the sample question below are math names entered as formulas.

- Click d=rt and then click sin/cos , or use **Insert + Formula** followed by **Insert + Math Name**.

- Turn on **Helper Lines** and look for background color to check that each of the entries a1, b1 or b1 / a1 is entered in a formula.

The sample question has the following appearance with neither of the variables a1 or b1 defined:

For which values of the variable x is a1 x − b1 < 0?

    a. $x <$ b1 / a1
    b. $x >$ b1 / a1
    c. $x >$ b1
    d. $x <$ b1
    e. None of these

Now define a1 $= 2$ and b1 $= 5$ by placing the insertion point in each equation and choosing **Definitions + New Definition**. You will obtain the following result:

For which values of the variable x is $2x - 5 < 0$?

    a. $x < 5/2$
    b. $x > 5/2$
    c. $x > 5$
    d. $x < 5$
    e. None of these

After printing a quiz, make different definitions for all the variables such as a1 and b1 to obtain variations of the quiz.

External Functions

You can access functions available to the computation engine that do not appear as menu items. These can be either functions from one of the libraries of the computation engine or user-defined functions.

Functions defined with the **Definitions + Define MuPAD Name** dialog can be saved with and restored to a document with **Save Definitions** and **Restore Definitions** as described previously for defined functions (see page 121). These functions, with their

MuPAD name correspondences, appear in the Show Definitions window but they are *not* removed by Clear Definitions. To remove a MuPAD function, select the function name and choose Compute + Definitions + Undefine.

Accessing Functions in MuPAD Libraries

The following example defines the function "divisors," which computes the divisors of a positive integer.

▶ **To access the MuPAD function divisors and to name it D**

1. From the Definitions submenu, choose Define MuPAD Name.

2. Respond to the dialog as follows:
   - MuPAD Name: **numlib::divisors(x)**
   - Scientific Notebook (WorkPlace) Name: $D(x) = x$
   - Check That is built in to MuPAD or is automatically loaded

3. Choose OK.

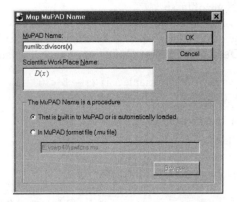

▶ Evaluate

$$D(24) = \{1, 2, 3, 4, 6, 8, 12, 24\}$$

Multiple notations for vectors are possible in *Scientific WorkPlace* and *Scientific Notebook*, including row or column matrices, and n-tuples enclosed by either parentheses or square brackets. However, to work with a function defined from the MuPAD libraries, *you must use the MuPAD syntax for the function arguments.* Consult MuPAD documentation for details of syntax needed with these functions.

An extensive MuPAD library is included with your system. Here is a short list from the many examples that are available using the Define MuPAD Name dialog.

MuPAD Name	Sample SNB Name
nextprime(x)	$p(n)$
ithprime(x)	$I(n)$
isprime(n)	$q(n)$
numlib::phi(x)	$\varphi(n)$
numlib::legendre(a,b)	$L(a, b)$
numlib::divisors(x)	$d(x)$
polylib::resultant(a,b,x)	$r(a, b, x)$
lllint(a)	$L(a)$

The following example defines the function "ithprime," which produces the ith member of the sequence of prime integers.

▶ **To access the MuPAD function ithprime and to name it I**

1. Choose Definitions + Define MuPAD Name.

2. In the MuPAD Name box, type **ithprime(x)**.

3. In the Scientific WorkPlace (Notebook) Name box, type $I(x)$.

4. Check That is built in to MuPAD or is automatically loaded, and choose OK.

Now apply Evaluate for the following result.

$$I(100) = 541$$
$$I(1000000000) = 22\,801\,763\,489$$

See page 456 for another example. In that section, the MuPAD function "**nextprime**" is used.

The guidelines for valid function and expression names (see page 109) apply to the names that can be entered in the Define MuPAD Name dialog box. You can give a multicharacter name to a MuPAD function as follows. With the Define MuPAD Name dialog box open and the insertion point in the Scientific WorkPlace (Notebook) Name box, click the Math Name button [sin cos], enter the desired function name, and click OK.

User-Defined MuPAD Functions

You can access user-defined functions written in the MuPAD language. Save the function to a file `filename.mu` in a MuPAD session. While in a document, from the Definitions submenu, choose Define MuPAD Name.

▶ **To access the function "myfunc" and name it "M"**

1. From the Definitions submenu, choose Define MuPAD Name.

2. Respond to the dialog box as follows.

- MuPAD Name: **myfunc(x)**
- Scientific Notebook (WorkPlace) Name: $M(x)$
- The MuPAD Name is a Procedure
    a. Check In MuPAD format file (.mu file) for MuPAD file
    b. Enter: **/dirname/subdirname/myfunc.mu**

3. Check OK.

This procedure defines a function $M(x)$ that behaves according to your MuPAD program. The guidelines on page 109 for valid function and expression names apply to the names that can be entered in the Define MuPAD Name dialog box.

MuPAD functions accessed through the Define MuPAD Name dialog can be saved with and restored to a document with Save Definitions and Restore Definitions, as described earlier for defined functions. The MuPAD functions accessed through the Define MuPAD Name dialog appear in the Show Definitions window, but they are *not* removed by Clear Definitions. To remove such a function definition, select the function name and choose Compute + Definitions + Undefine.

Tables of Equivalents

Constants and functions are available either as items on the Compute menu or through evaluating mathematical expressions.

Constants

The common constants can be expressed in ordinary mathematical notation. The constant gamma does not automatically gray when typed in mathematics. Use the Insert + Math Name dialog to enter this name. It is then interpreted by the system as the constant gamma.

SNB/SW	MuPAD	Comments
e	**exp(1) or E**	base of natural logs
i or j (see page 38)	**I**	imaginary unit: $\sqrt{-1}$
π	**PI**	circular constant
gamma	**EULER**	$\lim\limits_{n\to\infty}\left(\sum\limits_{m=1}^{n}\frac{1}{m}-\ln n\right)$
∞	**infinity**	positive real infinity
true	**True**	Boolean true
false	**False**	Boolean false
FAIL, undecidable	**FAIL**	Answer cannot be determined or non-existent function used

Compute Menu Items

Following is a summary of equivalents for some of the common MuPAD functions and procedures together with the equivalent **Compute** menu item. Items marked with π are programmed, generally using several MuPAD functions and procedures.

Other MuPAD functions can be implemented using **Definitions + Define MuPAD Name** (see page 127).

Algebra

Compute Menu	MuPAD
Evaluate	**eval**
Evaluate Numerically	**float**
Simplify	**simplify**
Factor	**factor**
Factor	**ifactor**
Expand	**expand**
Check Equality	**is**

Compute + Combine Menu	MuPAD
Exponentials	**combine**
Logs	**combine**
Powers	**simplify**
Trig Functions	**simplify**

Compute + Rewrite Menu	MuPAD
Rational	**numeric::rationalize**
Float	**float**
Exponential	**rewrite**
Factorial	**rewrite**
Gamma	**rewrite**
Logarithm	**rewrite**
Sin and Cos	**rewrite**
Sinh and Cosh	**rewrite**
Tan	**rewrite**
Polar	π
Rectangular	**rectform**
Equations as Matrix	π
Matrix as Equations	π

Compute + Solve Menu	MuPAD
Exact	**solve**
Numeric	**numeric::fsolve**
Integer	**Dom::Integer + solve**
Recursion	**rec + solve**

Compute + Polynomials Menu	MuPAD
Collect	**collect**
Divide	**div**
Partial Fractions	**parfrac**
Roots	**solve**
Sort	**polylib::sortMonomials**
Companion Matrix	**linalg::companion**

Calculus

Compute + Calculus Menu	MuPAD
Integrate by Parts	**intlib::by parts**
Change Variables	**intlib::changevar**
Partial Fractions	**parfrac**
Approximate Integral	**student::trapezoid**
Approximate Integral	**student::simpson**
Approximate Integral	**student::riemann**
Plot Approximate Integral	π
Plot Approximate Integral	π
Plot Approximate Integral	π
Find Extrema	π
Iterate	π
Implicit Differentiation	**diff**

Compute Menu	MuPAD
Power Series	**series**
Power Series	**taylor**

Differential Equations

Compute Menu	MuPAD
Solve ODE + Exact	**ode + solve**
Solve ODE + Laplace	**ode::laplace**
Solve ODE + Numeric	**numeric::odesolve2**
Solve ODE + Series	
Transforms + Fourier	**transform::fourier**
Transforms + Inverse Fourier	**transform::invfourier**
Transforms + Laplace	**transform::laplace**
Transforms + Inverse Laplace	**transform::invlaplace**

Vector Calculus

Compute + Vector Calculus Menu	MuPAD
Jacobian	**linalg::jacobian**
Hessian	**linalg::hessian**
Scalar Potential	**scalarpot**
Vector Potential	**linalg::vectorPotential**
Set Basis Variables	π

Matrices

Compute + Matrices Menu	MuPAD
Adjugate	linalg::adjoint
Characteristic Polynomial	linalg::charPolynomial
Cholesky Decomposition	linalg::cholesky
Column Basis	linalg::basis
Concatenate	linalg::concatMatrix
Condition Number	norm
Definiteness Tests	linalg::isPosDef
Determinant	linalg::det
Eigenvalues	linalg::eigenvalues
Eigenvectors	linalg::eigenvectors
Fill Matrix	𝝅
Fraction-free Gaussian Elimination	linalg::gaussElim
Gaussian Elimination	linalg::gaussElim
Hermite Normal Form	linalg::hermiteForm
Hermitian Transpose	conjugate + linalg::transpose

Compute + Matrices Menu	MuPAD
Inverse	numeric::inverse
Jordan Form	linalg::jordanForm
Minimum Polynomial	linalg::minpoly
Norm	norm
Nullspace Basis	linalg::nullspace
Orthogonality Test	linalg::isUnitary
Permanent	linalg::permanent
PLU Decomposition	numeric::factorLU
QR Decomposition	numeric::factorQR
Random Matrix	linalg::randomMatrix
Rank	linalg::rank

Compute + Matrices Menu	MuPAD
Rational Canonical Form	linalg::rationalForm
Reduced Row Echelon Form	linalg::GaussJordan
Reshape	𝝅
Row Basis	linalg::basis
Singular Values	numeric::singularvalues
Smith Normal Form	linalg::HermiteForm
Spectral Radius	𝝅
Stack	linalg::stackMatrix
SVD	numeric::singularvectors
Trace	linalg::tr
Transpose	linalg::transpose

Simplex

Compute + Simplex Menu	MuPAD
Dual	π
Feasible	π
Maximize	**linopt::maximize**
Minimize	**linopt::minimize**
Standardize	π

Statistics

Compute + Statistics Menu	MuPAD
Fit Curve to Data + Multiple Regression	**stats::reg**
Fit Curve to Data + Multiple Regression	**stats::reg**
Fit Curve to Data + Multiple Regression	**stats::linReg**
Fit Curve to Data + Polynomial of Degree	**stats::reg**

Compute + Statistics Menu	MuPAD
Random Numbers + Beta	π
Random Numbers + Binomial	π
Random Numbers + Cauchy	π
Random Numbers + Chi-Square	π
Random Numbers + Exponential	π
Random Numbers + F	π
Random Numbers + Gamma	π
Random Numbers + Normal	π
Random Numbers + Poisson	π
Random Numbers + Student's t	π
Random Numbers + Uniform	π
Random Numbers + Weibull	π

Compute + Statistics Menu	MuPAD
Mean	**stats::mean**
Median	**stats::median**
Mode	**stats::modal**
Correlation	π
Covariance	π
Geometric Mean	**stats::geometric**
Harmonic Mean	**stats::harmonic**
Mean Deviation	π
Moment	π
Quantile	**stats::a_quantil**
Standard Deviation	**stats::stdev**
Variance	**stats::variance**

Plot 2D

Compute + Plot 2D Menu	MuPAD
Rectangular	**plotfunc**
Rectangular	**plot::Polygon**
Polar	**plot::polar**
Implicit	**plot::contour**
Parametric	🔲
Conformal	🔲
Gradient	🔲
Vector Field	**plot::vectorfield**
ODE	**plot::ode**
Phase Plane	🔲
Rectangular, Edit + Properties, Axes Scaling	**plot2d**
Rectangular, Edit +Properties, Axes Scaling	**plot2d**

Plot 3D

Compute + Plot 3D Menu	MuPAD
Rectangular	**plot3d**
Rectangular	**plot3d**
Rectangular	**plot3d**
Cylindrical	**plotlib::cylindrical plot**
Spherical	**plotlib::sphericalplot**
Tube	**plot3d**
Gradient	**plot3d**
Vector Field	**plot3d**

Equivalents for Functions and Expressions

There are a number of built-in functions that can be evaluated with **Compute + Evaluate** or CTRL + E. Some are entered directly in mathematics and some use a **Math Name**. If a function name does not automatically turn gray when typed in mathematics (many do), choose **Insert + Math Name** and enter the function name in the **Name** box. The following lists show the MuPAD function names that have been used to implement the function.

Algebra

SN/SWP	MuPAD
$\sqrt{x}$ or $x^{1/2}$	sqrt(x)
$\sqrt[n]{x}$	x^(1/n)
$\|x\|$ or abs(x)	abs(x)
$\max(a,b,c)$ or $a \vee b \vee c$	max(a,b,c)
$\min(a,b,c)$ or $a \wedge b \wedge c$	min(a,b,c)
$\gcd(x^2+1, x+1)$	gcd(x^2+1,x+1)
$\mathrm{lcm}(x^2+1, x+1)$	lcm(x^2+1,x+1)
$\lfloor \frac{123}{34} \rfloor$	floor(123/34)
$\lceil \frac{123}{34} \rceil$	ceil(123/34)

SN/SWP	MuPAD
$\binom{6}{2}$	binomial(6,2)
$x!$	x!
$123 \bmod 17$	123 mod 17
$a^n \bmod m$	powermod(a,n,m)
$3x^3 + 2x \bmod x^2 + 1$	divide + Rem
$\{a,b\} \cup \{b,c\}$	{a,b}union{b,c}
$\{a,b\} \cap \{b,c\}$	{a,b}intersect{b,c}
signum (x)	sign(x)

Trigonometry *

SN/SWP	MuPAD
$\sin x$ or $\sin(x)$	sin(x)
$\cos x$	cos(x)
$\tan x$	tan(x)
$\cot x$	cot(x)
$\sec x$	sec(x)
$\csc x$	csc(x)
$\arcsin x$ or $\sin^{-1} x$	arcsin(x)
$\arccos x$ or $\cos^{-1} x$	arccos(x)
$\arctan x$ or $\tan^{-1} x$	arctan(x)
$\mathrm{arccot}\, x$ or $\cot^{-1} x$	arccot(x)
$\mathrm{arcsec}\, x$ or $\sec^{-1} x$	arcsec(x)
$\mathrm{arccsc}\, x$ or $\csc^{-1} x$	arccsc(x)

* These are "trigtype" functions for which the parentheses around the argument are optional. See page 141.

Exponential, Logarithmic, and Hyperbolic Functions *

SN/SWP	MuPAD
e^x or $\exp(x)$	**exp(x)**
$\log x$ or $\ln x$ (See page 91)	**ln(x)**
$\log_{10} x$ (See page 82)	**ln(x)/ln(10)**
$\sinh x$	**sinh(x)**
$\cosh x$	**cosh(x)**
$\tanh x$	**tanh(x)**
$\coth x$	**coth(x)**
$\cosh^{-1} x$ or $\text{arccosh}(x)$	**arccosh(x)**
$\sinh^{-1} x$ or $\text{arcsinh}(x)$	**arcsinh(x)**
$\tanh^{-1} x$ or $\text{arctanh}(x)$	**arctanh(x)**

   * These are "trigtype" functions for which the parentheses around the argument are optional. See page 141.

Calculus

SN/SWP	MuPAD
$\frac{d}{dx}\left(x \sin x\right)$	**diff(x*sin(x),x)**
f', Df, D	**D(f)**
$f'(3)$	**D(f)(3)**
$\int x \sin x\, dx$	**int(x*sin(x),x)**
$\int_0^1 x \sin x\, dx$	**int(x*sin(x),x = 0..1)**
$\lim_{x \to 0} \frac{\sin x}{x}$	**limit(sin(x)/x,x=0)**
$\sum_{i=1}^{\infty} \frac{i^2}{2^i}$	**sum(i^2/2^i, i = 1..infinity)**

Complex Numbers

SN/SWP	MuPAD
$\text{Re}(z)$	**Re(z)**
$\text{Im}(z)$	**Im(z)**
$|z|$	**abs(z)**
$\text{csgn}(z)$	
$\text{signum}(z)$	**sign(z)**
z^*	**conjugate(z)**
$\arg(z)$	**arctan(Im(z)/Re(z))**

The complex sign function $\operatorname{csgn}(z)$ and the signum function $\operatorname{signum}(z)$ are defined by

$$\operatorname{csgn}(z) = \begin{cases} 1 & \text{if } \operatorname{Re}(z) > 0; \text{ or } \operatorname{Re}(z) = 0 \text{ and } \operatorname{Im}(z) \geq 0 \\ -1 & \text{if } \operatorname{Re}(z) < 0; \text{ or } \operatorname{Re}(z) = 0 \text{ and } \operatorname{Im}(z) < 0 \end{cases}$$

$$\operatorname{signum}(z) = \begin{cases} \dfrac{z}{|z|} & \text{if } z \neq 0 \\ 0 & \text{if } z = 0 \end{cases}$$

Linear Algebra

SN/SWP	MuPAD
$\begin{pmatrix} 1 & 2 & 3 \\ 4 & 5 & 6 \end{pmatrix}$	**SWPmatrix(2,3,[[1,2,3],[4,5,6]])**
AB	**A*B**
A^{-1}	**A^(-1)**
A^T	**linalg::transpose(A)**
$A \bmod 17$	**map(A, x -> x mod 17)**
A^H	**conjugate + linalg::transpose**
AB^{-1}	**A*B^(-1)**
$A^{-1} \bmod 17$	**map(A^(-1), x -> x mod 17)**
$\|x\|_n$	**norm(x,n)**
$\|x\|_F$	**norm(x,Frobenius)**
$\|x\|_\infty$	**norm(x,Infinity)**

Vector Calculus

SN/SWP	MuPAD
∇xyz	**linalg::grad(x*y*z,[x,y,z])**
$\|(1,-3,4)\|_p$	**norm(SWPmatrix(1,3,[[1,-3,4]]),p)**
$S \times T$	**linalg::crossProduct(S,T)**
$S \cdot T$	**linalg::scalarProduct(S,T)**
$\nabla \cdot S$	**linalg::divergence(S,v)**
$\nabla \times S$	**linalg::curl(S,v)**
$\nabla^2 \left(x^2 y z^3 \right)$	**linalg::divergence(linalg::grad(x^2*y*z^3,[x,y,z]),[x,y,z])**

Differential Equations

SN/SWP	MuPAD
$\mathcal{F}\left(f\left(t\right),t,w\right)$	**transform::fourier(expr,t,w)**
$\mathcal{F}^{-1}\left(f\left(t\right),t,w\right)$	**transform::ifourier(expr,t,w)**
$\mathcal{L}\left(f\left(t\right),t,s\right)$	**transform::laplace(expr,t,s)**
$\mathcal{L}^{-1}\left(f\left(s\right),s,t\right)$	**transform::ilaplace(expr,s,t)**
$\mathrm{Dirac}\left(x\right)$	**dirac(x)**
$\mathrm{Dirac}\left(x,n\right)$	
$\mathrm{Heaviside}\left(x\right)$	**heaviside(x)**

Statistics

The following distribution and density function names may automatically gray when typed in mathematics mode. If they do not, use **Insert + Math Name** to create these function names. They are then interpreted by the system as the named function.

SN/SWP	Comment
NormalDist, NormalDen	Normal distribution
TDist, TDen, TInv	Student's t distribution
ChiSquareDist, ChiSquareDen, ChiSquareInv	Chi Square distribution
FDist, FDen, FInv	F distribution
ExponentialDist, ExponentialDen, ExponentialInv	Exponential distribution
WeibullDist, WeibullDen, WeibullInv	Weibull distribution
GammaDist, GammaDen	Gamma distribution
BetaDist, BetaDen	Beta distribution
CauchyDist, CauchyDen	Cauchy distribution
UniformDist, UniformDen	Uniform distribution
BinomialDist, BinomialDen	Binomial distribution
PoissonDist, PoissonDen	Poisson distribution
HypergeomDist, HypergeomDen	Hypergeometric distribution

Special Functions

Some of these function names do not automatically gray when typed in mathematics. Use the **Insert + Math Name** dialog to enter these names. They are then interpreted by the system as the named function.

SN/SWP	MuPAD	Comments
bernoulli (n)	**bernoulli(n)**	nth Bernoulli number: $\frac{t}{e^t-1} = \sum_{n=1}^{\infty} \text{bernoulli}\,(n)\,\frac{t^n}{n!}$
bernoulli (n, x)	**bernoulli(n,x)**	nth Bernoulli polynomial: $\frac{te^{xt}}{e^t-1} = \sum_{n=1}^{\infty} \text{bernoulli}\,(n, x)\,\frac{t^n}{n!}$
BesselI$_v$ (z) or $I_v(z)$	**besselI(v,z)**	Bessel function
BesselK$_v$ (z) or $K_v(z)$	**besselK(v,z)**	Bessel function
BesselJ$_v$ (z) or $J_v(z)$	**besselJ(v,z)**	Bessel function
BesselY$_v$ (z) or $Y_v(z)$	**besselY(v,z)**	Bessel function
Beta (x, y)	**beta(x,y)**	beta function: $\dfrac{\Gamma(x) + \Gamma(y)}{\Gamma(x+y)}$
Chi (z)		hyperbolic cosine integral: $\text{gamma} + \ln z - \int_0^z \frac{1-\cosh t}{t}dt \quad (\arg(z) < \pi)$
Ci(x)	**Ci(x)**	cosine integral: $\text{gamma} + \ln x - \int_0^x \frac{1-\cos t}{t}dt$
dilog (x)	**dilog(x)**	dilogarithm function: $\int_1^x \frac{\ln t}{1-t}dt$
Ei(x)	**eint(x)**	exponential integral: $\int_{-\infty}^x \frac{e^t}{t}dt$
erf(x)	**erf(x)**	error function: $\frac{2}{\sqrt{\pi}} \int_0^x e^{-t^2}\,dt$
$1 - \text{erf}(x)$	**erfc(x)**	complementary error function
$\Gamma(z)$	**igamma(z,0)**	Gamma function: $\int_0^\infty e^{-t}t^{z-1}dt$
$\Gamma(a, z)$	**igamma(a,z)**	incomplete Gamma function: $\int_z^\infty e^{-t}t^{a-1}dt$
LambertW(x)	**lambertW(x)**	LambertW $(x)\, e^{\text{LambertW}(x)} = x$
polylog (k, x)		polylogarithm function: polylog $(k, x) = \sum_{n=1}^{\infty} \frac{x^n}{n^k}$
Psi(x)	**psi(x)**	Psi function: $\psi(x) = \frac{d}{dx}\ln\Gamma(x)$
Psi(n, x)	**psi(x,n)**	nth derivative of Psi function
Shi (x)		hyperbolic sine integral: $\int_0^x \frac{\sinh t}{t}dt$
Si(x)	**Si(x)**	sine integral: $\int_0^x \frac{\sin t}{t}dt$
$\zeta(s)$	**zeta(x)**	zeta function: $\zeta(s) = \sum_{n=1}^{\infty} \frac{1}{n^s}$ for $s > 1$

Trigtype Functions

Your system recognizes two types of functions—ordinary functions and *trigtype* functions. The Gamma and exponential functions $\Gamma(x)$ and $\exp(x)$ are examples of ordinary functions, and $\sin x$ and $\ln x$ are examples of trigtype functions. The distinction is that the argument of an ordinary function is always enclosed in parentheses and the argument of a trigtype function often is not.

Twenty six functions are interpreted as trigtype functions: the six trig functions, the corresponding hyperbolic functions, the inverses of these functions written as "arc" functions (e.g. $\arctan(x)$), and the functions log and ln. These functions were identified as trigtype functions because they are commonly printed differently from ordinary functions in books and journal articles.

There is no ambiguity in determining the argument of an ordinary function because it is always enclosed in parentheses. Consider $\Gamma(a+b)x$ for example. It is clear that the writer intends that Γ be evaluated at $a+b$ and then the result multiplied by x. However, with the similar construction $\sin(a+b)x$, it is quite likely that the sine function is intended to be evaluated at the product $(a+b)x$. If this is not what is intended, the expression is normally written as $x\sin(a+b)$, or as $(\sin(a+b))x$.

To ascertain how an expression you enter will be interpreted, place the insertion point in the expression and press CTRL + ?.

▶ CTRL + ?

$$\sin x/2 = \sin \tfrac{x}{2}$$

You can reset your system to require that all functions be written with parentheses around the argument.

▶ **To disable the trigtype function option in all documents**

1. In the Tools + Computation Setup dialog, choose the General page.

2. Check Convert Trigtype to Ordinary.

3. Choose OK.

▶ **To disable the trigtype function option in one document**

1. In the Compute + Settings dialog, choose the General page.

2. Check Set Document Values and check Convert Trigtype to Ordinary.

3. Choose OK.

Your system will then *not* interpret $\sin x$ as a function with argument x, but will still recognize $\sin(x)$.

Determining the Argument of a Trigtype Function

Roughly speaking, the algorithm that decides when the end of the argument of a trigtype function has been reached stops when it finds a + or − sign, but tends to keep going as long as things are still being multiplied together. There many exceptions, some of which are shown in the following examples.

▶ CTRL + ?

- $\sin x + 5 = \sin x + 5$ It didn't write $\sin(x+5)$, so x is the argument of sin.
- $\sin(a+b)x = \sin(a+b)x$ It didn't write $(\sin(a+b))x$, so $(a+b)x$ is the argument.
- $\sin x(a+b) = \sin x(a+b)$ It didn't write $(\sin x)(a+b)$, so $x(a+b)$ is the argument.
- $\sin x \cos x = \sin x \cos x$ It didn't write $\sin(x\cos x)$, so x is the argument of sin.
- $\sin x(\cos x + \tan x) = \sin x(\cos x + \tan x)$ It didn't write $(\sin x)(\cos x + \tan x)$, so $x(\cos x + \tan x)$ is the argument of sin.
- $(\sin x)(\cos x + \tan x) = (\sin x)(\cos x + \tan x)$ Here x is the argument of sin.
- $\sin(x)(a + \cos b) = (\sin x)(a + \cos b)$ Here x is the argument of sin.

The algorithm stops parsing the argument of one trigtype function when it comes to another

$$\sin x \cos(ax + b) = (\sin x)(\cos(ax + b))$$

except when the second trigtype function is part of an expression inside expanding parentheses:

$$\sin x(\cos(ax + b)) = \sin(x(\cos(ax + b)))$$

Compare the following examples with the ordinary function $\exp(x) = e^x$.

- $\sin x \cos(ax + b) = \sin x \cos(ax + b)$ In this case, $\cos(ax+b)$ is not part of the argument.
- $\sin xe^x = (\sin x)e^x$ In this case, e^x is not part of the argument.
- $\sin x(\cos(ax + b)) = \sin x(\cos(ax + b))$ In this case, $\cos(ax+b)$ is part of the argument.
- $(\sin x)(\cos(ax + b)) = (\sin x)(\cos(ax + b))$ In this case, $\cos(ax+b)$ is not part of the argument.
- $\sin(x)(\cos(ax + b)) = (\sin x)(\cos(ax + b))$ In this case, $\cos(ax+b)$ is not part of the argument.
- $\sin x(a + e^b) = \sin x(a + e^b)$ In this case, $a + e^b$ is part of the argument.
- $(\sin x)(a + e^b) = (\sin x)(a + e^b)$ In this case, $a+e^b$ is not part of the argument.
- $\sin(x)(a + e^b) = (\sin x)(a + e^b)$ In this case, $a+e^b$ is not part of the argument.

Division using / is treated much like multiplication.
- $\sin x/2 = \sin \frac{x}{2}$ but $\sin(x)/2 = \frac{\sin x}{2}$ and $(\sin x)/2 = \frac{\sin x}{2}$

- $\sin x / \cos x = \frac{\sin x}{\cos x}$ and $(\sin x) / \cos x = \frac{\sin x}{\cos x}$ but $\sin (x/\cos x) = \sin \frac{x}{\cos x}$
- $\sin xy/2 = \sin x \frac{y}{2}$ and $\sin(xy)/2 = \sin \frac{xy}{2}$ but $(\sin xy) /2 = \frac{\sin xy}{2}$

As the examples above show, parentheses enclosing both the function and its argument will remove any ambiguity. If you write $(\sin xy) z$, the product xy will be taken as the argument of sin.

Exercises

1. Define $a = 5$. Define $b = a^2$. Evaluate b. Now Define $a = \sqrt{2}$. Guess the value of b and check your answer by evaluation.

2. Define $f(x) = x^2 + 3x + 2$. Evaluate $\frac{f(x+h)-f(x)}{h}$ and Simplify the result. Do computations in place to show intermediate steps in the simplification.

3. Rewrite the function $f(x) = \max\left(x^2 - 1, 7 - x^2\right)$ as a piecewise-defined function.

4. Experiment with the Euler phi function $\varphi(n)$, which counts the number of positive integers $k \le n$ such that $\gcd(k, n) = 1$. Use Definitions + Define MuPAD Name to open a dialog box. Type **numlib::phi(x)** as the MuPAD name, $\varphi(n)$ as the Scientific WorkPlace/Notebook Name. Test the statement "If $\gcd(n, m) = 1$ then $\varphi(nm) = \varphi(n)\varphi(m)$" for several specific choices of n and m.

5. Define $d(n)$ by typing **numlib::divisors(n)** as the MuPAD name, $d(n)$ as the Scientific WorkPlace/Notebook Name. Explain what the function $d(n)$ produces. (This is an example of a *set-valued function*, since the function values are sets instead of numbers.)

Solutions

1. If $a = 5$ then defining $b = a^2$ produces $b = 25$. Now define $a = \sqrt{2}$. The value of b is now $b = 2$.

2. Evaluate followed by Simplify yields
$$\frac{f(x+h) - f(x)}{h} = \frac{1}{h}\left(3h - x^2 + (h+x)^2\right) = h + 2x + 3$$
Select the expression $= \frac{1}{h}\left(3h - x^2 + (h+x)^2\right)$ and with the CTRL key down, drag the expression to create a copy. Select the expression $(h+x)^2$ and with the CTRL key down, choose Expand. Add similar steps (use Factor to rewrite $2xh + h^2 + 3h$) until you have the following:
$$\frac{f(x+h) - f(x)}{h} = \frac{1}{h}\left(3h - x^2 + (h+x)^2\right) = \frac{1}{h}\left(3h - x^2 + 2hx + h^2 + x^2\right)$$
$$= \frac{1}{h}\left(h\left(h + 2x + 3\right) + x^2 - x^2\right) = \frac{1}{h}\left(h\left(h + 2x + 3\right)\right) = h + 2x + 3$$

3. To rewrite $f(x) = \max\left(x^2 - 1, 7 - x^2\right)$ as a piecewise-defined function, first note that the equation $x^2 - 1 = 7 - x^2$ has the solutions $x = -2$ and $x = 2$. The function f is given by

$$g(x) = \begin{cases} x^2 - 1 & \text{if} \quad x < -2 \\ 7 - x^2 & \text{if} \quad -2 \le x \le 2 \\ x^2 - 1 & \text{if} \quad x > 2 \end{cases}$$

Check: $f(-5) = 24$, $g(-5) = 24$, $f(1) = 6$, $g(1) = 6$, $f(3) = 8$, and $g(3) = 8$.

4. Construct the following table:

n	$\varphi(n)$	n	$\varphi(n)$	n	$\varphi(n)$
1	1	11	10	21	12
2	1	12	4	22	10
3	2	13	12	23	22
4	2	14	6	24	8
5	4	15	8	25	20
6	2	16	8	26	12
7	6	17	16	27	18
8	4	18	6	28	12
9	6	19	18	29	28
10	4	20	8	30	8

Notice, for example, that

$$\begin{aligned} \varphi(4 \cdot 5) &= 8 = \varphi(4)\varphi(5) \\ \varphi(4 \cdot 7) &= 12 = \varphi(4)\varphi(7) \\ \varphi(3 \cdot 8) &= 8 = \varphi(3)\varphi(8) \end{aligned}$$

5. The output of $d(n)$ for $1 \le n \le 30$ is listed in the following table:

n	$d(n)$	n	$d(n)$	n	$d(n)$
1	1	11	$[1, 11]$	21	$[1, 3, 7, 21]$
2	$[1, 2]$	12	$[1, 2, 3, 4, 6, 12]$	22	$[1, 2, 11, 22]$
3	$[1, 3]$	13	$[1, 13]$	23	$[1, 23]$
4	$[1, 2, 4]$	14	$[1, 2, 7, 14]$	24	$[1, 2, 3, 4, 6, 8, 12, 24]$
5	$[1, 5]$	15	$[1, 3, 5, 15]$	25	$[1, 5, 25]$
6	$[1, 2, 3, 6]$	16	$[1, 2, 4, 8, 16]$	26	$[1, 2, 13, 26]$
7	$[1, 7]$	17	$[1, 17]$	27	$[1, 3, 9, 27]$
8	$[1, 2, 4, 8]$	18	$[1, 2, 3, 6, 9, 18]$	28	$[1, 2, 4, 7, 14, 28]$
9	$[1, 3, 9]$	19	$[1, 19]$	29	$[1, 29]$
10	$[1, 2, 5, 10]$	20	$[1, 2, 4, 5, 10, 20]$	30	$[1, 2, 3, 5, 6, 10, 15, 30]$

Observe that $d(n)$ consists of all the divisors of n.

6 Plotting Curves and Surfaces

The plotting capabilities of symbolic computation systems are among their most powerful features. With the system you are using, you can carry out operations interactively. You can plot functions and expressions; examine the results; revise the plot and examine the results of the revision; add multiple functions to the plot; and carry out a variety of other plotting procedures. This adds an experimental dimension to problem solving. In the preceding chapter, several plots were provided to illustrate properties of functions. You will find yourself creating plots in many situations to help answer questions about the behavior of different functions or families of functions.

In this chapter, you will find techniques for creating plots and examples showing how to plot lines and curves in the Euclidean plane, and lines, curves, and surfaces in three-dimensional Euclidean space using the basic routines Rectangular, Polar, Implicit, and Parametric from the Plot 2D submenu, and Rectangular, Cylindrical, Spherical, and Tube from the Plot 3D submenu. The submenus of Plot 2D, Plot 3D, and Calculus also contain a variety of specialized plotting routines for advanced topics in calculus, vector calculus, and differential equations. Those plotting options are introduced and discussed in later chapters, along with the related mathematics.

Getting Started With Plots

You can plot an expression or function in several ways, as described in the ensuing sections. Most of these are variations on the following basic procedure.

▶ **To plot an expression involving one variable**

1. Place the insertion point in the expression.

2. Click ⊞ or, from the Plot 2D submenu, choose Rectangular.

A frame containing a plot of the expression appears after the expression, either displayed or in line (that is, with the lower edge resting on the text baseline) and the insertion point appears at the right of the plot. In the plot layout section, starting on page 146, you will find information on repositioning and resizing the frame. Following that is information on revising plots.

The first attempt at a plot uses the default parameters that are set on the 2D Plots page of the Computation Setup or Document Computation Settings dialog (see page 210). There are many settings you can adjust to obtain the view you prefer.

The following plot shows the function $y = x \sin x$ with the default plot interval $-5 < x < 5$ and the default view intervals on x and y. To make this plot, leave the insertion point in the expression $x \sin x$, and click ⊞ or, from the **Plot 2D** submenu, choose **Rectangular**.

▶ Plot 2D + Rectangular

$x \sin x$

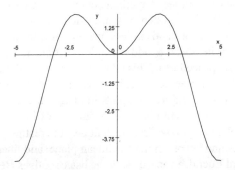

There are other ways to enter plot information, to change both the appearance and the position of the plot, and to plot multiple items on the same axes. These points are explained one at a time in the following sections. First, we discuss terminology and general properties pertaining to all plots.

The Frame, the View, and the Plot Properties Dialog

When you choose any of the options from either the **Plot 2D** or **Plot 3D** submenu to create a plot, a rectangular *frame* appears on the screen. The region of Euclidean space shown inside the frame is referred to as the *view*.

- To use the mouse to make changes to the size of the plot and to its placement in your document, you must select the frame.

- To use the mouse to make changes to a plot, including changes to the domain intervals, or any other attributes, you must select the view.

- To use the **Plot Properties** dialog to make these changes, you may select either the frame or the view.

▶ **To select the frame**

- Click the frame.

You can click anywhere within the rectangle, including in any part of the plot that is showing in the frame. As shown on the next page, eight black handles (solid boxes) appear around the edges of the frame and one tool appears in the lower-right corner, indicating that the frame is selected. *You select the frame to make changes to the size, shape, and placement of the plot.*

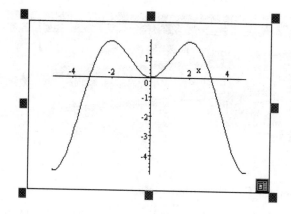

▶ **To select the view**

- Double-click the frame.

You can double-click anywhere within the rectangle, including in any part of the plot that is showing in the frame.

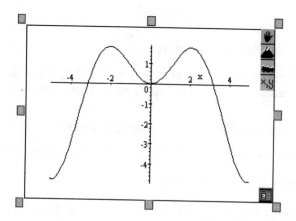

Eight gray handles appear around the edges of the frame. For 2D plots, four tools appear along the right edge of the frame; for 3D plots, one tool; and for a picture rather than a plot, three tools. When the mouse pointer is over the frame, it has the shape of a hand. In this state, the view is selected. *You select the view to make changes to the functions that are being plotted, to the view intervals, and to many other attributes of the actual picture that appears in the frame.*

From the Plot Properties dialog, you can change properties of the Layout, Labeling, Items Plotted, Axes, and View.

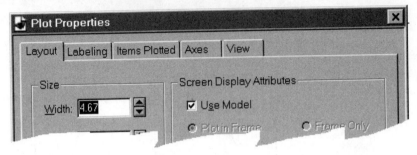

▶ **To open the** Plot Properties **dialog**

1. Click to select the frame, or double click to select the view.

2. Click the dialog tool in the lower-right corner of the frame or view.

or

Click and release the right mouse button and choose Properties from the menu that appears.

or

From the Edit menu, choose Properties.

or

On the Standard toolbar, click .

There is a keyboard shortcut to make the Plot Properties dialog open when a plot is created, so that you can customize settings before generating the plot.

▶ **To open the** Plot Properties **dialog while creating the plot**

1. Place the insertion point to the right of an expression or function name.

2. Press CTRL while choosing the plot command.

This opens the Plot Properties dialog with the tabbed pages Layout, Labeling, Items Plotted, Axes, and View. If you want this behavior to occur every time you generate a plot, change the default on the Plot Behavior page under Tools + Computation Setup (or under Compute + Settings for individual documents) as described on page 210.

Layout

Plot layout properties include the size of a graphic, its placement within your document, and the print and screen display attributes. The defaults for the plot layout can be changed on the Layout page of the Tools + Computation Setup dialog or the Compute + Settings dialog (see page 210).

Resizing the Frame

All plots have an attribute known as "fit to frame." When you resize the frame, the plot is resized with it. You can resize the frame either with the mouse or with the Plot Properties dialog.

▶ **To resize the frame with the mouse**

1. Click the plot to select the frame.

2. Click and drag one of the handles to enlarge or reduce the frame.

When the frame is selected, eight black handles are visible and you can resize the frame by dragging one of the handles. The corner handles leave the opposite vertex fixed while moving the two sides adjacent to the handle, creating a frame that has edges proportional to the original frame . The edge handles move only the corresponding edge in or out. Either type of change stretches or shrinks a plot in the view, along with the frame. Resizing the frame retains the same plot and view intervals. For example, use one of the side handles to create a tall and narrow frame or use one of the handles on the top or bottom to create a short and fat frame . The examples in this paragraph illustrate the use of *in-line plots*, one of the two placement options described in the next section.

Use the Plot Properties dialog to specify frame dimensions precisely.

▶ **To resize the frame with the Plot Properties dialog**

1. Select the plot and open the Plot Properties dialog.

2. Click the Layout tab.

3. In the Size boxes for Width and Height, set the sizing options you want.

4. Choose OK.

Frame Placement

With *Scientific Notebook*, there are two choices for frame placement—In Line and Displayed. With *Scientific WorkPlace* you can also choose Floating. These choices are explained in the following paragraphs.

Open the Plot Properties dialog to see how a frame is placed in your document.

▶ **To verify and/or change the frame placement**

1. Select the plot and open the Plot Properties dialog.

2. Click the Layout tab and check your choice for Placement.

In-Line Frames

When a frame is in line, you can move it up or down within the line.

▶ **To move an in-line frame up or down with the mouse**

1. Select the frame.

2. Drag the frame up or down.

You can drag the frame such that its lower edge is resting on the text baseline, is centered on the line, hangs with the upper edge at the text baseline, or rests anywhere in between. An in-line frame behaves like a word in the text, in the sense that, when you enter additional items to the left of it, the frame is pushed along in the line.

You can also adjust the offset on the Layout page of the Plot Properties dialog. After choosing In Line, specify a value for Baseline, Offset from Bottom of Frame. A drop-down list allows you to choose points, inches, centimeters, or picas as units of measurement.

Displayed Frames

Displayed frames appear on the screen centered on a separate line.

▶ **To display a plot or graphic**

1. Select the frame.

2. Click 🔍 on the Standard toolbar.

   or

   Click and release the right mouse button and choose Properties from the menu that appears.

   or

   Click the dialog tool ▥ in the lower-right corner of the frame.

   or

   From the Edit menu, choose Properties.

3. Click the Layout tab.

4. In the Placement area of the Layout page, choose Displayed.

Note   To center a plot, choose Displayed from the Layout page of the Plot Properties dialog; or choose In Line from the Layout page, then choose Center Text from the Body Tags popup list. To minimize vertical white space above and/or below the plot, choose Displayed and use the backspace or delete key to remove any new paragraph symbols ¶ that occur immediately before/after the plot. (To see these symbols, click

¶ on the Standard toolbar, or turn on Invisibles on the View menu.)

The use of the mathematics display, which treats the frame like mathematics, can lead to unpredictable results when you preview or print your document. If the frame appears red on your screen, and if it is not in a mathematics display, you can change the frame to text mode by selecting it with the mouse and clicking the Math/Text button.

Floating Frames

Floating placement is a typesetting option, available in *Scientific WorkPlace* or *Scientific Word* only. Floating frames containing plots are not anchored to a precise location in your document. Instead, they are positioned when you typeset the document, according to the options you choose for placement: Here, On a Page of Floats, Top of Page, or Bottom of Page.

Floating frames can carry figure numbers, captions, and keys. The figure number is created automatically by LaTeX unless you suppress it. With File + Preview or File + Print, floating frames behave like displayed frames. See Help + Search + floating objects for further information on floating frames.

Screen Display and Print Attributes

There are several choices for the appearance of graphics, both on the screen and in print. The choices are

- Plot in Frame: displays both the picture and the surrounding frame.
- Plot Only: displays only the plot.
- Frame Only: displays only the frame.
- Iconified: minimizes the plot so that it appears as an icon.
- Use Model: uses the default set in the Plot Layout page of the Tools + Computation Setup or Compute + Settings dialog.

to save the time required for generating plots while working on a document, you may wish to use the Iconified option for Screen Display Attributes.

Plot Intervals and View Intervals for 2D Plots

The view of a plot depends on both the Plot Intervals and the View Intervals.
- The Plot Intervals determine the sampling points for evaluation.
- The View Intervals determine the coordinates that are visible.

The following views of the plot $y = \sin x$ demonstrate the relationship between the Plot and View intervals on the x-axis for Rectangular plots.

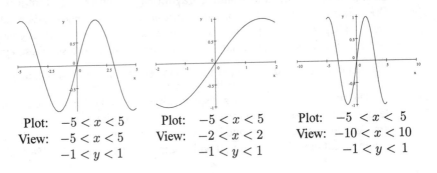

Plot: $-5 < x < 5$	Plot: $-5 < x < 5$	Plot: $-5 < x < 5$
View: $-5 < x < 5$	View: $-2 < x < 2$	View: $-10 < x < 10$
$-1 < y < 1$	$-1 < y < 1$	$-1 < y < 1$

Note   When the View Interval is smaller than the Plot Interval, the system is plotting points in the domain outside the view, and consequently plotting fewer points inside the view than would normally be the case. Large differences in this direction can lead to some distorted graphs and you will need to change the Plot Intervals accordingly.

You can change the Plot Intervals from the Items Plotted page of the Plot Properties dialog (see page 159), and you can change the View Intervals from the View page of the Plot Properties dialog (see below).

A default is preset for bounds on the Plot Intervals for each of the choices on the Plot 2D submenu. The default bounds are somewhat arbitrary, and you will often want to change them. See page 210 for instructions on changing these default settings. The factory default is $-5 \leq x \leq 5$ for rectangular coordinates and $-5 \leq x \leq 5, -5 \leq y \leq 5$ for polar coordinates and implicit plots.

The default View Intervals are determined by the Plot Intervals and the graph. They are determined automatically to maximize the information contained in the plot, given the Plot Intervals that are set from the Items Plotted page. The x- and y-axes can have radically different scales, and the curve is normally not truncated at the top or bottom unless there are vertical asymptotes.

From the View page of the Plot Properties dialog you can set the view intervals, and generate or remove a snapshot of the plot. See page 208 for an explanation of plot snapshots.

▶ **To change the view for a 2D plot**

1. Select the plot and open the Plot Properties dialog.

2. Click the View tab.

3. Click to remove the check mark from Default, and set the View Intervals.

4. Choose OK to make the indicated changes and refresh the function plot. Choose Cancel to leave the dialog without taking any action.

▶ **To generate a plot snapshot**

1. Select the plot and choose Edit + Properties.

2. Choose the View page and click Generate Snapshot.

▶ **To remove a plot snapshot**

1. Select the plot and choose Edit + Properties.

2. Choose the View page and click Remove Snapshot.

In the following sections, you will see examples of the default plot and view for different types of 2D plots.

Rectangular Coordinates

When you choose Rectangular from the Plot 2D submenu, the view that appears is determined by inequalities of the form $a \leq x \leq b$ and $c \leq y \leq d$. The default for the view is the region bounded by $-5 \leq x \leq 5$ and $c \leq y \leq d$, where c and d are chosen by the underlying computational system and depend on the shape of the function plot.

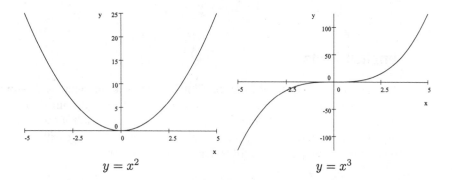

$$y = x^2 \qquad\qquad y = x^3$$

Polar Coordinates

When you choose Polar from the Plot 2D submenu, the view that appears is determined by the inequality $-\pi \leq \theta \leq \pi$ on the angle. The view intervals are chosen by the underlying computational system and depend on the shape of the function plot. The following are default views.

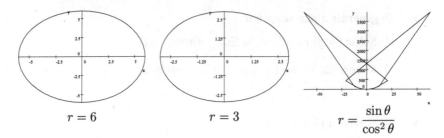

$$r = 6 \qquad\qquad r = 3 \qquad\qquad r = \frac{\sin\theta}{\cos^2\theta}$$

Changing the view for the graph of $r = \dfrac{\sin\theta}{\cos^2\theta}$ to $-5 < x < 5$ and $0 < y < 25$, or changing the plot interval to $-1.5 < \theta < 1.5$, gives the following views of the same plot.

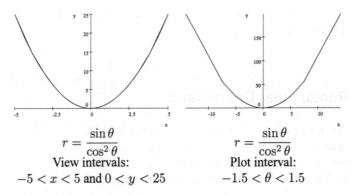

$$r = \frac{\sin\theta}{\cos^2\theta}$$
View intervals:
$-5 < x < 5$ and $0 < y < 25$

$$r = \frac{\sin\theta}{\cos^2\theta}$$
Plot interval:
$-1.5 < \theta < 1.5$

Implicit Plots

When you choose Implicit from the Plot 2D submenu, the view is determined by inequalities of the form $a \leq x \leq b$ and $c \leq y \leq d$. The default values for the Plot Intervals are $-5 \leq x \leq 5$ and $-5 \leq y \leq 5$, and the default View Intervals are determined by the underlying computer algebra system. Plotting with Implicit assumes rectangular coordinates. The following are default views.

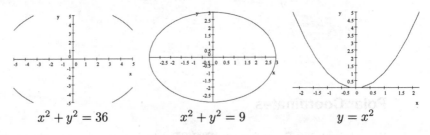

$$x^2 + y^2 = 36 \qquad\qquad x^2 + y^2 = 9 \qquad\qquad y = x^2$$

Parametric Plots

When you choose Parametric from the Plot 2D submenu, the default values for the Plot Interval are $-5 \leq t \leq 5$ and the View Intervals $a \leq x \leq b$ and $c \leq y \leq d$ are determined by the underlying computer algebra system. The following are default views.

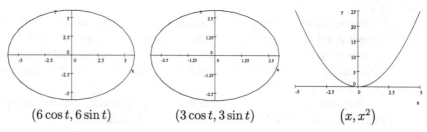

$(6 \cos t, 6 \sin t)$ $(3 \cos t, 3 \sin t)$ (x, x^2)

Plotting Tools for 2D Plots

Several plotting tools designed for investigating plots—collecting data, or changing the view by zooming and panning—make this process efficient and enjoyable.

► **To display plotting tools**

- Select the view by double-clicking a plot.

  Four plotting tools appear at the upper-right edge of the frame for translating the view, zooming in, zooming out, and reading and storing plot coordinates;

eight gray handles appear on the frame; and the mouse pointer takes the shape of a hand when it is over the view.

Zooming In and Out

The icons for the two zoom tools were designed to look like pictures of mountains. Click the large mountain range ![mountain] to zoom in for a closer look; click the small mountain range ![mountain] to zoom out to see additional parts of the graph. You will find these features to be very useful when investigating properties of functions that you plot.

▶ **To zoom in or out**

1. Click one of the two zoom tools.

2. Move the mouse pointer over the view.

3. Click and drag to create a small rectangle inside the frame.

If you zoom in (with the large mountain range), the contents of the small rectangle expand to fill the entire frame. Parts of the plot that were outside the small rectangle are now outside the view. If you zoom out (with the small mountain range), the original view shrinks to fit inside the small rectangle. Additional parts of the function plot appear in the new, expanded view.

There are two types of behavior that occur, depending on the setting for the View Intervals in the View page of the Plot Properties dialog.

- To zoom in or out depending on only the x-coordinate, Default should be checked in the View page of the Plot Properties dialog. (This is the default setting.)

- To control both the x- and y-intervals when zooming in or out, click the Default box to clear it.

After zooming in or out, you can return the view to approximately where it was previously. Click the opposite icon, and create a small rectangle with approximately the same size and position as before, relative to the frame.

Translating the View

▶ **To translate the view**

1. Select the view by double-clicking. (If the view has already been selected, click the hand icon [🤚] in the upper-right corner of the frame.)

2. Move the mouse pointer over the view, so that it takes on the shape of a hand.

3. Click and drag the frame.

An outline of the frame moves as you drag, while the plot remains fixed. When you release the mouse button, the frame is redrawn at its original position with a translated view. Use this feature to "pan" across the Cartesian plane in any direction to capture different portions of the plot.

Note Zooming and translating change the view without altering the size or position of the frame. Resizing and moving the frame change the frame size and placement but do not change the view.

Plot Coordinates Dialog

Double-click a plot and click the 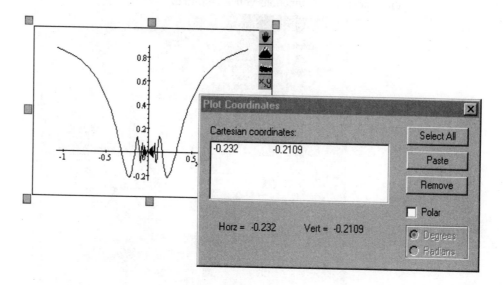 tool. A dialog will appear on your screen that displays the rectangular coordinates of the mouse cursor whenever the plot is selected and the mouse cursor is over the plot area. Check Polar on the Plot Coordinates dialog to see polar coordinates of points at the cursor.

▶ **To record selected points**

- Select the ✕،ᵧ tool, then click on points in the graph.

The rectangular coordinates of each points that you click will appear in the Plot Coordinates dialog box.

▶ **To create a matrix of coordinates**

1. Click the coordinate pairs you want in the matrix or choose Select All.

2. Place the insertion point at the position where you want the matrix.

3. From the Plot Coordinates dialog, choose Paste.

▶ **To plot selected points in a polygonal plot**

- Drag the matrix to a 2D plot.

    or

- Create the matrix with the plot selected, or with the insertion point immediately to the right of the matrix (Step 2 above).

    If you wish to plot only the points, first create the polygonal plot as described above.

Then select the plot, choose the Items Plotted page of the Plot Properties dialog, and change Plot Style to Point.

▶ **To remove coordinate pairs from the** Plot Coordinates **dialog box**

1. Click the coordinate pairs you want to remove or choose Select All.

2. From the Plot Coordinates dialog, choose Remove.

Items Plotted

From the Items Plotted page of the Plot Properties dialog you can edit, add, and delete expressions to be plotted, choose the plot style, change the interval to be plotted, and set the number of sampling points.

Expressions and Relations

The Expressions and Relations box in the upper-left corner of the dialog shows one of the expressions or functions you are plotting.

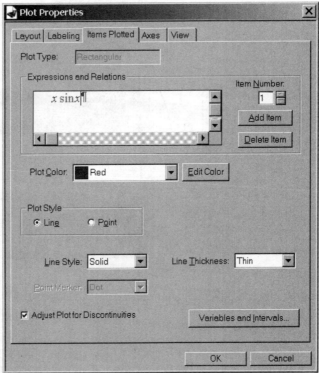

These expressions or functions are referred to as "items" and listed by Item Number. You can view all current functions or expressions by clicking the up-down arrow buttons

by the Item Number box. You can change the curves that are plotted by editing the
displayed item, by adding a new item, or by removing an item from the list.

- To change an item, edit it in the box.

- To add an item, click Add Item, and then type or paste the item in the box.

- To delete an item from the list, click an Item Number arrow button until the item is
  displayed in the box, and then click Delete Item.

Intervals and Sample Size

For each item, you can reset the interval to be plotted.

▶ **To change** Plot Intervals

- Choose Variables and Intervals and edit the Plot Intervals.

  The setting for Plot Intervals specifies the interval containing the sampling points
  that generate the plot. The standard default for Rectangular plots is the interval $-5 \leq x \leq 5$.

▶ **To increase the accuracy of the graph**

- Choose Variables and Intervals and increase the setting for Points Sampled.

  Because increasing the Points Sampled slows down the plotting process, you may
  want a relatively small number of data points for real-time demonstrations and a relatively large number of data points for printed documents. If a plot does not appear
  smooth, it may be that the plot requires a larger number of data points for an accurate
  representation.

Plot Color and Style

You can specify a color for each item to be plotted.

▶ **To specify** Plot Color

- Set Plot Color to one of the twenty named colors: Black, Blue, Blue Green,
  Brown, Cyan, Dark Gray, Green, Light Blue, Light Gray, Light Green, Light
  Red, LtBlueGreen, Magenta, Maroon, Navy, Purple, Red, Sienna, White, or
  Yellow.
  or

- Choose Edit Color and choose from among additional unnamed colors or define
  your own custom colors.

  You can specify the Plot Style for each item to be plotted. In a plot, the data points
  computed either are connected by line segments or are displayed as points, depending
  on your choice of Plot Style as Line or Point. The default is Line, which connects the
  points with lines. See the section beginning on page 170 for examples of Line and Point
  plots.

If the Plot Style is set to Line, you can change the appearance of a plot by changing Line Style and Line Thickness.

▶ **To specify** Line Style **and** Line Thickness

- Set Line Style to Solid, Dash, Dots, DotDash, or DotDotDash.
- Set Thickness to Thin, Medium, or Thick.

If several expressions appear as different items in the same view, you can set different parameters (such as Plot Style, Line Style, Thickness, and Plot Color) to distinguish visually between the different expressions. You may prefer to distinguish between expressions by setting Plot Color to a different color for each expression in screen-based documents, and to Black (the default) with a variety of Line Style and/or Thickness for black and white printed documents.

You can also choose Plot Style as Point, which displays only the computed points. You can choose the Point Marker (Box, Circle, Cross, Dot, or Diamond) used to plot the points.

Adjust Plot for Discontinuities

With Adjust Plot for Discontinuities you can change the appearance of a discontinuous graph. See page 166 for a discussion of this option.

Axes and Axis Scaling

From the Axes page of the Plot Properties dialog you can choose the type of axes displayed and the scaling for the axes. You can change the appearance of the axes and add axis labels.

▶ **To specify properties of axes**

- Click Equal Scaling Along Each Axis to force the y-axis to use the same scale as the x-axis.
- Choose Axis Scaling from among Linear, Log, and Log-Log.
- Customize the x-axis and y-axis by setting a specified number of tick marks for each.
- Add Custom labels to the x-axis and y-axis.
- Choose Disable tick labeling to suppress the numeric labels on the axes.
- Specify the appearance of the axes by choosing an Axes Type from among Normal, Boxed, Framed, or None. The choices Normal and None are self-explanatory. The Boxed appearance shows the plot inside a rectangular frame, and Framed displays the left and lower edges of the box. All of these choices (except None) display numerical labels on the axes or box edges.

The four choices for axes are shown below.

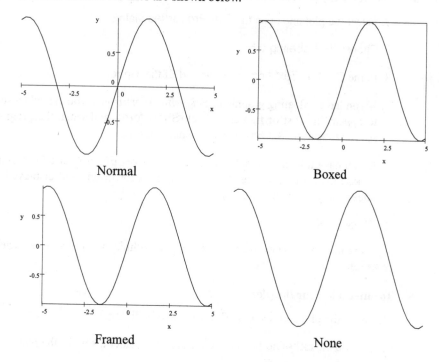

Use the UP ARROW to locate hidden choices, should any of these fail to appear in the dialog box. Choose OK to make the indicated changes and refresh the function plot. Choose Cancel to leave the dialog without taking any action.

The default for axis labels is that the independent variable will be labeled according to the variable name you use in the expression for the plot (or x if you do not use a variable name). In the case of an implicit plot, both axes will be labeled according to the variable names you use in the expression for the plot.

Plot Labeling

You can provide a caption for a plot. Captions appear under in-line and displayed frames and, in *Scientific WorkPlace* only, either over or under floating frames.

- If you produce the document without typesetting it, approximately as much of the caption as can fit in the width of the frame will appear in print. You can alter the width of the frame as necessary.

- If you typeset the document in *Scientific WorkPlace*, the caption appears in its entirety.

If the typesetting specifications call for a list of figures, the caption for a floating figure or a special short form of the caption can be used to generate the list. See Help + Search + list of figures for more information about this option.

▶ **To enter a caption for a plot**

1. Select the plot and open the **Plot Properties** dialog.

2. Choose the **Labeling** tab.

3. In the **Caption Text** box, enter the text of the caption.

4. If you have a floating graphic in *Scientific WorkPlace*, you can add a short caption to appear in a list of figures. Check **Short form**, and when the program opens a second **Caption Text** box, enter the short caption.

5. If you have a floating graphic in *Scientific WorkPlace*, check **Above** or **Below** to choose the position of the caption. (Captions for in-line and displayed frames can appear only below the frame.)

6. Choose **OK**.

You can provide a key as a target for hypertext links or, in *Scientific WorkPlace*, for cross references to the plot.

▶ **To enter a key for the plot**

1. Select the plot and open the **Plot Properties** dialog.

2. Choose the **Labeling** tab and, in the **Key** box, enter a key for the plot.

3. Choose **OK**.

Plots can also have a name that appears on screen when the plot is iconified.

▶ **To enter a name for an iconified plot**

1. Select the plot and open the **Plot Properties** dialog.

2. Choose the **Labeling** tab.

3. In the **Name** box, enter the name you want to appear on the screen when the plot is iconified.

4. Choose **OK**.

2D Plots of Functions and Expressions

In the equation $f(x) = x \sin x$, each of the two sides—$f(x)$ and $x \sin x$—is an *expression* while f is a *function*. The function f is a rule that assigns to each number the product of that number with the sine of that number. Thus the function f defined by the equation $f(x) = x \sin x$ is the same function as the function g defined by the equation $g(t) = t \sin t$. The expression $x \sin x$ (or $f(x)$) is different from the expression $t \sin t$ (or $f(t)$), since $x \sin x$ is tied to the variable x, and $t \sin t$ is tied to the variable t.

Expressions

▶ **To plot an expression involving a single variable**

1. Enter the expression in your document.

2. With the insertion point in the expression, click 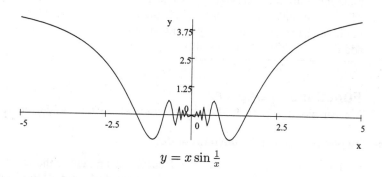, or from the **Plot 2D** submenu choose **Rectangular**.

▶ **Plot 2D + Rectangular**

$$x \sin \frac{5}{x}$$

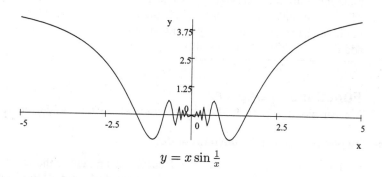

$$y = x \sin \tfrac{1}{x}$$

Add the Label $y = x \sin \frac{1}{x}$ by typing it into the **Plot Properties + Labeling** dialog in the **Caption Text** entry field.

▶ **To add an expression to a plot**

- Select the expression with the mouse and drag it onto the plot.
  or

- Choose **Add Item** in the **Items Plotted** page of **Plot Properties** dialog and type or paste the expression in the **Expressions and Relations** box.

Example   To plot $x \sin \frac{1}{x}$ together with its envelope x and $-x$,

1. With the insertion point in $x \sin \frac{1}{x}$, choose **Plot 2D + Rectangular**.

2. Select and drag to the frame x.

3. Select and drag to the frame $-x$.

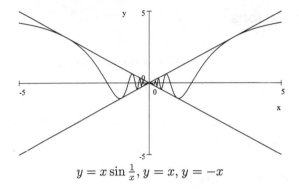

$$y = x \sin \frac{1}{x}, y = x, y = -x$$

Note   If the *Reference Library* is installed on your system, you can find equations and plots for many more curves. Look under *Tables, reference: Curves in the Plane*. (A "Typical" installation does not include the *Reference Library* for space considerations, but if you have space on your system you can copy the files from the installation CD or add them with a "Custom" install.)

Functions of Degrees

You can plot trigonometric functions written as functions of degrees rather than radians.

▶ **To plot trigonometric functions of degrees**

1. Enter the expression(s) in your document, using either the red degree symbol in a superscript or the green degree symbol from the Insert + Unit Names dialog.

2. With the insertion point in an expression, click [+] or, from the Plot 2D submenu, choose Rectangular.

3. With the plot selected, click [Q] or choose Edit + Properties.

4. Click the Items Plotted tab and choose Variables and Intervals.

5. Change Plot Intervals to $-180 < x < 180$ (or other limits as appropriate).

6. Choose OK.

7. Click and drag additional expressions onto the plot, as desired.

▶ Plot 2D + Rectangular

$\sin x°$

- Edit + Properties, Items Plotted: choose Variables and Intervals and change Plot Interval to $-180 < x < 180$.

- Select and drag to the plot $\cos 2x°$.

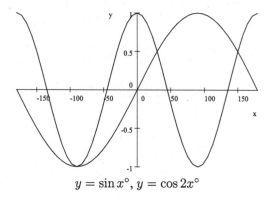

$$y = \sin x°, \, y = \cos 2x°$$

► Plot 2D + Rectangular

$$\sin 2x° + \cos 3x°$$

- Edit + Properties, Items Plotted: choose Variables and Intervals and change Plot Intervals to $-360 < x < 360$.

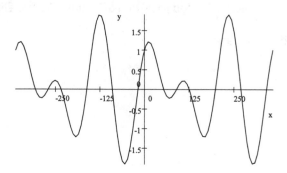

Defined Functions

You can plot a defined function in two different ways. Recall that you define a function such as $f(x) = x \sin x$ by placing the insertion point in the expression and choosing New Definition from the Definitions submenu.

► **To plot a defined function f of one variable**

1. Select the function name f or select the expression $f(x)$.

2. From the Plot 2D submenu, choose Rectangular.

Example   To plot $g(x) = \tan \sin(x^2)$, define the function with **Definitions + New Definition** and plot the function name g.

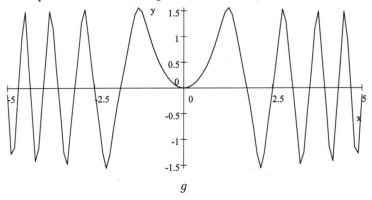

g

Continuous and Discontinuous Plots

Plotting an expression or a function name gives a (possibly) discontinuous plot.

▶ **To display vertical asymptotes**

- From the **Items Plotted** page, turn off **Adjust Plot for Discontinuities**.

▶ **Plot 2D + Rectangular**

$$\frac{x+1}{x-1}$$

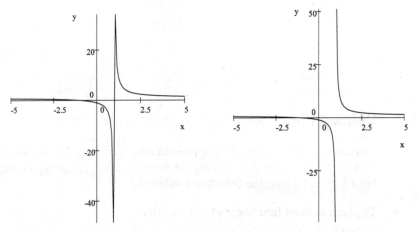

Adjust Plot for Discontinuities unchecked   Adjust Plot for Discontinuities checked

This setting applies to individual items so it is possible to plot together two functions that require opposite settings.

There can be expressions that do not plot with the setting Adjust Plot for Discontinuities, but that *will* plot with that setting unchecked. The expression

$$10^5 \left(x^{10^{-5}} - 1 \right) / \ln(x)$$

is one example. If you know you will need to change this setting to obtain the plot, hold down the CTRL key while applying the plot command. The Plot Properties dialog will open for you to edit before the system attempts to generate the plot.

For more examples of continuous and discontinuous piecewise-defined functions, see the following section.

Plotting Piecewise-Defined Functions

A piecewise-defined function must be entered in a two- or three-column matrix enclosed in expanding brackets—a left brace and right null bracket. See page 117 for detailed instructions on entering such expressions.

▶ Plot 2D + Rectangular

$$\begin{cases} x^2 - 1 & \text{if} & x < -1 \\ 10 - 10x^2 & \text{if} & -1 \leq x \leq 1 \\ x^2 - 1 & \text{if} & 1 < x \end{cases}$$

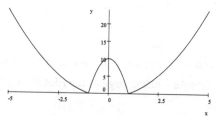

You can also plot a continuous graph from a discontinuous expression $g(x)$ (or directly from the defining matrix) by unchecking Adjust Plot for Discontinuities on the Items Plotted page of the Plot Properties dialog, as described in the previous section.

▶ Plot 2D + Rectangular

$$\begin{cases} x^2 - 1 & \text{if} & x < -1 \\ 20 - x^2 & \text{if} & -1 \leq x \leq 1 \\ x^2 - 1 & \text{if} & 1 < x \end{cases}$$

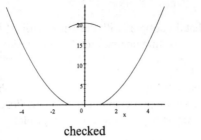

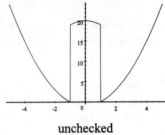

checked   unchecked

Special Functions

You can enter the *greatest integer function* or *floor function* $f(x) = \lfloor x \rfloor$ by clicking the

brackets icon and choosing ⌊⌋ (see *floor function* on page 35).

▶ Plot 2D + Rectangular

$\lfloor x \rfloor$

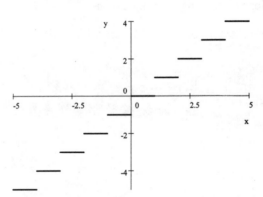

You can get a continuous plot of this function by unchecking **Adjust Plot for Disconti-nuities** on the **Items Plotted** page of the **Plot Properties** dialog.

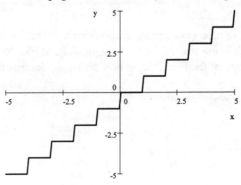

In a similar fashion, you can enter the *absolute value function* $f(x) = |x|$ by choosing vertical brackets from ▨. The following shows the graph of $f(x) = |\sin x|$.

▶ Plot 2D + Rectangular

$|\sin x|$

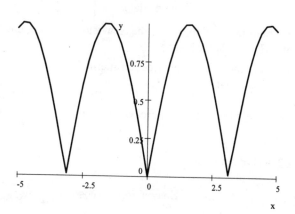

The *Gamma function* $\Gamma(x)$ extends the factorial function in the sense that for each nonnegative integer n the identity $\Gamma(n+1) = n!$ holds. The plot of the Gamma function displays the vertical asymptotes with the graph.

▶ Plot 2D + Rectangular

$\Gamma(x)$

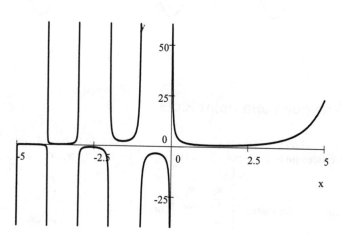

The *Heaviside function* has the following graph:

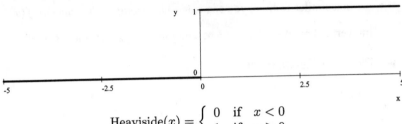

$$\text{Heaviside}(x) = \begin{cases} 0 & \text{if} \quad x < 0 \\ 1 & \text{if} \quad x \geq 0 \end{cases}$$

This is a built-in function: to use the Heaviside function, choose **Insert + Math Name** and enter Heaviside for **Name**.

The Heaviside function provides an alternative method for creating piecewise-defined functions. Note that

$$\text{Heaviside}(x - 2)\sin(x) + \text{Heaviside}(-x)\cos x = \begin{cases} \sin x & \text{if} & x \geq 2 \\ 0 & \text{if} & 0 \leq x \leq 2 \\ \cos x & \text{if} & x \leq 0 \end{cases}$$

▶ **Plot 2D + Rectangular**

$\text{Heaviside}(x - 2)\sin x + \text{Heaviside}(-x)\cos x$

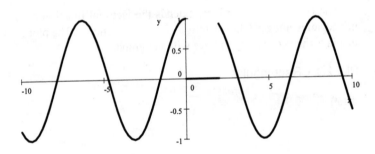

Polygons and Point Plots

You can plot the points $\{(x_1, y_1), (x_2, y_2), (x_3, y_3), \ldots, (x_n, y_n)\}$, or a polygon whose vertices lie at these points, by typing the vector $(x_1, y_1, x_2, y_2, x_3, y_3, \ldots, x_n, y_n)$ or by

entering the matrix $\begin{bmatrix} x_1 & y_1 \\ x_2 & y_2 \\ x_3 & y_3 \\ \vdots & \vdots \\ x_n & y_n \end{bmatrix}$ or $\begin{bmatrix} x_1 & x_2 & x_3 & \vdots & x_n \\ y_1 & y_2 & y_3 & \vdots & y_n \end{bmatrix}$ and choosing **Plot 2D**

+ Rectangular.

▶ Plot 2D + Rectangular

$(1, 1, 2, 1, 2, 2, 1, 2, 1, 1)$, **View Intervals** $0 < x < 3, 0 < y < 3$

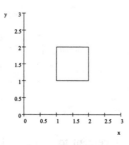

The default is to connect the points with straight-line segments. To plot points alone, in the **Plot Properties** dialog box under **Plot Style** choose **Point**.

▶ Plot 2D + Rectangular, Edit + Properties, Items Plotted, Point, Circle

$$\begin{bmatrix} 1 & 1 \\ 2 & 1 \\ 2 & 2 \\ 1 & 2 \end{bmatrix}, \text{View Intervals } 0 < x < 3, 0 < y < 3$$

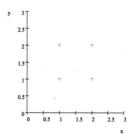

Example You can generate a regular pentagon with an enclosed five-point star by using the points $\left(\cos \frac{2\pi k}{5}, \sin \frac{2\pi k}{5}\right)$ as k ranges from 0 to 5.

1. Place the insertion point in the vector

$$\left[1, 0, \cos \frac{2\pi}{5}, \sin \frac{2\pi}{5}, \cos \frac{4\pi}{5}, \sin \frac{4\pi}{5}, \cos \frac{6\pi}{5}, \sin \frac{6\pi}{5}, \cos \frac{8\pi}{5}, \sin \frac{8\pi}{5}, 1, 0 \right]$$

2. Choose **Plot 2D + Rectangular.**

3. Choose **Edit + Plot Properties** or double-click the frame and click the dialog box tool.

4. Choose **Equal Scaling Along Each Axis.**

5. Select the vector
$$\left[1, 0, \cos\frac{4\pi}{5}, \sin\frac{4\pi}{5}, \cos\frac{8\pi}{5}, \sin\frac{8\pi}{5}, \cos\frac{2\pi}{5}, \sin\frac{2\pi}{5}, \cos\frac{6\pi}{5}, \sin\frac{6\pi}{5}, 1, 0\right]$$
with the mouse and drag it to the frame.

You may find it convenient to combine **Line** and **Point** styles, as in the following plot that combines a data cloud with a line of best fit. (See page 439 for information on curves of best fit.)

▶ **Plot 2D + Rectangular**

$$\begin{bmatrix} 1 & 3 & 4 & 6 & 7 & 7 & 10 & 11 \\ 8 & 7 & 9 & 12 & 15 & 16 & 19 & 21 \end{bmatrix}$$

- **Edit + Properties**, choose **Point**, **Circle**, and select and drag $\frac{2792}{647} + \frac{957}{647}x$ to the view.

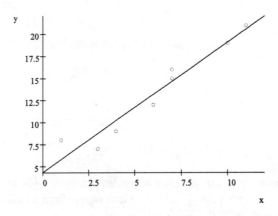

You can create line graphs and bar charts with polygonal plots, as demonstrated in the following two examples. The first example is a line graph depicting the data

1	2	3	4	5	6
0.11	0.24	0.28	0.21	0.1	0.031

▶ **Plot 2D + Rectangular**

$(1, 0, 1, 0.11)$

- Select and drag to the plot each of $(2, 0, 2, 0.24)$, $(3, 0, 3, 0.28)$, $(4, 0, 4, 0.21)$, $(5, 0, 5, 0.1)$, $(6, 0, 6, 0.031)$

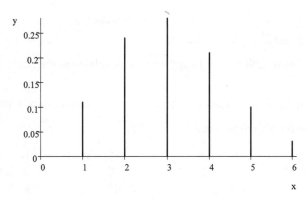

Following is a bar chart, or histogram, depicting the data

1-2	2-3	3-4	4-5	5-6	6-7
0.11	0.24	0.28	0.21	0.1	0.031

▶ **Plot 2D + Rectangular**

$(1, 0, 1, 0.11, 2, 0.11)$

- Select and drag to the plot each of $(2, 0, 2, 0.24, 3, 0.24)$, $(3, 0, 3, 0.28, 4, 0.28, 4, 0)$, $(4, 0.21, 5, 0.21, 5, 0)$, $(5, 0.1, 6, 0.1, 6, 0)$, $(6, 0.031, 7, 0.031, 7, 0)$

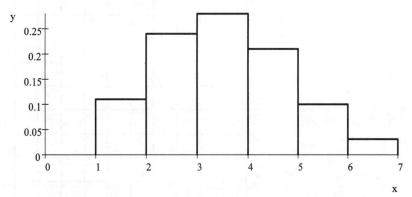

For information on importing data from an external source, see page 412.

Plotting a Grid
You can create a grid using point plots. You may wish to add a grid similar to the following to your list of fragments.

Example   To make a grid with a unit mesh for $-5 < x < 5$, $-5 < y < 5$

1. Place the insertion point in the vector

$$[-5, 0, 5, 0]$$

2. Choose **Plot 2D + Rectangular.**

3. Choose **Edit + Plot Properties** or double-click the frame and click the dialog box tool.

4. Choose **Equal Scaling Along Each Axis** and set **Axis Type** to **Boxed** or **Framed.**

5. Select each of the vectors

$$[-5, 5, -5, -5] \qquad [-4, 5, -4, -5] \qquad [-3, 5, -3, -5]$$
$$[-2, 5, -2, -5] \qquad [-1, 5, -1, -5] \qquad [0, 5, 0, -5]$$
$$[1, 5, 1, -5] \qquad [2, 5, 2, -5] \qquad [3, 5, 3, -5]$$
$$[4, 5, 4, -5] \qquad [5, 5, 5, -5] \qquad [-5, -5, 5, -5]$$
$$[-5, -4, 5, -4] \qquad [-5, -3, 5, -3] \qquad [-5, -2, 5, -2]$$
$$[-5, -1, 5, -1] \qquad [-5, 1, 5, 1] \qquad [-5, 2, 5, 2]$$
$$[-5, 3, 5, 3] \qquad [-5, 4, 5, 4] \qquad [-5, 5, 5, 5]$$

with the mouse and drag them to the frame.

6. On the **Axes** page of the **Plot Properties** dialog, change **Axes Type** to **Framed.**

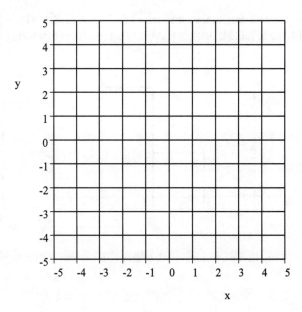

You can drag an expression to this grid to create a plot on the grid.

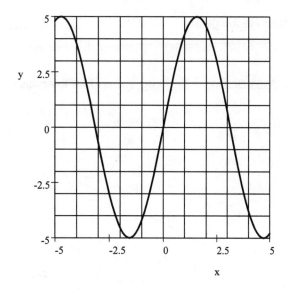

Log and Log-Log Plots

Options for both Log and Log-Log plots are listed on the Axes page of the Plot Properties dialog.

A log plot is a two-dimensional plot with the vertical axis given in a log scale. Exponential functions $f(x) = cb^x$ plot as straight lines on a Log coordinate system.

▶ Plot 2D + Rectangular, set Axes Scaling to Log

$5\left(2^x\right)$

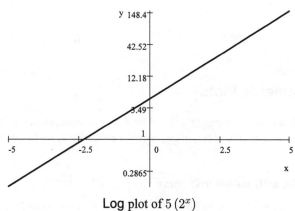

Log plot of $5\left(2^x\right)$

A log-log plot is a two-dimensional plot with both the vertical and horizontal axes given in a logarithmic scale. Power functions $f(x) = ax^n$ plot as straight lines on a Log-Log coordinate system.

▶ Plot 2D + Rectangular

$$5x^{2/3}$$

- Set Axes Scaling to Log-Log, and set Plot Intervals to $0.1 < x < 1.1$

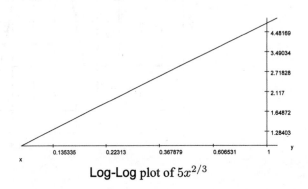

Log-Log plot of $5x^{2/3}$

You can plot expressions such as 2^x with Linear, Log, or Log-Log scaling.

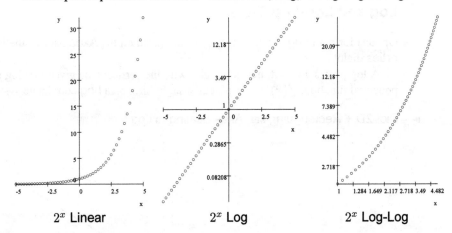

2^x Linear 2^x Log 2^x Log-Log

Parametric Plots

A 2D parametric curve is defined by a pair of equations $x = f(t)$, $y = g(t)$. The curve is the set of points $(f(t), g(t))$, where t ranges over an interval.

▶ **To plot a 2D parametric curve**

1. Make the two defining expressions the components of a vector. You can use any of the standard notations for a vector, including the forms $[\sin 2t, \cos 3t]$, $(\sin 2t, \cos 3t)$, $\begin{bmatrix} \sin 2t & \cos 3t \end{bmatrix}$, $(\ \sin 2t \quad \cos 3t\ )$, $\begin{bmatrix} \sin 2t \\ \cos 3t \end{bmatrix}$, or $\begin{pmatrix} \sin 2t \\ \cos 3t \end{pmatrix}$.

(The last four vectors are 1×2 and 2×1 matrices, respectively.)

2. Place the insertion point in the vector.

3. Choose **Plot 2D + Parametric** or **Plot 2D + Rectangular**.

The following plot shows the parametric curve defined by $x = \sin 2t$, $y = \cos 3t$ as the parametric plot of the vector $[\sin 2t, \cos 3t]$ with $0 \le t \le 2\pi$ and **Equal Scaling Along Each Axis**.

▶ Plot 2D + Parametric

$(\sin 2t, \cos 3t)$

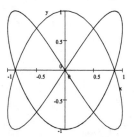

inverse relation of a function $y = f(x)$. For example, to plot the cube root function $y = x^{\frac{1}{3}}$, observe that it is the inverse function to $y = x^3$ and create a parametric plot. You can make a parametric plot of the pair $(f(x), x)$ to plot the inverse function or

▶ Plot 2D + Parametric

$\left(x^3, x\right)$

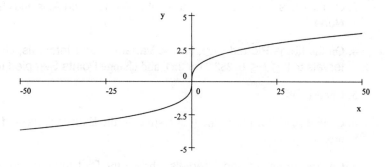

The inverse relation of $\sin x$ follows. You can adjust the view to get the plot of the inverse function $\sin^{-1} x$.

▶ **Plot 2D + Parametric**

$(\sin x, x)$

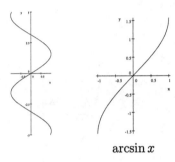

$\arcsin x$

Example   You can generate a regular pentagon with an enclosed five-point star by creating two parametric plots of $(\cos t, \sin t)$ and changing the **Plot Intervals** and number of **Points Sampled**.

1. With the insertion point inside $(\cos t, \sin t)$, choose **Plot 2D + Parametric**.

2. Choose **Edit + Plot Properties**, or double-click the frame and click the dialog box tool, or click the right mouse button and choose **Properties** from the menu that appears.

3. On the **Axes** page, check **Equal Scaling Along Each Axis** and choose **Axes type: None**.

4. On the **Items Plotted** page, choose **Variables and Intervals**, change the **Plot Intervals** to $0 \le t \le 6.2832$ ($\approx 2\pi$), and change **Points Sampled** to 6.

5. Choose **OK**.

6. Choose **Add Item** and type $(\cos t, \sin t)$ in the **Expressions and Relations** input box.

7. Choose **Variables and Intervals**, change the **Plot Intervals** to $0 \le t \le 12.566$ ($\approx 4\pi$), and change **Points Sampled** to 6.

8. Choose **OK** (twice).

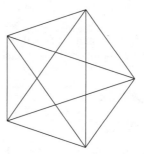

Envelopes

An interesting phenomenon occurs when simple curves are displayed as the "envelope" of a more complicated function. Such things happen in practice when low-frequency waves (say, frequencies in the audible range for the human ear) ride carrier waves broadcast from a radio station.

The following example shows the curve $y = 4\sin x + 3\cos 3x$ "riding" on top of the carrier $y = \sin 30x$. To get an accurate plot, increase the point sampled—from the **Items Plotted** page of the **Plot Properties** dialog, raise **Item Number** to 2, choose **Variables and Intervals**, and raise the number of **Points Sampled** to 150 or higher.

▶ Plot 2D + Rectangular

$4\sin x + 3\cos 3x$

- Select and drag to the frame $(4\sin x + 3\cos 3x)\sin 30x$

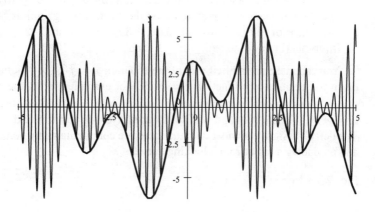

Implicit Plots

The equation of a circle cannot be rewritten as a function of one variable. You can, however, plot the set of points satisfying such an equation using the 2D implicit plot feature.

▶ **To plot an equation**

1. Enter the equation.

2. With the insertion point in the equation, choose **Plot 2D + Implicit**.

   The **Plot Intervals** for the following plot were set to $-3 \le x \le 7$ and $-2 \le y \le 8$.

▶ **Plot 2D + Implicit**

$$(x-2)^2 + (y-3)^2 = 25$$

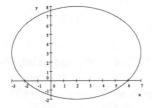 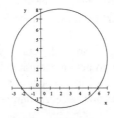 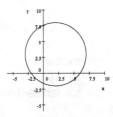

- Unequal scales on the x- and y-axes can cause distortion of geometric figures. Click the **Equal Scaling Along Each Axis** box to make the figure on the left look like the circle (middle figure).

- With the **Plot Intervals** unchanged, click to remove the check mark from the **View Intervals + Default** and edit the **View Intervals** to $-6 \le x \le 10$ and $-6 \le y \le 10$ in order to generate the plot above on the right. This again illustrates the difference between **Plot Intervals** (in this case, $-3 \le x \le 7$ and $-2 \le y \le 8$) and **View Intervals** (in this case, $-6 \le x \le 10$ and $-6 \le y \le 10$).

You can make an implicit plot of the equation $x = f(y)$ to plot the inverse function or inverse relation of a function $y = f(x)$. For example, to plot the cube root function $y = x^{\frac{1}{3}}$, observe that it is the inverse function to $y = x^3$ and create an implicit plot of $x = y^3$. Revise the plot and set **Plot Intervals** to $-5 < x < 5$ and $-1.75 < y < 1.75$. The default assigns y a wider domain, therefore computing many points outside the view and producing a rather rough looking curve.

▶ **Plot 2D + Implicit**

$$y^3 = x$$

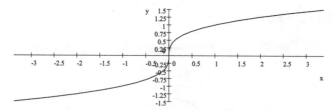

For the inverse relation of the sine function, do an implicit plot of $x = \sin y$. Changing the view appropriately will give the plot of the inverse sine function. For a smooth curve, revise the plot and set the Plot Intervals to match the view that appears on your screen.

▶ Plot 2D + Implicit

$x = \sin y$

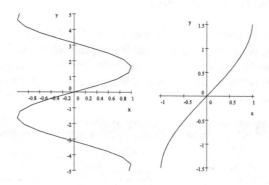

Polar Coordinates

In polar coordinates, you specify a point P by giving the angle θ that the ray from the origin to the point P makes with the polar axis and the distance r from the origin. The equations that relate rectangular coordinates to polar coordinates are given by

$$x = r\cos\theta, \qquad y = r\sin\theta$$

or equivalently,

$$x^2 + y^2 = r^2, \qquad \tan\theta = \frac{y}{x}$$

▶ **To make a plot in polar coordinates**

1. Type an expression for the radius in terms of the angle θ.

2. With the insertion point in the expression, choose Plot 2D + Polar.

To obtain the views shown in the following plots, select the frame and choose Edit + Properties. Click the Axes tab, check Equal Scaling on Both Axes, and choose OK.

▶ Plot 2D + Polar

$$\sin 2\theta \qquad 1 - \cos\theta \qquad 1 - \sin\theta + 2\sin 3\theta \qquad \cos 6\theta$$

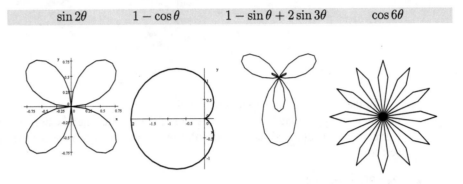

Parametric Polar Plots

The polar plot of $\theta = r^2$ is obtained as the 2D polar plot of the vector (r, r^2). To get the following plots, revise the first plot that appears, choosing **Polar** and setting the **Plot Intervals** to 0 to 5 and -5 to 5, respectively. For the second plot, set **Thickness** to **Thick** and **Line Color** to **Gray**. Choose **Variables and Intervals** and raise the number for **Point Sampled** to 200 or higher.

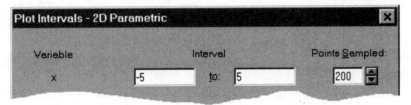

▶ Plot 2D + Parametric, Edit + Properties, Polar

$$(r, r^2)$$

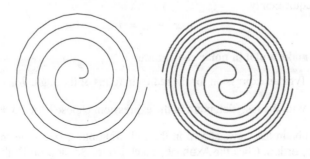

Both the radius r and the angle θ may be defined in terms of some third variable t. You can make the polar plot of the parametric curve defined by the equations $r = 1 - \sin t$, $\theta = \cos t$ as the polar plot of the vector $(1 - \sin t, \cos t)$, using **Parametric** from the **Plot 2D** submenu. Revise the first plot, choosing **Polar** and setting the **Plot Intervals** to $0 \le t \le 2\pi$.

▶ Plot 2D + Parametric, Edit + Properties, Polar

$(1 - \sin t, \cos t)$

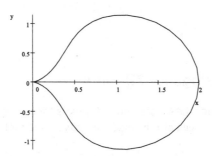

The View for 3D Plots

The environment for plotting curves and surfaces in space is similar to the environment for plotting in the plane. The view is a box, a rectangular solid determined by inequalities of the form $x_0 \le x \le x_1$, $y_0 \le y \le y_1$, and $z_0 \le z \le z_1$. The frame is a rectangular region of the computer screen.

The default view has the **Plot Intervals** $-5 \le x \le 5$, $-5 \le y \le 5$, with the z-coordinates determined automatically from properties of the plot. If you use other variable names, the order is determined alphabetically.

▶ **To plot an expression involving two variables**

1. Place the insertion point in the expression.

2. Click the **3D Plot** button on the **Compute** toolbar; or from the **Plot 3D** submenu, choose **Rectangular**.

The following plot shows the surface $z = x^3 - 3xy^2$ with the default **Plot Intervals** for x and y, and the default **View Intervals** for x, y, and z. To make this plot, leave the insertion point in the expression $x^3 - 3xy^2$ and click the **3D Plot** button on the **Compute** toolbar or, from the **Plot 3D** submenu, choose **Rectangular**.

▶ Plot 3D + Rectangular

$x^3 - 3xy^2$

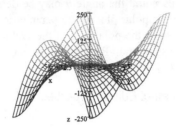

Plotting Tools and Dialogs for 3D Plots

The Plot Properties dialog provides tools for a variety of plot changes. Changes in the plot orientation can also be made directly on the plot using the plot orientation tool.

The Plot Orientation Tool

The plot orientation determines the angle from which you view the 3D space. You can change the plot orientation with the 3D plot orientation tool or from the View page of the Plot Properties dialog.

▶ **To display the 3D plot orientation tool**

- Select the view by double-clicking the plot.

    The 3D plot orientation tool ⬛ appears at the upper-right edge of the frame. The frame also has eight gray handles, and the mouse pointer has the shape of a hand when it is over the view.

▶ **To change the orientation with the plot orientation tool**

1. Double-click to select the view.

2. Move the mouse pointer over the plot so that the pointer changes to a hand.

3. Press the left mouse button and move the mouse.

4. Release the left mouse button.

    As you move the mouse while holding the button down, a 3-dimensional box indicates the orientation of the axes. When you release the button, the plot is redrawn with the new orientation.

    You can also change plot orientation from the View page of the Plot Properties dialog, as described in the next section.

The 3D Plot Properties Dialog

▶ **To open the** Plot Properties **dialog for a 3D plot**

1. Select the frame by clicking the frame.

   or

   Select the view by double-clicking the frame.

2. Click the plot-properties icon in the lower right corner of the frame.

   or

   From the Edit menu, choose Properties.

   or

   Press and release the right mouse button and choose Properties from the menu that appears.

   or

   Click ![magnifier icon] on the Standard toolbar.

Make your changes, and then choose OK to save your changes and to refresh the function plot, or choose Cancel to exit the dialog box without taking any action.

View Page

Orientation   You can change the plot orientation on the View page of the Plot Properties dialog by setting the Tilt and Turn. The Tilt setting indicates the number of degrees the z-axis is tilted—the angle normally denoted by φ. The Turn setting indicates the number of degrees the xy-plane is rotated about the z-axis—the angle normally denoted by θ.

Tilt can be set between -180 and 180, and Turn between -360 and 360. With Tilt set to 0, or ± 180, the z-axis is horizontal. With Tilt set to ± 90, the z-axis is vertical. Here are three views of the cylinder $r = 3$ illustrating Turn and Tilt settings.

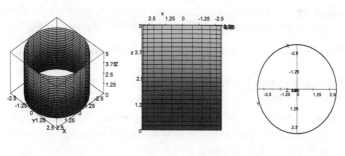

Turn 45, Tilt 45   Turn 90, Tilt 90   Turn 0, Tilt 0

View Intervals   The View Intervals determine the coordinates that are visible. The default View Intervals are determined by the Plot Intervals—the z-axis can have quite a different scale from the x- and y-axes, and a bounded surface is generally not truncated at the top or bottom. The View Intervals appear in the View page of the Plot Properties dialog and the Plot Intervals, which determine the sampling points for evaluation for each expression plotted, are found from the Items Plotted page. The plot may have different Plot Intervals for each expression being plotted, but it can have only one set of View Intervals.

To change the View Intervals, remove the check from Default and set the View Intervals by editing the numbers in the boxes.

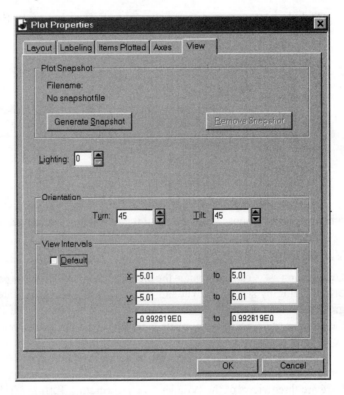

Lighting   You can also change the Lighting option on the View page. The choices—0, 1, 2, 3, 4—are various predetermined color schemes.

Plot Snapshots   See page 208 for a discussion of Plot Snapshots.

Items Plotted page

The Items Plotted page of the Plot Properties dialog contains settings that can be chosen individually for each item to be plotted. You can change the scale and translate each item in the plot from the Items Plotted page. You can individualize the color and surface or line style for each expression.

Item List   The Expressions and Relations box in the upper-left corner of the Items Plotted page contains one of the expressions or functions that you are plotting. These expressions are referred to as "items" and are listed by Item Number. You can view all of the current functions or expressions by clicking the up-down arrow buttons to the right of the Item Number box.

To change the curves or surfaces that are plotted, you can change, delete, or add expressions or functions in the box. To change an item, edit it in the box. To add an item, click the Add Item button, and then type the item in the box that opens. To delete an item from the list, click the arrow buttons until the item is displayed in the box, and then click the Delete Item button.

Plot Intervals and Points Sampled   On the Items Plotted page of the Plot Properties dialog choose Variables and Intervals to get a Plot Intervals dialog. These intervals determine the region within which the program chooses the sampling points that determine the plot.

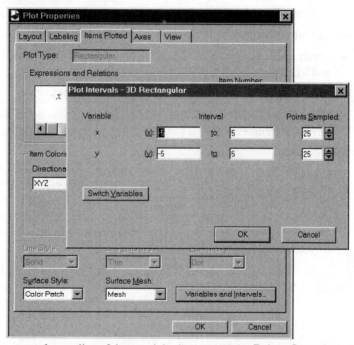

You can increase the quality of the graph by increasing the Points Sampled. Increasing the Points Sampled also slows down the plotting process, of course. This is especially critical for Plot 3D because the number of data points computed is the product of the two numbers entered under Points Sampled. Thus the default number of data points computed is $25^2 = 625$. So you may want to specify a relatively small sample size for real-time demonstrations and a relatively large sample size for printed documents.

Surface Style   In a plot, the grid points computed are connected by polynomial surfaces or curves, or are simply displayed as points, depending on your choice of Surface Style. The default style is WireFrame, which is generated relatively quickly on the screen. You may wish to switch to Color Patch, which takes much longer but can be much prettier, for a final printed document. Examples illustrating the following styles begin on page 188.

- Color Patch connects the points by small patches of polynomial surfaces.
- WireFrame is the default setting; it connects the points by straight line segments on a transparent surface.
- Hidden Line is the same as WireFrame, except that the surface is opaque.
- Points displays only the computed points.

Surface Mesh   The Surface Mesh controls the type of curves that are drawn on the surface.

- None with Color Patch draws a surface by using color only. (None has no effect on other surface styles.)
- ULines adds curves with constant x-coordinates onto the surface.
- VLines adds curves with constant y-coordinates onto the surface.
- Mesh adds curves in both directions onto the surface; that is, it plots both ULines and VLines.
- With MuPAD, Contour has the same effect as VLines.

If you do not see all these choices in the dialog, use the up arrow and down arrow to display the complete list.

Item Coloring   You can select the Directional Shading from among XYZ, XY, Z, Z Hue, Z Grayscale, and Flat. Choice Z means that the color choice depends on the z-coordinate only, whereas XYZ means that color is a function of all three coordinates.

- For each choice of Directional Shading, you can select both the Base Color and Secondary Color from among 20 named colors, or from additional unnamed colors, or you can customize your own colors.

Following are some examples of different Surface Style, Surface Mesh, and Directional Shading.

Surface Style: Color Patch
Surface Mesh: Mesh
Directional Shading: Z[Greyscale]

Surface Style: Color Patch
Surface Mesh: None
Directional Shading: Z[Greyscale]

Surface Style: Color Patch
Surface Mesh: Contour (VLines)
Directional Shading: Z[Greyscale]

Surface Style: Wire Frame
Surface Mesh: Mesh
Directional Shading: Flat

Surface Style: Hidden Line
Surface Mesh: Mesh
Directional Shading: Flat

Surface Style: Hidden Line
Surface Mesh: Contour (VLines)
Directional Shading: Flat

Surface Style: Point (Cross)
Surface Mesh: None
Directional Shading: Flat

Axes Page

You can specify the appearance of the axes as **Normal, None, Boxed,** or **Frame.**

- **Normal** gives the usual x-, y-, and z-axes if the origin is contained inside the view box; otherwise, it draws three axes as close as possible to the origin.

- **Boxed** shows the plot inside a box frame.

- **Frame** displays the left edge and two lower edges of the box.

    All of these choices (except **None**) display numerical labels on the axes or box edges.

The axes of the independent variables will be labeled according to the variable names you use. You also have options to

- Customize the x-axis, y-axis and z-axis by setting a specified number of tick marks for each or disable tick labeling.

- Add custom labels to the x-axis, y-axis and z-axis.

- Set **Equal Scaling Along Each Axis.**

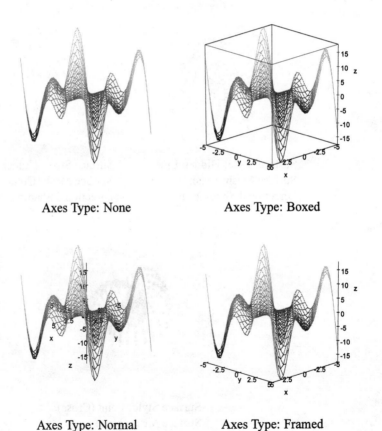

Axes Type: None   Axes Type: Boxed

Axes Type: Normal   Axes Type: Framed

Labeling Page

The options on the Labeling page of the Plot Properties dialog are basically the same for all plots and graphics. See page 161 for a discussion of these options, or choose Help + Search and choose one of the graphics topics under Labels.

3D Plots of Functions and Expressions

You can plot a wide variety of surfaces with Plot 3D. This section provides examples of Rectangular, Cylindrical, Spherical, Implicit, and Tube plots. The details of Gradient and Vector Field plots are provided in Vector Calculus (see page 370).

If you are using *Scientific Notebook*, you may see a reduced Compute menu that shows only Rectangular on the Plot 3D submenu. To use the other 3D plotting techniques (except Plot 3D Implicit), go to Tools + Engine Setup, General page, and uncheck Display Simplified Compute Menu.

Expressions

▶ **To plot an expression involving two variables**

1. Enter the expression in your document.

2. Place the insertion point in the expression.

3. Choose Plot 3D + Rectangular.

▶ Plot 3D + Rectangular

$\sin x + \cos y$

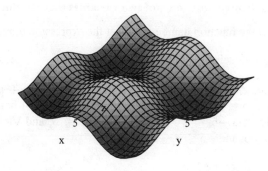

▶ **To add an expression involving two variables to an existing 3D plot**

1. Select the expression.

2. Drag the expression onto the plot.

or

1. Choose Plot Properties + Items Plotted.

2. Click Add Item.

3. Type or Paste the expression in the Item box.

See the next section for an example with defined functions.

Defined Functions

You can plot a defined function of two variables in two different ways. Recall that you define a function such as

$$f(x,y) = \frac{xy}{(x^2 + y^2)^2}$$

by placing the insertion point in the expression and choosing New Definition from the Definitions submenu.

▶ **To plot a defined function f of two variables**

1. Select the function name f or select the expression $f(x,y)$.

2. From the Plot 3D submenu, choose Rectangular.

▶ **To add a defined function g of two variables to a 3D plot**

1. Select the function name g or select the expression $g(x,y)$.

2. Drag your selection onto the plot.

For the example that follows, define $f(x,y) = x^2 + y^2$ and $g(x,y) = -5$. This example shows 3D rectangular plots of $f(x,y)$ and of both $f(x,y)$ and $g(x,y)$, with Plot Intervals $-5 \le x \le 5$ and $-5 \le y \le 5$, and View Intervals $-5 \le x \le 5$, $-5 \le y \le 5$, and $-5 \le z \le 50$.

▶ Plot 3D + Rectangular

$f(x,y)$ Add $g(x,y)$ for the second plot.

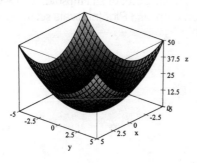 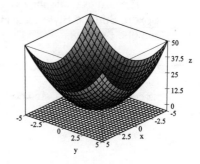

Parametric Plots

Parameterized surfaces in rectangular coordinates are given by equations of the form $x = f(s,t)$, $y = g(s,t)$, and $z = h(s,t)$. These are very general and allow you to generate a wide variety of interesting plots.

▶ **To plot a parameterized surface**

1. Enter expressions in a vector, making each expression a separate component.

2. Place the insertion point in the vector and choose **Plot 3D + Rectangular**.

The parameterized surface shown in the following examples, which is defined by the equations

$$x = s \sin s \cos t$$
$$y = s \cos s \cos t$$
$$z = s \sin t$$

can be created as the 3D rectangular plot of the vector $[s \sin s \cos t, s \cos s \cos t, s \sin t]$. In the following plot, $0 \le s \le 2\pi$ and $0 \le t \le \pi$.

▶ Plot 3D + Rectangular

$[s \sin s \cos t, s \cos s \cos t, s \sin t]$

The next example shows a 3D parametric plot of an ellipsoid, with **Axes Type** set to **Boxed**, **Surface Style** set to **Hidden Line**, and **Plot Intervals** set to $-1.57 \leq s \leq 1.57$ and $0 \leq t \leq 6.28$.

▶ **Plot 3D + Rectangular**

$[2 \cos t \cos s, 3 \sin s, \sin t \cos s]$

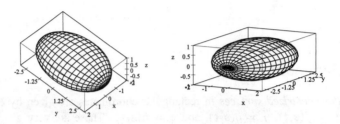

The next example shows a 3D parametric plot of a hyperboloid of one sheet, with **Axes Type** set to **None**, **Surface Style** set to **Hidden Line**, and **Plot Intervals** $-1 \leq s \leq 1$, $-3.14 \leq t \leq 3.14$.

▶ **Plot 3D + Rectangular**

$[2 \sec s \sin t, 3 \sec s \cos t, \tan s]$

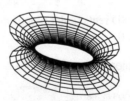

The next example shows a 3D parametric plot of a hyperboloid of two sheets, with **Boxed** axes, **Hidden Line** style, and **Plot Intervals** $0 \leq s \leq 1.4$, $-3.1416 \leq t \leq 3.1416$.

▶ **Plot 3D + Rectangular**

$[2 \tan s \sin t, 3 \tan s \cos t, \sec s]$

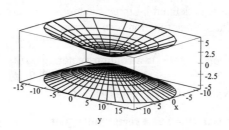

The next example shows a 3D parametric plot of two planes, with **Hidden Line** style and **Plot Intervals** $-5 \leq x \leq 5$, $-5 \leq y \leq 5$.

► Plot 3D + Rectangular

(x, x, z)

Drag $(x, -x, z)$ to the plot

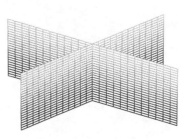

Note If the *Reference Library* is installed on your system, you can find equations and plots for many more parametric plots. Look under *Tables, reference: Surfaces and Curves in Space*. (A "Typical" installation does not include the *Reference Library* for space considerations, but if you have space on your system you can copy the files from the installation CD or add them with a "Custom" install.)

Curves in Space

A space curve is defined by three functions $x = f(t)$, $y = g(t)$, $z = h(t)$ of a single variable. These three functions can be presented as a row vector: $\begin{bmatrix} f(t) & g(t) & h(t) \end{bmatrix}$ or $\begin{pmatrix} f(t) & g(t) & h(t) \end{pmatrix}$; a column vector: $\begin{bmatrix} f(t) \\ g(t) \\ h(t) \end{bmatrix}$ or $\begin{pmatrix} f(t) \\ g(t) \\ h(t) \end{pmatrix}$; or as a fenced list: $(f(t), g(t), h(t))$ or $[f(t), g(t), h(t)]$.

▶ **To plot a space curve as a rectangular plot**

1. Enter the three defining expressions as the components of a three-element vector.

2. With the insertion point in the vector, choose Plot 3D + Rectangular.

To change the appearance, open the Plot Properties dialog, go to the appropriate page, and change the settings.

▶ Plot 3D + Rectangular

$$\begin{bmatrix} t & 2\sin t & t^2 \end{bmatrix}$$

For a smooth plot, you may need to increase the number of points plotted. In the following plot, Points Sampled is set to 200.

▶ Plot 3D + Rectangular

$$\begin{bmatrix} -10\cos t - 2\cos(5t) + 15\sin(2t) \\ -15\cos(2t) + 10\sin t - 2\sin(5t) \\ 10\cos(3t) \end{bmatrix}$$

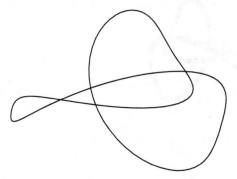

You can create a "fat curve" by using **Plot 3D Tube** and specifying a radius for the curve in the **Plot Properties** dialog box. This radius can be constant or can be a function of t. The **Sample Size** is the number of computed points along the curve; the **Number of Tube Points** is the number of computed points in a cross section of the tube. **Ranges** refers to the range of computed values for the parameter t. The **View Intervals** include intervals for x, y, and z of the form $x_0 \leq x \leq x_1$, $y_0 \leq y \leq y_1$, $z_0 \leq z \leq z_1$.

▶ **To plot a space curve as a tube plot**

1. Enter the three defining expressions as the components of a three-element vector.

2. With the insertion point in the vector, from the **Plot 3D** submenu, choose **Tube**.

3. To change the radius, open the **Plot Properties** dialog and change the setting on the **Items Plotted** page.

The "fat curve" is designed to show which parts of the curve are close to the observer and which are far away. Otherwise, a curve in space is difficult to visualize. In the following example, **Radius** is set to 1, the **Plot Interval** is set to $0 \leq t \leq 6.28$ ($\approx 2\pi$), and **Surface Style** is set to **Hidden Line**. To draw the "thin curve" as a tube plot, set **Radius** to 0.

▶ Plot 3D + Tube

$$
\begin{bmatrix}
-10\cos t - 2\cos(5t) + 15\sin(2t) \\
-15\cos(2t) + 10\sin t - 2\sin(5t) \\
10\cos(3t)
\end{bmatrix}
$$

By typing an expression in t for the radius and choosing the curve to be a straight line, you can get surfaces of revolution. In the following example, the radius is set to $1 - \sin t$, the range for t is $-2\pi \leq t \leq 2\pi$, and the setting for **Points per Cross Section** is 30.

▶ Plot 3D + Tube

$[t, 0, 0]$

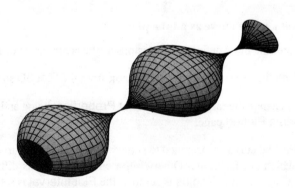

The spine of the surface of revolution can be any line, as illustrated by the next example plotted with **Radius:** $4 + \sin 3t + 2\cos 5t$, **Axes: Frame, Style: Hidden Line**, and **Plot Interval:** $-5 \leq t \leq 5$.

▶ Plot 3D + Tube

$(2t, -3t, t)$

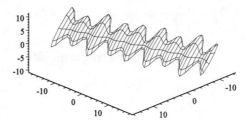

Polygonal Paths

You can plot the polygon whose vertices lie at the points
$$\{(x_1, y_1, z_1), (x_2, y_2, z_2), (x_3, y_3, z_3), \ldots, (x_n, y_n, z_n)\}$$
by entering the three-column or three-row matrix

$$\begin{bmatrix} x_1 & y_1 & z_1 \\ x_2 & y_2 & z_2 \\ x_3 & y_3 & z_3 \\ \vdots & \vdots & \vdots \\ x_n & y_n & z_n \end{bmatrix} \text{ or } \begin{bmatrix} x_1 & x_2 & x_3 & \vdots & x_n \\ y_1 & y_2 & y_3 & \vdots & y_n \\ z_1 & z_2 & z_3 & \vdots & z_n \end{bmatrix}$$

or by entering the fenced list
$$(x_1, y_1, z_1, x_2, y_2, z_1, x_3, y_3, z_1, \ldots, x_n, y_n, z_n)$$
and choosing **Plot 3D + Rectangular**. The points are connected with straight-line segments as in the following box.

▶ Plot 3D + Rectangular

$(0, 0, 0, 0, 1, 0, 1, 1, 0, 1, 0, 0, 0, 0, 0)$

- Select and drag to the plot each of $(0, 0, 1, 0, 1, 1, 1, 1, 1, 1, 0, 1, 0, 0, 1)$, $(1, 1, 0, 1, 1, 1, 0, 1, 1, 0, 1, 0)$, and $(1, 0, 0, 1, 0, 1, 0, 0, 1, 0, 0, 0)$

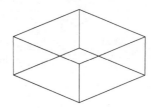

▶   Plot 2D + Rectangular

$$\begin{bmatrix} 1 & 0 & 0 \\ \cos\frac{4\pi}{5} & \sin\frac{4\pi}{5} & 0 \\ \cos\frac{8\pi}{5} & \sin\frac{8\pi}{5} & 0 \\ \cos\frac{2\pi}{5} & \sin\frac{2\pi}{5} & 0 \\ \cos\frac{6\pi}{5} & \sin\frac{6\pi}{5} & 0 \\ 1 & 0 & 0 \end{bmatrix}, \text{Select and drag to the plot} \begin{bmatrix} 1 & 0 & 1 \\ \cos\frac{4\pi}{5} & \sin\frac{4\pi}{5} & 1 \\ \cos\frac{8\pi}{5} & \sin\frac{8\pi}{5} & 1 \\ \cos\frac{2\pi}{5} & \sin\frac{2\pi}{5} & 1 \\ \cos\frac{6\pi}{5} & \sin\frac{6\pi}{5} & 1 \\ 1 & 0 & 1 \end{bmatrix}$$

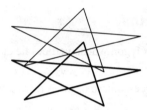

You can plot polygonal paths by adjusting the setting for **Points Sampled**.

▶   Plot 2D + Rectangular

$(\cos t, \sin t, 0)$

Select and drag to the frame $(\cos t, \sin t, 1)$

For **Items** 1 and 2: **Plot Intervals:** $0 \le t \le 12.566$, **Points Sampled:** 6.

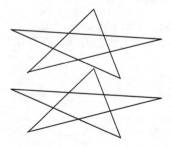

Cylindrical Coordinates

In the *cylindrical coordinate system*, a point P is represented by a triple (r, θ, z), where (r, θ) represents a point in polar coordinates and z is the usual rectangular third coordinate. Thus, to convert from cylindrical to rectangular coordinates, we use the equations

$$x = r \cos \theta \qquad y = r \sin \theta \qquad z = z$$

To go from rectangular to cylindrical coordinates, we use the equations

$$r^2 = x^2 + y^2 \qquad \tan \theta = \frac{y}{x} \qquad z = z$$

The default assumption is that r is a function of θ and z. As usual, you can plot several surfaces on the same axes by dragging expressions onto a plot.

Expressions

▶ **To make a cylindrical plot of an expression**

1. Enter the expression in your document.

2. With the insertion point in the expression, from the Plot 3D submenu choose Cylindrical.

The following examples show a plot of the cylinder $r = 1$ and the cone $r = 1 - z$, obtained as the 3D cylindrical plot of the expressions 1 and $1 - z$, with $0 \leq \theta \leq 2\pi$, and $0 \leq z \leq 1$.

▶ Plot 3D + Cylindrical

1

$1 - z$

1 $1 - z$

The following example shows a plot of the cylinders $r = 1$ and $r = 1 - z$ obtained as the 3D cylindrical plot of the expressions 1 and $1 - z$, with $0 \leq \theta \leq 2\pi$, and $0 \leq z \leq 1$. To get this plot, either plot both expressions together (as indicated), or drag one of the expressions onto a plot of the other. The Surface Style for the first view is WireFrame and for the second is Hidden Line.

▶ Plot 3D + Cylindrical

> 1

- Select and drag to the frame $1 - z$

 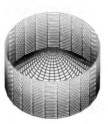

Surface Style: Wire Frame   Surface Style: Hidden Line

Defined Functions

For the following cylindrical plot, define $r(\theta, z) = (z + \sin \theta)^2$ with **New Definition** from the **Definitions** submenu. Use the limits $-3.1416 \leq \theta \leq 3.1416$ and $-5 \leq z \leq 5$. Choose **Surface Style: Color Patch** and **Surface Mesh: Mesh**.

▶ Plot 3D + Cylindrical

> $(r(\theta, z))^2$

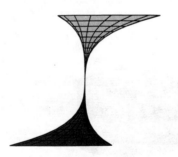

Parameterized Surfaces in Cylindrical Coordinates

You can create cylindrical plots of parameterized surfaces defined by functions such as $z(r, \theta) = r + \cos \theta$. You plot the parameterized surface $r = f(s, t)$, $\theta = g(s, t)$, $z = h(s, t)$ in cylindrical coordinates by entering the expressions for r, θ, and z into a vector $(f(s, t), g(s, t), h(s, t))$ and choosing **Cylindrical** from the **Plot 3D** submenu.

▶ **To create a parameterized cylindrical plot**

1. Enter the three defining expressions for r, θ, and z as the components of a vector.

2. With the insertion point in the vector, choose **Plot 3D + Cylindrical**.

The following example shows the "spiral staircase" $z = \theta$, a 3D cylindrical plot of the vector $[r, \theta, \theta]$, with $0 \leq r \leq 1$, $0 \leq \theta \leq 4\pi$, and.Surface Style set to Hidden Line.

▶ Plot 3D + Cylindrical

$[r, \theta, \theta]$

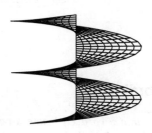

The following examples show a sphere of radius 1, and a sphere together with a cylinder as 3D cylindrical plots of the vector $\left[\sqrt{1 - z^2}, \theta, z\right]$, with $0 \leq \theta \leq 2\pi$, **Surface Style** set to **Hidden Line**, and **Equal Scaling Along Each Axis**. The first plot uses $-1 \leq z \leq 1$ and the second $-\frac{\sqrt{3}}{2} \leq z \leq \frac{\sqrt{3}}{2}$ for both items.

▶ Plot 3D + Cylindrical

$\left[\sqrt{1 - z^2}, \theta, z\right]$

Select and drag to the plot 0.5

A hyperboloid of one sheet appears in this example as a 3D cylindrical plot of the vector $[1 + z^2, \theta, z]$, with $-1 \leq z \leq 1$, $0 \leq \theta \leq 2\pi$, **Surface Style** set to **Hidden Line**, and **Equal Scaling Along Each Axis**.

▶ Plot 3D + Cylindrical

$$\left[\sqrt{1-z^2},\theta,z\right]$$

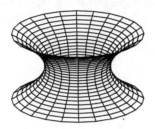

Spherical Coordinates

The *spherical coordinates* (ρ,θ,ϕ) locate a point P in space by giving the distance ρ from the origin, the angle θ projected onto the xy-plane (the *polar angle*), and the angle ϕ with the positive z-axis (the *vertical angle*). The conversion into rectangular coordinates is given by

$$x = \rho\sin\phi\cos\theta \qquad y = \rho\sin\phi\sin\theta \qquad z = \rho\cos\phi$$

and the distance formula implies

$$\rho^2 = x^2 + y^2 + z^2$$

The default assumption is that ρ is a function of ϕ and θ. You can use other names for the polar and vertical angles. Any two variables you give will be interpreted as the polar and vertical angles. Even when you use the standard notation, however, the roles of the variables may be reversed in the default interpretation from what you intended. You can correct this interpretation with the **Switch Variables** option in the **Plot Properties** dialog box.

You can plot more than one surface on the same axes by dragging additional expressions to the plot or by adding additional items on the **Items Plotted** page of the **Plot Properties** dialog.

Expressions

▶ **To make a spherical plot**

1. Enter an expression involving θ and ϕ in your document.

2. Place the insertion point in the expression.

3. From the **Plot 3D** submenu, choose **Spherical**.

A sphere and a cylinder can each be plotted as a function of the radius. Below is a sphere of radius 2, and a sphere of radius 4 with a cylinder of radius 3 removed from its center. For the latter, set $-2.65 \leq z \leq 2.65$ for the View. For both, take $0 \leq \theta \leq 2\pi$ and $0 \leq \phi \leq \pi$, set Equal Scaling on Each Axis, and choose Hidden Line for Surface Style.

▶ Plot 3D + Spherical

2

4

Select and drag to the plot $3 \csc \phi$

Changing the setting for Points Sampled of θ to 4 creates a solid with a triangular cross section. In the following example, take $0 \leq \theta \leq 2\pi$, $-1 \leq z \leq 1$, and $0 \leq \phi \leq \pi$.

▶ Plot 3D + Spherical

2

Defined Functions

You can create a plot of a function defined in spherical coordinates $\rho = \rho(\theta, \phi)$.

▶ **To make a spherical plot of a defined function ρ of θ and ϕ**

1. Define ρ as a function of θ and ϕ using New Definition on the Definitions submenu.

2. Select the function name ρ or select the expression $\rho(\theta, \phi)$.

3. From the Plot 3D submenu, choose Spherical.

Example   To plot the nautilus determined by the expression $(1.2)^\phi \sin(\theta)$, you can do any one of the following:

- Plot the expression $(1.2)^\phi \sin(\theta)$ and then choose Switch Variables on the Items Plotted page of the Plot Properties dialog. Use the ranges $-1 \le \phi \le 2\pi$ and $0 \le \theta \le \pi$ to get the view of the nautilus shown below.

- Define $\rho(\theta, \phi) = (1.2)^\phi \sin(\theta)$, plot the expression $\rho(\theta, \phi)$, and then choose Switch Variables on the Items Plotted page of the Plot Properties dialog. Use the ranges $-1 \le \phi \le 2\pi$ and $0 \le \theta \le \pi$ to get the view of the nautilus shown below.

- Define the function $\rho(\phi, \theta) = (1.2)^\phi \sin(\theta)$ and plot the function name ρ. (Note the variables are already switched here.) Use the ranges $-1 \le \theta \le 2\pi$ and $0 \le \phi \le \pi$ to get the view of the nautilus shown below.

▶ Plot 3D + Spherical

ρ

Note   The Switch Variables option is not available for a plot that depends only on the function name.

Parameterized Surfaces in Spherical Coordinates

Parameterized surfaces in spherical coordinates are given by equations of the form $\rho = f(s,t)$, $\theta = g(s,t)$, and $\varphi = h(s,t)$. This approach is very general and allows you to generate a wide variety of interesting plots.

▶ **To plot a parameterized surface**

1. Enter the defining expressions as the three components of a vector.

2. With the insertion point in the vector, from the Plot 3D submenu, choose Spherical.

The 3D spherical plot of the vector $[\rho, \theta, 1]$ gives the cone $\varphi = 1$. For the following plot, the view is set with $-1 \le \rho \le 1$ and $0 \le \theta \le 2\pi$.

▶ Plot 3D + Spherical

$[\rho, \theta, 1]$

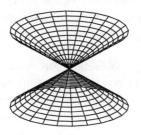

You can plot the surface defined by $\rho = s$, $\theta = s^2 + t^2$, $\varphi = t$ by entering the three expressions as coordinates of a vector. For the following plot, take $0 \le s \le 1$ and $-1 \le t \le 1$.

▶ Plot 3D + Spherical

$$\left(\begin{array}{ccc} s & s^2 + t^2 & t \end{array} \right)$$

Plot Snapshots

When you preview or print with File + Preview or File + Print, the plots you see on the screen are also generated for preview and printing. However, a plot snapshot is required when you preview, print, or compile with any of the Typeset options available with *Scientific Word* or *Scientific WorkPlace*, use any LaTeX compiler to produce a DVI file of your document, or export to an RTF document. *Plot snapshots* are windows metafiles, computer graphics files of type WMF. They are generated with a random file name of the form EWHCUS00.WMF and stored in the same directory as your document. Plot snapshots are generated either by automatic snapshot generation or from the View page of the Plot Properties dialog.

Plot snapshots are not required for viewing plots on your screen or for printing or previewing from the File menu. However, the program for displaying plots on the screen uses the plot snapshot, if it exists, rather than recomputing the plot, enabling you to scroll more quickly through a document. You can also import these snapshot files as pictures (see page 209).

Automatic Snapshot Generation, which creates a corresponding snapshot file any time it creates a new plot, has the following advantages and disadvantages.

- *Advantage:* Plot snapshots are required for typesetting documents and for viewing plots from programs without an active compute engine.

- *Advantage:* Plot snapshots can be imported as pictures.

- *Advantage:* Having snapshots for all plots speeds viewing on screen and speeds previewing and printing from the File menu.

- *Disadvantage:* Having snapshots for all plots adds (possibly numerous) WMF files to your disk that you may not need.

Your system may or may not be set for automatic snapshot generation. You can change the default for automatic snapshot generation as follows.

▶ **To change the default for automatic snapshot generation**

1. From the Tools menu, choose Computation Setup.

2. Click the Plot Behavior tab and change Automatic Snapshot Generation by checking or unchecking the appropriate box.

3. Choose OK.

For individual documents, you can override the global default with a similar dialog under Compute + Settings.

For individual plots, go to the View page of the Plot Properties dialog to generate or to delete snapshot files.

▶ **To generate a plot snapshot**

1. Select the plot and open the Plot Properties dialog.

2. Choose the View page and click Generate Snapshot.

3. Choose OK.

▶ **To remove a plot snapshot**

1. Select the plot and open the Plot Properties dialog.

2. Choose the View page and click Remove Snapshot.

3. Choose OK.

Plots imported from the plot snapshot metafile and viewed on screen as pictures will generate more rapidly than active plots.

▶ **To import a plot snapshot as a picture**

1. Generate a snapshot as described above, and record the Filename that appears on the View page of the Plot Properties dialog.

2. With a file manager, rename the file with an appropriate descriptive name, retaining the WMF extension, e.g. myplot.wmf.

3. Open the document and, with the insertion point in the location for the picture, choose File + Import Picture. Go to the appropriate directory and select myplot.wmf.

This will no longer be an active plot, but will have the same appearance as one. Pictures made from plot snapshots are sometimes preferable for on screen viewing as they do not take any significant time to generate. Documents with snapshots or pictures can also be viewed with systems that do not have an active computing engine, such as *Scientific Word* and *Scientific Viewer*, the free browser version of the program. Both the .tex file and the picture files must be present.

Snapshots imported as pictures can be given to students as on-line puzzles—although the plots themselves contain all the information in the Plot Properties dialog that is needed to replicate them when viewed in *Scientific Notebook*, the snapshot pictures contain no such information. Students can recover an approximate formula for the plot by knowing the appropriate form for the formula and solving appropriate systems of equations.

Setting Plot Default Options

You can set defaults for plot generation and display. These defaults can either apply to all documents (global defaults) or only to the active document (local defaults).

Universal Default Options For Plots

Global defaults can be set from the Tools menu.

▶ **To set global defaults for** Plot Layout **settings**

1. From the Tools menu, choose Computation Setup.

2. Click the Plot Layout tab.

3. Change Size, Screen Display Attributes, Units, Print Attributes, and Placement by checking appropriate boxes or making an appropriate selection from a list.

▶ **To set global defaults for** Plot Behavior **settings**

1. From the Tools menu, choose Computation Setup.

2. Click the Plot Behavior tab.

3. Check or uncheck each of the listed behaviors and choose OK.

The behaviors are the following.

- Generate Plot Snapshots Automatically
  See page 208 for details on Automatic Snapshot Generation.

- Recompute Plot When Definitions Change
  If functions or parameters determining the plot are defined, you can redefine them to change the graph in the plot. This is an interactive way to do experiments on line.

- Display Plot Properties Dialog Before Plotting
  This behavior gives you the opportunity to reset Plot Intervals or other features before the system makes the first attempt to plot your expression.

Remark   If you do not have the option Display Plot Properties Dialog Before Plotting checked in Computation Setup, you can evoke this behavior for individual plots by holding down the CTRL key when starting the plot.

▶ **To set global defaults for** 2D Plots

1. From the Tools menu, choose Computation Setup.

2. Click the 2D Plots tab.

3. Change Default Plot Style, Line Style, Line Thickness, Point Marker, Axes

Type, Plot Color, and Intervals Plotted by checking the appropriate box or making an appropriate selection from a list.

4. Under Intervals Plotted, choose any plot type and specify default settings for Plot Intervals and Sample Size.

5. Choose OK.

▶ **To set global defaults for** 3D Plots

1. From the Tools menu, choose Computation Setup.

2. Click the 3D Plots tab.

3. Change Line Style, Line Thickness, Point Marker, Axes Type, Surface Style, Surface Mesh, Item Coloring, and Intervals Plotted by checking the appropriate box or making an appropriate selection from a list.

4. Under Intervals Plotted, choose any plot type and specify default settings for Plot Intervals and Points Plotted.

5. Choose OK.

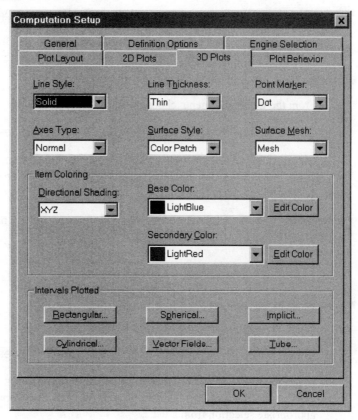

Default Plot Options for a Document

The first attempt at a plot uses the default parameters that are set on the Plot Layout, Plot Behavior, and either 2D Plots or 3D Plots pages of the Tools + Computation Setup dialog. You can change these defaults for a single document in the Compute + Settings dialog.

▶ **To set local defaults for Plot Layout settings**

1. From the Compute menu, choose Settings.

2. Choose the Plot Layout page and check Set Document Values.

3. Change Size, Screen Display Attributes, Units, Print Attributes, and Placement by checking appropriate boxes.

4. Choose OK.

▶ **To set local defaults for Plot Behavior settings**

1. From the Compute menu, choose Settings.

2. Choose the Plot Behavior page and check Set Document Values.

3. Check or uncheck each of the listed behaviors and choose OK.

▶ **To set local defaults for 2D Plots**

1. From the Compute menu, choose Settings.

2. Choose the 2D Plots page and check Set Document Values

3. Change Default Plot Style, Line Style, Line Thickness, Point Marker, Axes Type, and Plot Color by checking the appropriate box or making an appropriate selection from a list.

4. Under Intervals Plotted, choose any plot type and specify default settings for Plot Intervals and Sample Size.

5. Choose OK.

▶ **To set local defaults for 3D Plots**

1. From the Compute menu, choose Settings.

2. Choose the 3D Plots page and check Set Document Values.

3. Change Line Style, Line Thickness, Point Marker, Axes Type, Surface Style, Surface Mesh, and Item Coloring by checking the appropriate box or making an appropriate selection from a list.

4. Under Intervals Plotted, choose any plot type and specify default settings for Plot Intervals and Sample Size.

5. Choose OK.

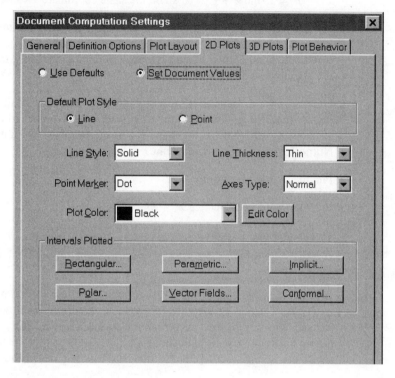

▶ **To set local defaults for** Plot Behavior

1. From the Compute menu, choose Settings.

2. Choose the Plot Behavior page and check Set Document Values.
   - Check Generate Plot Snapshots Automatically if you wish this behavior.
   - Check Recompute Plot when Definitions Change if you wish this behavior.
   - Check Display Plot Properties Dialog Before Plotting if you wish to set plotting options each time you make a plot, rather than using default settings.

3. Choose OK.

Remark   If you do not have the option Display Plot Properties Dialog Before Plotting checked in Compute + Settings, you can evoke this behavior for individual plots by holding down the CTRL key when starting the plot.

Exercises

1. Use Implicit under the Plot 2D submenu to plot the conic sections $x^2 + y^2 = 1$, $x^2 - y^2 = 1$, and $x + y^2 = 0$ all on the same coordinate axes.

2. Use Implicit under the Plot 2D submenu to plot the conic sections $(x - 1)^2 + (y + 2)^2 = 1$, $(x - 1)^2 - (y + 2)^2 = 1$, and $(x - 1) + (y + 2)^2 = 0$ on one pair of coordinate axes. With the hand symbol visible over the view, translate the view so that the curves match the curves in Exercise 1. In which direction did the axes move?

3. Plot $x^2 + y^2 = 4$ and $x^2 - y^2 = 1$ together. How many intersection points are there? Zoom in on the one in the first quadrant to estimate where the curves cross each other. Verify your estimate by typing the formulas into a matrix and choosing Numeric from the Solve submenu.

4. Plot the astroid $x^{2/3} + y^{2/3} = 1$.

5. Plot the folium of Descartes $x^3 + y^3 = 6xy$.

6. Plot the surface $z = \sin xy$, with $-4 \le x \le 4$ and $-4 \le y \le 4$. Compare the location of the ridges with the implicit plot of the three curves $xy = \frac{\pi}{2}$, $xy = \frac{3\pi}{2}$, and $xy = \frac{5\pi}{2}$.

7. A standard calculus problem involves finding the intersection of two right circular cylinders of radius 1. View this problem by choosing Rectangular from the Plot 3D submenu to plot the two parametric surfaces $[s, \cos t, \sin t]$ and $[\cos t, s, \sin t]$. Create a second view by choosing Plot 3D + Tube to plot the line segment $[0, 0, t]$ where $-1 \le t \le 1$, setting the Radius to $\sqrt{2}\sqrt{1 - t^2}$ and Tube Points to 5.

8. Do the two space curves
$$[(2 + \sin t)10\cos t, \ (2 + \cos t)10\sin t, 3\sin 3t]$$
and
$$[20\cos t, 20\sin t, -3\sin 3t]$$
intersect? Use Tube from the Plot 3D submenu and rotate the curves to find out.

9. View the intersection of the sphere $x^2 + y^2 + z^2 = 1$ and the plane $x + y + z = \frac{1}{2}$ by expressing these equations in parametric form and choosing Rectangular from the Plot 3D submenu. Verify that the points of intersection lie on an ellipse (it is actually a circle) by solving $x + y + z = \frac{1}{2}$ for z, substituting this value into the equation $x^2 + y^2 + z^2 = 1$, and calculating the discriminant (see page 85) of the resulting equation.

10. Explore the meaning of horizontal and vertical lines by plotting the surface $z = xy$. Choose Color Patch and VLines as Surface Style and Surface Mesh. Rotate the surface until only the top face of the cube is visible, and interpret the meaning of the curves that you see. Rotate the cube until the top face just disappears, and interpret the meaning of the contours that appear.

Solutions

1. Plot 2D + Implicit: $x^2 + y^2 = 1, x^2 - y^2 = 1, x + y^2 = 0$

   (Take $-2 \leq x \leq 2$ and $-2 \leq y \leq 2$. Choose **Equal Scaling Along Each Axis.**)

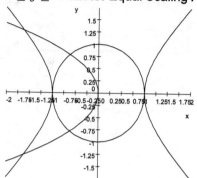

2. Plot 2D + Implicit: $(x-1)^2 + (y+2)^2 = 1, (x-1)^2 - (y+2)^2 = 1, (x-1) + (y+2)^2 = 0$

   (Take $-1 \leq x \leq 3$ and $-4 \leq y \leq 0$. Choose **Equal Scaling Along Each Axis.**)

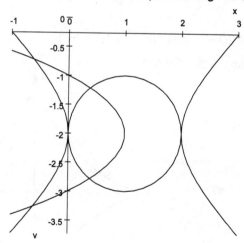

3. Plot 2D + Implicit: $x^2 + y^2 = 4, x^2 - y^2 = 1$

   (Take $-5 \leq x \leq 5$ and $-5 \leq y \leq 5$. Choose **Equal Scaling Along Each Axis.**)

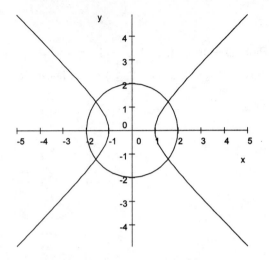

Solve + Numeric: $\begin{aligned} x^2 + y^2 &= 4 \\ x^2 - y^2 &= 1 \\ x &\in (1,2) \\ y &\in (1,2) \end{aligned}$, Solution is : $\{x = 1.58113883, y = 1.224744871\}$

4. Plot 2D + Implicit: $|x|^{2/3} + |y|^{2/3} = 1$

(Take $-1 \le x \le 1$ and $-1 \le y \le 1$.)

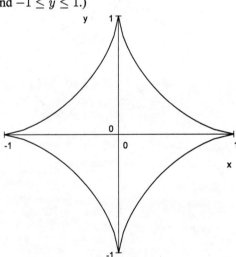

Without the absolute values, you obtain only the first quadrant portion of the graph.

5. Plot 2D + Implicit: $x^3 + y^3 = 6xy$

(Take $-5 \le x \le 5$ and $-5 \le y \le 5$ and set the grid to 50 by 50.)

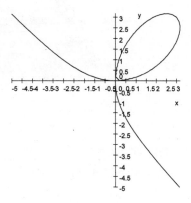

Notice how the folium of Descartes shows up as a level curve on the surface $z = x^3 + y^3 - 6xy$.

Plot 3D + Rectangular: $x^3 + y^3 - 6xy$

Plot 3D + Rectangular: 0

(Take $-5 \le x \le 5$, $-5 \le y \le 5$. Set **Turn** to 16 and **Tilt** to 1. For the surface, use **Hidden Line** and **Mesh**, and for the plane, **Color Patch** and **None**.)

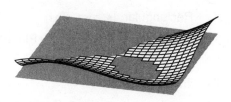

6. Plot 3D + Rectangular: $\sin xy$

(Choose **Patch** and **VLines** and take $-4 \le x \le 4$, $-4 \le y \le 4$. Set **Turn** to 108 and **Tilt** to 17.)

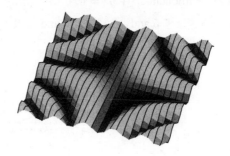

Plot 2D + Implicit: $xy = \pi/2, xy = 5\pi/2, xy = 3\pi/2$

(Take $-4 \le x \le 4$ and $-4 \le y \le 4$.)

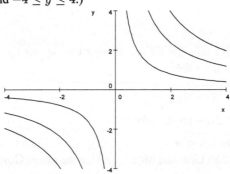

7. **Plot 3D + Rectangular:** $[s, \cos t, \sin t]$

Drag $[\cos t, s, \sin t]$ onto the plot, and set **Plot Intervals** at $-2 \le s \le 2$ and $0 \le t \le 2\pi$.

Plot 3D + Tube: $[0, 0, t]$

Set **Plot Intervals** at $-1 \le t \le 1$, **Radius** to $\sqrt{2}\sqrt{1 - t^2}$, and **Points per Cross Section** to 4.

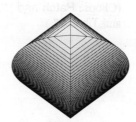

8. Plot 3D + Tube:
$$\begin{bmatrix} (2+\sin t)10\cos t \\ (2+\cos t)10\sin t \\ 3\sin 3t \end{bmatrix}$$

Drag $\begin{bmatrix} 20\cos t \\ 20\sin t \\ -3\sin 3t \end{bmatrix}$ onto the plot. Set $0 \le t \le 2\pi$. Set **Radius** for both items to 1.

Solve + Exact:
$$\begin{bmatrix} (2+\sin t)10\cos t = 20\cos s \\ (2+\cos t)10\sin t = 20\sin s \\ 3\sin 3t = -3\sin 3s \end{bmatrix}, \text{ Solution is} : \{t=0, s=0\}, \{t=\pi, s=\pi\}$$

9. Plot 3D + Rectangular: $\left(\sqrt{1-s^2}\cos t, \sqrt{1-s^2}\sin t, s\right)$

Take $-1 \le s \le 1$ and $0 \le t \le 6.283$ (2π). Set style to **Hidden Line** and check Set Equal Scaling Along Each Axis.

Drag to the plot: $\left(s, t, \frac{1}{2} - s - t\right)$

Take $-1 \le s \le 1$, $-1 \le t \le 1$, and set style to **Patch & Contour**.

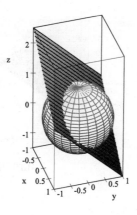

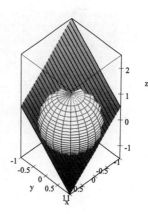

Solving for z on the plane gives $z = \frac{1}{2} - x - y$, giving the equation for points on the intersection of the plane and sphere:

$$x^2 + y^2 + \left(\frac{1}{2} - x - y\right)^2 = 1$$

Expanding this expression yields the equation

$$2x^2 + 2xy + 2y^2 - x - y - \frac{3}{4} = 0$$

for the curve of intersection. The discriminant $B^2 - 4AC$ is

$$2^2 - 4\,(2)\,(2) = -12 < 0$$

which indicates that the curve of intersection is an ellipse (see page 85).

10. Plot 3D + Rectangular: xy

(This is the default plot with settings $-5 \leq x \leq 5$, $-5 \leq y \leq 5$, and style Wire-Frame.)

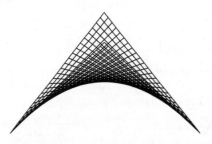

(For the next plots, take $-1 \leq x \leq 1$, $-1 \leq y \leq 1$, and choose the style Patch & VLines.)

Turn 90, Tilt 0   Turn 90, Tilt 90

7 Calculus

This chapter covers the standard topics from differential and integral calculus, including limits, sequences, and series. The notion of a function is fundamental to the study of calculus. Functions were introduced in Chapter 3, Algebra, with a description of procedures for naming expression and functions. Basic information on working with functions and expressions is summarized in Chapter 5, Function Definitions, along with additional information on storing and retrieving definitions. In this chapter we assume that you have read and understand how to define and manipulate functions. We give several examples in this chapter that illustrate connections between calculus and the function plots introduced in Chapter 6, Plotting Curves and Surfaces.

Evaluating Calculus Expressions

You can evaluate calculus expressions in the same manner as expressions from algebra or trigonometry.

▶ **To calculate a derivative or an integral**

1. Enter the derivative or integral using standard mathematical notation.

2. Leave the insertion point in the expression.

3. Click ⬛ ; or from the **Compute** menu, choose **Evaluate**; or press CTRL + E.

▶ **To calculate the derivative $\frac{d}{dx}x\sin x$**

1. Click ⬛ for a fraction, type d in the numerator, and press TAB to take the insertion point to the denominator.

2. Type dx and press SPACEBAR to put the insertion point back in line, then type $x\sin x$.

3. Leave the insertion point in the expression $\frac{d}{dx}x\sin x$ and click ⬛ ; or from the **Compute** menu, choose **Evaluate**; or press CTRL + E.

▶ Evaluate

$$\frac{d}{dx}x\sin x = \sin x + x\cos x$$

▶ **To calculate the integral $\int_0^\pi x \sin x \, dx$**

1. Click $\int$ on the **Math Templates** toolbar, or press CTRL + I.

2. Click N_x on the **Math Templates** toolbar, or press CTRL + DOWN ARROW, and enter the lower limit 0 in the subscript box.

3. Press TAB to take the insertion point to a superscript position and enter the upper limit π.

4. Press SPACEBAR to put the insertion point back in line, and type $x \sin x \, dx$.

5. Leave the insertion point in the expression $\int_0^\pi x \sin x \, dx$ and click $\boxed{=?}$; or, from the **Compute** menu, choose **Evaluate**; or press CTRL + E.

▶ Evaluate

$$\int_0^\pi x \sin x \, dx = \pi$$

Compare the area under the curve $y = x \sin x$ between 0 and π and the area of the rectangle having sides of length 1 and π.

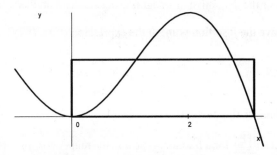

Limits

The concept of a limit is fundamental to the study of calculus. It is the central idea of the subject and is what distinguishes calculus from earlier mathematics. The notion, which encompasses subtle concepts such as instantaneous velocity, can be fully understood only through experience and experimentation. With *Scientific WorkPlace* and *Scientific Notebook*, you have a variety of tools for computing and experimenting with limits.

Notation for Limits

The *limit* of f as x approaches a is L, written $\lim_{x \to a} f(x) = L$, if for each number $\varepsilon > 0$ there exists a number $\delta > 0$ such that $|f(x) - L| < \epsilon$ whenever $0 < |x - a| < \delta$.

▶ **To find a limit of the form** $\lim_{x \to a} f(x)$

1. Type **lim** while in mathematics mode, or click the Math Name button [sin cos] and choose lim from the **Name** list.

2. Click [Nx] and enter the subscript $x \to a$.

3. Press SPACEBAR to put the insertion point back in line, then enter a mathematical expression $f(x)$.

4. Click [=?], or choose **Evaluate**, or press CTRL + E.

▶ Evaluate

$$\lim_{x \to 1} \frac{x^2 - 1}{x - 1} = 2$$

This result is reasonable since $x \neq 1$ implies $\frac{x^2-1}{x-1} = x+1$, which is close to 2 when x is close to 1.

Limits of rational functions are not always apparent. You cannot evaluate the following expression at $x = -3/2$, because the denominator is 0 at this point. The expression does, however, have a limit at $-3/2$.

▶ Evaluate

$$\lim_{x \to -3/2} \frac{4x^4 + 6x^2 + 19x + 6x^3 + 15}{2x^3 + 5x^2 + 5x + 3} = -\frac{25}{7}$$

Factoring the numerator and denominator suggests a method for evaluating this limit by direct substitution.

▶ Factor

$$4x^4 + 6x^2 + 19x + 6x^3 + 15 = (2x + 3)(x + 1)(2x^2 - 2x + 5)$$

$$2x^3 + 5x^2 + 5x + 3 = (2x + 3)(x + x^2 + 1)$$

If an expression has a removable singularity, factoring in place may allow you to fill in the steps leading to evaluation by direct substitution. This is illustrated in the following example, where the second step removes the singularity from the expression. For the first two lines, copy the entire expression after an equals sign, and carry out in-place operations. Then substitute $x = -3/2$ into the expression and **Evaluate**.

$$\lim_{x \to -3/2} \frac{4x^4 + 6x^2 + 19x + 6x^3 + 15}{2x^3 + 5x^2 + 5x + 3} = \lim_{x \to -3/2} \frac{(2x + 3)(x + 1)(2x^2 - 2x + 5)}{(2x + 3)(x + x^2 + 1)}$$

$$= \lim_{x \to -3/2} \frac{(x + 1)(2x^2 - 2x + 5)}{(x + x^2 + 1)}$$

$$= \left[\frac{(x + 1)(2x^2 - 2x + 5)}{(x + x^2 + 1)} \right]_{x = -3/2}$$

$$= -\frac{25}{7}$$

You can carry out the substitution (see page 79) in the preceding example as follows.

▶ **To substitute a value into an expression**

1. Select the expression $\frac{(x+1)(2x^2 - 2x + 5)}{x^2 + x + 1}$ with the mouse, or place the insertion point

   to the left of the expression and press SHIFT + RIGHT ARROW, and click $\boxed{\text{▣}}$.

3. Click $\boxed{\text{N}_\text{x}}$ and enter the subscript $x = -3/2$.

4. Choose **Evaluate**.

▶ **Evaluate**

$$\left[\frac{(x + 1)(2x^2 - 2x + 5)}{x^2 + x + 1} \right]_{x = -3/2} = -\frac{25}{7}$$

You can also carry out a replacement using the editing features.

▶ **To do an automatic replacement of mathematics**

1. Select the expression $\frac{(x+1)(2x^2 - 2x + 5)}{x^2 + x + 1}$ with the mouse, or place the insertion point
   to the left of the expression and press SHIFT + RIGHT ARROW.

2. From the **Edit** menu, choose **Replace**.

3. Fill in the choices in the dialog box in mathematics mode:
   - Search for: x
   - Replace with: $(-3/2)$

The result is the expression

$$\frac{((-3/2) + 1)\left(2(-3/2)^2 - 2(-3/2) + 5\right)}{(-3/2)^2 + (-3/2) + 1}$$

Special Limits

You can compute one-sided limits, limits at infinity, and infinite limits. Define $a = 0$.

▶ Evaluate

$$\lim_{x \to 0+} \frac{x}{|x|} = 1 \qquad\qquad \lim_{x \to 0-} \frac{x}{|x|} = -1$$

$$\lim_{x \to 2+} \frac{x+2}{x-2} = \infty \qquad\qquad \lim_{x \to 2} \frac{x+2}{x-2} = \text{undefined}$$

$$\lim_{x \to 0} \sin\left(\tfrac{1}{x}\right) = [-1, 1] \qquad\qquad \lim_{x \to a+} \frac{\sin x}{x} = 1$$

The notation $[-1, 1]$ produced for one of the preceding limits indicates the interval $-1 \le x \le 1$. Although this output is somewhat unusual, it is explained by the graph of $y = \sin \frac{1}{x}$. As x approaches 0, the function values of $\sin \frac{1}{x}$ continue to include every number in the interval $-1 \le x \le 1$ infinitely often. To obtain the appearance of the following plot, click the plot, choose **Edit + Properties**, and choose the **Items Plotted** page. Click **Variables and Intervals** and increase **Points Sampled** to 500.

▶ Plot 2D + Rectangular

$\sin \frac{1}{x}$

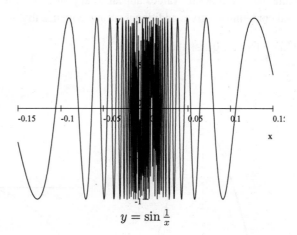

$$y = \sin \tfrac{1}{x}$$

Tables of Values and Plots

You can generate a table of values by applying a function to a vector of domain values and then concatenating matrices, or you can do it in one step by defining appropriate auxiliary functions. The limit $\lim_{x \to 0} \frac{\sin x}{x} = 1$ is of special interest. After evaluating this limit, the following paragraphs examine the behavior of the function $f(x) = \frac{\sin x}{x}$ near the origin, first by looking at numerical evidence and then at plots containing the origin. Two methods are then illustrated for constructing a table of values for this function.

▶ Evaluate

$$\lim_{x \to 0} \frac{\sin x}{x} = 1$$

To see numerical evidence that $\lim_{x \to 0} \frac{\sin x}{x} = 1$, you can evaluate the expression $\frac{\sin x}{x}$ for several values of x near 0. First define a function f to be equal to this expression so that it can be evaluated easily, then evaluate numerically at several points near zero.

▶ Definitions + New Definition

$$f(x) = \frac{\sin x}{x}$$

▶ Evaluate Numerically

$$\begin{aligned} f(0.1) &= 0.9983341665 \\ f(0.01) &= 0.9999833334 \\ f(-0.001) &= 0.9999998333 \end{aligned}$$

Note that the function values appear to approach 1. The graph of $y = \frac{\sin x}{x}$ on an interval containing 0 gives additional strong evidence that $\lim_{x \to 0} \frac{\sin x}{x} = 1$.

▶ Plot 2D + Rectangular$(0.5, 0.96, 0.5, 1)$

$$\frac{\sin x}{x}$$

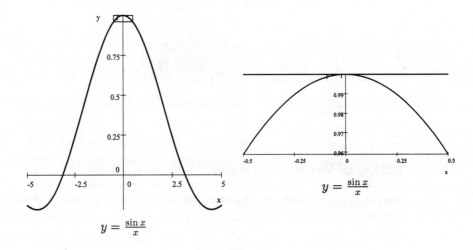

$$y = \frac{\sin x}{x}$$

$$y = \frac{\sin x}{x}$$

Creating a Table of Values Using Auxiliary Functions

The matrix feature called Fill Matrix is useful for creating tables of values.

▶ **To create a table of values for the function $y = f(x)$ by defining auxiliary functions**

1. Define the function $f(x)$.

2. Define a function $g(n)$ to provide a sample of values of the independent variable.

3. Define the function $h(i,j) = (2 - j)g(i) + (j - 1)f(g(i))$.

4. From the Matrices submenu, choose Fill Matrix.

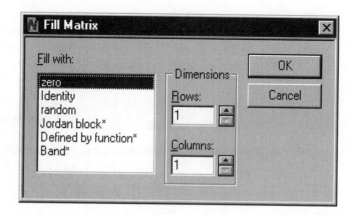

5. Set Columns to 2 and set Rows to match the size of your sample.

6. Under Fill with, choose Defined by function.

7. In the input box for function name, enter h.

8. Choose OK.

The following example illustrates this procedure for the function $f(x) = \dfrac{\sin x}{x}$, with a sample of 10 values for the independent variable.

▶ **To create a table of values for the function $y = \dfrac{\sin x}{x}$**

1. Enter the equation $f(x) = \dfrac{\sin x}{x}$ and, with the insertion point in the equation, choose Definitions + New Definition.

2. Enter the equation $g(i) = i * 10^{-2}$ and, with the insertion point in the equation, choose Definitions + New Definition.

3. Enter the equation $h(i,j) = (2 - j)g(i) + (j - 1)f(g(i))$ and, with the insertion point in the equation, choose Definitions + New Definition.

4. Click the expanding **Square Brackets** button ▣ on the **Math Templates** toolbar and leave the insertion point in the input box.

5. From the **Matrices** submenu, choose **Fill Matrix**.

6. Set **Rows** to 10 and set **Columns** to 2.

7. Under **Fill with**, choose **Defined by function**.

8. In the input box for function name, enter h.

9. Choose **OK**.

The result is the matrix below on the left, in which the numbers in the first column are values of the independent variable, and the numbers in the second column are the corresponding function values. To create the matrix on the right, place the insertion point in or on the right of the matrix and choose **Evaluate Numerically**.

$$
\begin{bmatrix}
\frac{1}{100} & 100\sin\frac{1}{100} \\
\frac{1}{50} & 50\sin\frac{1}{50} \\
\frac{3}{100} & \frac{100}{3}\sin\frac{3}{100} \\
\frac{1}{25} & 25\sin\frac{1}{25} \\
\frac{1}{20} & 20\sin\frac{1}{20} \\
\frac{3}{50} & \frac{50}{3}\sin\frac{3}{50} \\
\frac{7}{100} & \frac{100}{7}\sin\frac{7}{100} \\
\frac{2}{25} & \frac{25}{2}\sin\frac{2}{25} \\
\frac{9}{100} & \frac{100}{9}\sin\frac{9}{100} \\
\frac{1}{10} & 10\sin\frac{1}{10}
\end{bmatrix}
=
\begin{bmatrix}
0.01 & 0.999\,98 \\
0.02 & 0.999\,93 \\
0.03 & 0.999\,85 \\
0.04 & 0.999\,73 \\
0.05 & 0.999\,58 \\
0.06 & 0.999\,4 \\
0.07 & 0.999\,18 \\
0.08 & 0.998\,93 \\
0.09 & 0.998\,65 \\
0.1 & 0.998\,33
\end{bmatrix}
$$

Creating a Table of Values by Concatenating Matrices

▶ **To generate a table of values by concatenating matrices**

1. Click the expanding **Square Brackets** button ▣ on the **Math Templates** toolbar and leave the insertion point in the input box.

2. Click ⊞ on the **Math Objects** toolbar, or choose **Insert + Matrix**.

3. Specify one column and some number of rows and choose **OK**.

4. Enter your choice of domain values in the matrix.

5. Evaluate f(your matrix).

With the same function as in the previous example, this gives

$$f \begin{bmatrix} \frac{1}{100} \\ \frac{1}{50} \\ \frac{3}{100} \end{bmatrix} = \begin{bmatrix} 100\sin\frac{1}{100} \\ 50\sin\frac{1}{50} \\ \frac{100}{3}\sin\frac{3}{100} \end{bmatrix}$$

and concatenating produces

$$\begin{bmatrix} \frac{1}{100} \\ \frac{1}{50} \\ \frac{3}{100} \end{bmatrix} \begin{bmatrix} 100\sin\frac{1}{100} \\ 50\sin\frac{1}{50} \\ \frac{100}{3}\sin\frac{3}{100} \end{bmatrix}, \text{ concatenate: } \begin{bmatrix} \frac{1}{100} & 100\sin\frac{1}{100} \\ \frac{1}{50} & 50\sin\frac{1}{50} \\ \frac{3}{100} & \frac{100}{3}\sin\frac{3}{100} \end{bmatrix}$$

Differentiation

The *derivative* f' of a function f is defined by the equation

$$f'(x) = \lim_{h\to 0} \frac{f(x+h) - f(x)}{h}$$

Notation for Derivative

You can use a variety of notations for the derivative, including the forms

$$\frac{d}{dx}, \frac{d^n}{dx^n}, D_x, D_{xx}, D_{x^2}, D_{xy}, D_{x^s y^t}, \frac{\partial}{\partial x}, \text{ and } \frac{\partial^n}{\partial x^s \partial y^t}$$

To compute a derivative, enter an expression with one of these forms and, with the insertion point in the expression, choose **Evaluate**.

▶ Evaluate

$$\frac{d}{dx}\left(x^3\right) = 3x^2 \qquad \frac{d^4}{dx^4}\left(3x^8\right) = 5040x^4 \qquad D_{x^5 y^2}\left(x^9 y^3\right) = 90\,720x^4 y$$

$$\frac{\partial}{\partial x}\left(\sin^2 x\right) = \sin 2x \qquad \frac{\partial^5}{\partial x^2 \partial y^3}\left(\sin x \cos y\right) = \tfrac{1}{2}\cos\left(x+y\right) - \tfrac{1}{2}\cos\left(x-y\right)$$

If f is defined as a function of one variable, then the forms $f'(x)$, $f''(x)$, ..., and $f^{(n)}(x)$ are recognized as first, second, and nth derivatives, respectively.

Note   The parentheses used to enclose the superscript on an nth derivative $f^{(n)}$ must be expanding parentheses (entered from **Insert + Brackets** or by clicking $\boxed{(\square)}$), not left and right parentheses on the keyboard. With keyboard parentheses, $f^{(n)}(x)$ is interpreted the same way as $f^n(x)$, namely, as $(f(x))^n$.

▶ Define + New Definition

$$f(x) = \sin x \cos x$$

▶ **Evaluate**

$$f(x) = \cos x \sin x \qquad f'(x) = \cos^2 x - \sin^2 x \qquad f''(x) = -4\cos x \sin x$$
$$f^{(4)}(x) = 16 \cos x \sin x \qquad f^4(x) = \cos^4 x \sin^4 x$$

The following examples include some time-saving steps for keyboard entry.

▶ **To enter a derivative of the form $\frac{d}{dx}x^2$**

1. Place the insertion point where you want the derivative to appear, even in an existing input box.

2. Click ⊟ or, choose **Insert + Fraction**, and type the numerator.

3. Move to the denominator by pressing DOWN ARROW, or pressing TAB, or clicking the denominator input box; and type the denominator (usually similar to dx).

4. Press RIGHT ARROW or SPACEBAR to leave the fraction, and type the mathematical expression.

▶ **To enter a derivative of the form $f^{(3)}(x)$**

1. Place the insertion point where you want the derivative to appear, even in an existing input box.

2. If the insertion point is not in mathematics, click ⊤ or, from the **Insert** menu, choose **Math**.

3. Type f.

4. Click N^x and then click (□), or choose **Insert + Superscript** and then choose **Insert + Brackets** and choose parentheses.

5. Type 3 in the input box.

6. Press RIGHT ARROW twice to leave the superscript.

7. Click (□), or choose **Insert + Brackets** and select parentheses.

8. Type x in the input box.

▶ **To find the derivative of x^2**

1. Place the insertion point in the expression $\frac{d}{dx}\left(x^2\right)$.

2. Click =? , or choose **Evaluate**, or press CTRL + E.

You obtain the same result from any of the following expressions.

$$\frac{dx^2}{dx} \qquad \frac{d}{dx}x^2 \qquad \frac{d}{dx}\left(x^2\right) \qquad \frac{\partial}{\partial x}\left(x^2\right)$$

$$D_x x^2 \qquad D_x\left(x^2\right) \qquad \frac{\partial x^2}{\partial x} \qquad \frac{\partial}{\partial x}x^2$$

Note that the "prime" notation works only for defined functions, not for expressions. For example, Evaluate applied to $(x + \sin x)'$ does *not* give the derivative:

▶ Evaluate

$$(x + \sin x)' = x + \sin x \qquad\qquad \tfrac{d}{dx}(x + \sin x) = 1 + \cos x$$

A derivative is applied to the term directly to the right of the operator, as illustrated in the following two examples.

▶ Evaluate

$$\frac{\partial^2}{\partial x^2}x^2 + 3x = 2 + 3x \qquad\qquad \frac{\partial^2}{\partial x^2}\left(x^2 + 3x\right) = 2$$

Using good notation is important. The program may accept ambiguous notation, but it may lead to an unexpected output. Experiment with expressions such as

$$\frac{\partial^2}{\partial x^2}\left((x^2 + 3x\right) \qquad \text{and} \qquad \frac{\partial^2}{\partial x^2}\{x^2 + 3x$$

to see examples of how ill-formed expressions are interpreted. Apply CTRL + ? or choose Compute + Interpret to observe the interpretation of an expression.

Tip Making good use of the parentheses button $\boxed{(\square)}$ eliminates many common types of ill-formed expressions.

The derivative of a piecewise-defined function is again a piecewise-defined function. (See page 117 for more information on piecewise-defined functions.)

▶ Definitions + New Definition

$$f(x) = \begin{cases} x & if \quad x < 0 \\ 3x^2 & if \quad x \geq 0 \end{cases}$$

▶ Evaluate

$$\tfrac{d}{dx}f(x) = \begin{cases} 1 & if \quad x < 0 \\ 6x & if \quad 0 < x \end{cases}$$

It is not necessary to name a piecewise function in order to take its derivative.

▶ Evaluate

$$\frac{d}{dx}\left(\left\{\begin{array}{lll} x+2 & \text{if} & x<0 \\ 2 & \text{if} & 0<x<1 \\ 2/x & \text{if} & 1<x \end{array}\right.\right) = \left\{\begin{array}{lll} 1 & \text{if} & x<0 \\ 0 & \text{if} & 0<x \wedge x<1 \\ -\frac{2}{x^2} & \text{if} & 1<x \end{array}\right.$$

Plotting Derivatives

You can plot several functions on the same graph. In particular, a function can be plotted together with one or more of its derivatives. Defining the function first is often convenient.

▶ Definitions + New Definition

$$f(x) = x^4 - 7x^3 + 14x^2 - 8x$$

▶ **To view the graph of f with its first and second derivatives**

1. Type $f(x)$ and, with the insertion point in $f(x)$, click ⊞ or choose **Plot 2D + Rectangular**.

2. Type $f'(x)$, select it and drag it to the frame

3. Type $f''(x)$, select it and drag it to the frame.

4. Click the plot and choose **Edit + Properties** to open the **Plot Properties** dialog.

5. Choose the **View** page, uncheck **Default**, and set the **View Intervals** to $-1 \le x \le 5$ and $-15 \le y \le 15$.

6. Choose the **Items Plotted** tab.
   - For **Item Number** 1, set **Thickness** to **Thick**. Choose **Variables and Intervals** and set the **Plot Interval** to $-1 \le x \le 5$. Choose **OK**.
   - For **Item Number** 2, set **Thickness** to **Medium**. Choose **Variables and Intervals** and set the **Plot Interval** to $-1 \le x \le 5$. Choose **OK**.
   - For **Item Number** 3, choose **Variables and Intervals** and set the **Plot Interval** to $-1 \le x \le 5$. Choose **OK**.

7. Choose **OK**.

You will see f as a thick curve, f' as a medium curve, and f'' as a thin curve.

▶   Plot 2D + Rectangular

$f(x)$

$f'(x)$

$f''(x)$

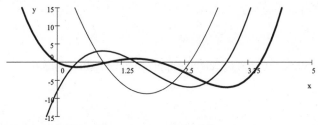

You can also change Line Color for each Item Number. Another way to distinguish the graphs is by determining the values at 0. Use Evaluate (or inspection) to find $f(0) = 0$, $f'(0) = -8$, and $f''(0) = 28$.

It is not necessary to define the functions. You can plot an expression and drag the first and second derivatives to the plot, as indicated below.

▶   Plot 2D + Rectangular

$\sin 2x$

$\frac{d}{dx}(\sin 2x)$

$\frac{d^2}{dx^2}(\sin 2x)$

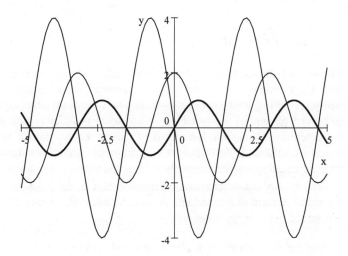

Generic Functions

It is possible to specify the symbol $f(x)$ to be an arbitrary, or "generic," function. Simply define $f(x)$ to be a function, without associating it with a formula.

▶ Definitions + New Definition

$f(x)$

$g(x)$

Standard rules of calculus apply to generic functions.

▶ Evaluate

$$\frac{d}{dx} f\left(g\left(x\right)\right) = f'\left(g\left(x\right)\right) \frac{\partial g(x)}{\partial x} \qquad D_x\left(f(x)g(x)\right) = \left(f\left(x\right) \frac{\partial g(x)}{\partial x} + g\left(x\right) \frac{\partial f(x)}{\partial x}\right)$$

$$D_x \frac{f(x)}{g(x)} = \frac{\frac{\partial f(x)}{\partial x} g\left(x\right) - f\left(x\right) \frac{\partial g(x)}{\partial x}}{g^2\left(x\right)} \qquad \frac{d}{dx}\left(f\left(x\right) + g(x)\right) = \frac{\partial f\left(x\right)}{\partial x} + \frac{\partial g\left(x\right)}{\partial x}$$

$$\frac{d}{dx} \int_0^x f(t)\, dt = f\left(x\right)$$

▶ Power Series, Expand in Powers of: x

$$f(x) = f\left(0\right) + xf'\left(0\right) + \tfrac{1}{2}x^2 f''\left(0\right) + \tfrac{1}{6}x^3 f^{(3)}\left(0\right) + \tfrac{1}{24}x^4 f^{(4)}\left(0\right) + O\left(x^5\right)$$

Implicit Differentiation

Variables can be linked to one another implicitly via an equation rather than in an explicit way. For example, $xy = 1$ implicitly determines y as a function of x. This example is easily solved to give the explicit formula $y = 1/x$. Many other equations cannot easily be solved for one of the variables. Also, some equations, such as $x^2 + y^2 = 1$, do not determine a function, but pieces of the curves determined by such equations are functions. Implicit Differentiation, an item on the Calculus submenu, finds derivatives from an equation without explicitly solving the equation for any one variable.

You specify the differentiation variable—that is, the independent variable. It is important to remember this variable in order to interpret the result, because the derivative is returned in the prime notation y'.

▶ **To find a derivative of an implicitly defined function**

1. Place the insertion point in the equation.

2. From the Calculus submenu, choose Implicit Differentiation.

3. Place the insertion point in the solution and solve for the derivative with **Solve + Exact**.

▶ Calculus + Implicit Differentiation

$xy + \sin x = y$ (Differentiation variable x), Solution: $y + xy' + \cos x = y'$

$xyz - x^2y = 0$ (Differentiation variable t), Solution: $xyz' - 2xyx' + xzy' + yzx' - x^2y' = 0$

Note that in the first example above, $y' = dy/dx$. In the second example above, $x' = dx/dt$, $y' = dy/dt$, and $z' = dz/dt$.

▶ Solve + Exact

$y + xy' + \cos x = y'$ (Variable(s) to Solve For: y'),

Solution is: $\begin{cases} \left\{ \frac{1}{x-1}(-y - \cos x) \right\} & \text{if} & x \neq 1 \\ \mathbb{C} & \text{if} & y \in \{-\cos 1\} \wedge x = 1 \\ \emptyset & \text{if} & x = 1 \wedge y \in \mathbb{C} \setminus \{-\cos 1\} \end{cases}$

▶ **To ignore special cases**

• Go to **Tools + Engine Setup**, choose the **General** page, and in the **Solve Options** area, check **Ignore Special Cases**.

▶ Solve + Exact

$y + xy' + \cos x = y'$ (Variable(s) to Solve For: y'), Solution is: $\frac{1}{x-1}(-y - \cos x)$

$xyz' - 2xyx' + xzy' + yzx' - x^2y' = 0$ (Variable(s) to Solve For: z'),
    Solution is: $\frac{1}{xy}\left(2xyx' - xzy' - yzx' + x^2y'\right)$

You can use **Definitions + New Definition** to declare any valid expression name to be a constant. Defined constant names are ignored under certain circumstances. For example, when identifying dependent and independent variables for implicit differentiation, a defined constant is not considered as a variable. Observe the difference below, where a is a defined constant and b is not.

▶ Definitions + New Definition

a

▶ Calculus + Implicit Differentiation

$axy = \sin y$ (Differentiation Variable x) Solution: $ay + axy' = (\cos(y)) y'$

$bxy = \sin y$ (Differentiation Variable x) Solution: $by + xby' + xyb' = (\cos(y)) y'$

▶ Solve + Exact

$ay + axy' = (\cos(y)) y'$ (Variable to Solve for y'), Solution is: $-a\frac{y}{ax - \cos y}$

$by + xby' + xyb' = (\cos(y)) y'$ (Variable to Solve for y'),

Solution is: $\frac{1}{bx - \cos y}(-by - xyb')$

Example   Use Implicit Differentiation combined with word processing editing features to find the second derivative y''.

1. Leave the insertion point in $y' = -\frac{y + \cos x}{x - 1}$, and from the Calculus submenu, choose Implicit Differentiation. Type x for the Differentiation Variable, and choose OK. This returns the equation

$$y'' = -\frac{y' - \sin x}{x - 1} + \frac{y + \cos x}{(x - 1)^2}$$

2. Use editing techniques to replace y' by $-\frac{y + \cos x}{x - 1}$.

1. Apply Simplify, and Factor the denominator in place, to obtain the following:

$$y'' \;=\; -\frac{-\frac{y + \cos x}{x - 1} - \sin x}{x - 1} + \frac{y + \cos x}{(x - 1)^2}$$

$$=\; \frac{2y + 2\cos x + (\sin x)\,x - \sin x}{x^2 - 2x + 1}$$

$$=\; \frac{2y + 2\cos x + (\sin x)\,x - \sin x}{(x - 1)^2}$$

Example   You can use Implicit Differentiation to find an equation of a tangent line. Find the derivative y', evaluate at a point on the curve to find the slope of the tangent at that point, and use the point-slope formula to find the equation for the tangent line. You can then plot the graph of the equation together with the tangent line.

1. Place the insertion point in the equation $x^3 + 3x^2 y = 2y^3 + 2$ and choose Calculus + Implicit Differentiation (Differentiation variable x) to obtain

Solution is: $6xy + 3x^2 + 3x^2 y' = 6y^2 y'$

2. Choose **Solve + Exact** (Variable(s) to solve for: y') for the result

$$\text{Solution is:} \begin{cases} \left\{\frac{-6xy-3x^2}{3x^2-6y^2}\right\} & \text{if} & 3x^2-6y^2 \neq 0 \\ \mathbb{C} & \text{if} & -6xy-3x^2 = 0 \wedge 3x^2-6y^2 = 0 \\ \emptyset & \text{if} & -6xy-3x^2 \neq 0 \wedge 3x^2-6y^2 = 0 \end{cases}$$

and simplify the principal solution to obtain

$$\frac{-6xy-3x^2}{3x^2-6y^2} = \frac{2xy+x^2}{2y^2-x^2}$$

3. For the slope at the point $(1,1)$ on the curve, enclose the expression in expanding brackets, add limits in a subscript, and choose **Evaluate**. This yields

$$\left[\frac{2xy+x^2}{2y^2-x^2}\right]_{x=1,y=1} = 3$$

4. Place the insertion point in the point-slope formula $y - 1 = 3(x - 1)$ and choose **Solve + Exact** (Variable(s) to solve for: y) to find the formula for the tangent line in standard form: $y = 3x - 2$.

5. Place the insertion point in the equation $x^3 + 3x^2y = 2y^3 + 2$ and choose **Plot 2D + Implicit** to plot the curve. Select and drag the equation for the tangent line to the plot.

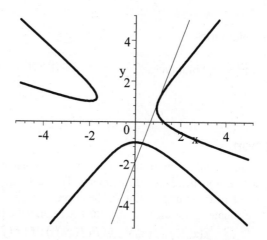

You can use **Definitions + New Definition** to declare a character to be a (generic) constant. Such a constant is recognized as an arbitrary constant by the implicit differentiation operation, as shown in the following example.

▶ **Definitions + New Definition**

a

▶ Calculus + Implicit Differentiation

$ax + y = 0$ (Differentiation Variable: x) Derivative: $a + y' = 0$

$bx + y = 0$ (Differentiation Variable: x) Derivative: $b'x + b + y' = 0$

Note that $a' = 0$ while b' is not automatically assumed to be 0.

Numerical Solutions to Equations

You can use both exact and numerical methods for solving equations, as illustrated in the following three examples.

▶ Solve + Exact

$5x^3 - 5x^2 = x$, Solution is: $0, \frac{1}{2} - \frac{3}{10}\sqrt{5}, \frac{3}{10}\sqrt{5} + \frac{1}{2}$

$5.0x^3 - 5.0x^2 = x$, Solution is: $0, -0.17082, 1.1708$

▶ Solve + Numeric

$5x^3 - 5x^2 = x$, Solution is: $\{x = 0\}, \{[x = 1.1708], [x = -0.17082], [x = 0.0]\}$

Iteration

You can also obtain numerical solutions for many equations of the form $f(x) = x$ by using Iterate from the Calculus submenu. This technique works for functions satisfying $|f'(x)| < 1$ near the intersection of the curve $y = f(x)$ and the line $y = x$. You start with an estimate x_0 for the root, and Iterate returns the list of values

$$f(x_0), f(f(x_0)), f(f(f(x_0))), f(f(f(f(x_0)))), \ldots$$

up to the number of iterations you specify. In appropriate situations, these values converge to a root of the equation $f(x) = x$.

You can use Iterate to solve the equation $\cos x = x$.

▶ Definitions + New Definition

$f(x) = \cos x$

Choosing Calculus + Iterate opens a dialog. In the boxes, enter f as the Iteration Function, select 1.0 as Starting Value, and select 10 as the Number of Iterations. With Digits Shown in Results set to 5, you receive the following vector of iterations.

▶ Calculus + Iterate

$$
\text{Iterates:} \quad
\begin{bmatrix}
1.0 \\
0.5403 \\
0.85755 \\
0.65429 \\
0.79348 \\
0.70137 \\
0.76396 \\
0.7221 \\
0.75042 \\
0.7314 \\
0.74424
\end{bmatrix}
$$

These entries are the initial value, followed by the values
$$f(1.0), f(f(1.0)), \ldots, f(f(f(f(f(f(f(f(f(f(1.0))))))))))$$
You can generate these numbers geometrically by starting at the point $(1,0)$ and moving vertically to the curve $y = \cos x$, then horizontally to the line $y = x$, then vertically to the curve $y = \cos x$, then horizontally to the line $y = x$, and so forth, as illustrated in the following figure.

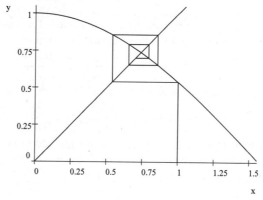

This figure can be generated by plotting $\cos x$ and x as usual, then selecting the matrix

$$
\begin{bmatrix}
1.0 & 0 \\
1.0 & 0.5403 \\
0.5403 & 0.5403 \\
0.5403 & 0.85755 \\
0.85755 & 0.85755 \\
0.85755 & 0.65429 \\
0.65429 & 0.65429 \\
0.65429 & 0.79348 \\
0.79348 & 0.79348 \\
0.79348 & 0.70137 \\
0.70137 & 0.70137
\end{bmatrix}
$$

and dragging it to the frame. This matrix can be created from two copies of the column computed previously, modified appropriately, using Matrices + Concatenate. (See page 310 for details on concatenating matrices.)

Newton's Method

The iteration method in the previous section can work very slowly. However, it provides the basis for Newton's method , which is usually much faster than direct iteration. Newton's method is based on the observation that the tangent line is a good local approximation to the graph of a function.

Let $(x_0, f(x_0))$ be a point on the graph of the function f. The tangent line is given by the equation

$$y - f(x_0) = f'(x_0)(x - x_0)$$

This line crosses the x-axis when $y = 0$. The corresponding value of x is given by

$$x = x_0 - \frac{f(x_0)}{f'(x_0)}$$

In general, given an approximation x_n to a zero of a function $f(x)$, the tangent line at the point $(x_n, f(x_n))$ crosses the x-axis at the point $(x_{n+1}, 0)$ where

$$x_{n+1} = x_n - \frac{f(x_n)}{f'(x_n)}$$

The *Newton iteration function* for a function f is the function g defined by

$$g(x) = x - \frac{f(x)}{f'(x)}$$

Given a first approximation x_0, Newton's method produces a list x_1, x_2, ..., x_n of approximations to a zero of f. In the following graph, $f(x) = x - x^3$, $x_0 = 0.44$, $x_1 \approx -0.41$, $x_2 \approx 0.27$, and $x_3 \approx -0.048$.

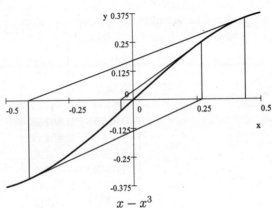

$$x - x^3$$

This figure can be generated by plotting $x - x^3$ as usual, zooming in to change the viewing rectangle, then selecting the matrix

$$
\begin{bmatrix}
0.44 & 0 \\
0.44 & 0.904\,75 \\
-0.41 & 0 \\
-0.41 & 0.917\,12 \\
0.27 & 0 \\
0.27 & 0.963\,77 \\
-0.048 & 0 \\
-0.048 & 0.998\,85
\end{bmatrix}
$$

and dragging it to the frame.

You can use Newton's method to solve the equation $x = \cos x$.

▶ **Definitions + New Definition**

$$f(x) = x - \cos x$$
$$g(x) = x - \frac{f(x)}{f'(x)}$$

In the dialog box, give g as the **Iteration Function**, enter 0.7 as the **Initial Value**, and select 5 as the **Number of Iterations**.

With **Digits Shown in Results** set at 20, you receive the vector of values shown below.

▶ **Calculus + Iterate**

0.7
0.73943649784805819543
0.7390851604651073986
0.73908513321516080562
0.73908513321516064166
0.73908513321516064166

These values converge to the display precision in four iterations. As a check, use **Evaluate** to verify that

$$\cos(0.73908513321516064166) = 0.73908513321516064165$$

A graph of $y = \cos x$ and $y = x$ displays the approximate solution to the equation $x = \cos x$.

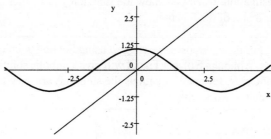

You can observe that there is only one solution, so you do not need to specify the interval for the solution. Enter the equation $\cos x = x$, leave the insertion point in the equation, and from the **Solve** submenu, choose **Numeric**.

▶ **Solve + Numeric**

$\cos x = x$, Solution is : $\{x = 0.73909\}$

This is a good place to use the **Plot Coordinates** plotting tool. Double-click the plot and click the ×,y tool. Move the cursor to the intersection point to see a good approximation of the solution.

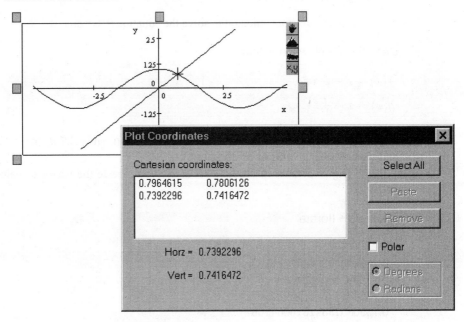

Optimization

Many of the applications of differentiation involve finding a value of x that yields a local maximum or local minimum value of some function $f(x)$. A good way to begin the investigation, when you know the function $f(x)$ either implicitly or explicitly, is to examine a plot of the function.

Tip   For most purposes, we suggest using floating point coefficients for optimization problems. Although **Solve + Exact** will give symbolic solutions to equations with rational coefficients, for many equations the solutions are very long, full of nested radicals, and difficult to work with.

Given the function $f(x) = \cos x + \sin 3x$, a plot suggests that there are numerous extreme values.

► **Plot 2D + Rectangular**

$\cos x + \sin 3x$

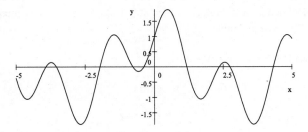

You can locate extreme values by solving $f'(x) = 0$ with **Solve + Numeric**, since the function $f(x) = \cos x + \sin 3x$ is everywhere differentiable.

Note   In this section, **Digits Shown in Results** is set at 5 on the **General** page of the **Computation Setup** dialog. See page 30 for details on changing this setting.

► **Solve + Numeric**

$\frac{d}{dx}(\cos x + \sin 3x) = 0$, Solution is: $\{[x = -82.267]\}$

This calculation yields only one critical number, although the graph indicates many more. You can specify the interval for a solution by placing the equation in a one-column matrix and entering a solution interval in the second row.

► **Solve + Numeric**

$\frac{d}{dx}(\cos x + \sin 3x) = 0$, Solution is: $[x = 0.4728]$
    $x \in (0, 2)$

Another strategy is to give the function a floating point coefficient and then use an exact method.

► **Definitions + New Definition**

$f(x) = 1.0 \cos x + \sin 3x$

▶ Solve + Exact

$f'(x) = 0$, Solution is: $\{6.2832X_{36} + 2.0\,(\arctan X_{34}) \mid X_{36} \in \mathbb{Z},$

$X_{34} \in \{-4.1510, -0.89335, -0.30132, 0.24091, 1.1194, 3.3187\}$

$\cap \mathbb{C} \setminus \{i\} \cap \mathbb{C} \setminus \{-i\}\}$

Indeed, the absolute minimum $f(-2.6688) = -1.8787$ occurs at $x = -2.6688$ (and at $-2.6688 + 2\pi n$ for any integer n), and the absolute maximum $f(0.4728) = 1.8787$ occurs at $x = 0.4728$ (and at $0.4728 + 2\pi n$ for any integer n).

The extreme values of $y = x^3 - 5x + 1$ can be found directly.

▶ Calculus + Find Extrema

$x^3 - 5x + 1$, Candidate(s) for extrema: $\left\{\frac{10}{9}\sqrt{15} + 1, -\frac{10}{9}\sqrt{15} + 1\right\}$,

at $\left\{\left[x = \frac{1}{3}\sqrt{15}\right], \left[x = -\frac{1}{3}\sqrt{15}\right]\right\}$

Floating-point coefficients produce floating-point approximations. Thus, applying **Find Extrema** to $x^3 - 5.0x + 1.0$ gives numerical approximations to the extreme values.

▶ Calculus + Find Extrema

$x^3 - 5.0x + 1.0$ Candidate(s) for extrema: $\{-3.3033, 5.3033\}$,

at $\{[x = -1.2910], [x = 1.2910]\}$

Geometrically, the points $(-1.291, 5.3033)$ and $(1.291, -3.3033)$ represent a high point and a low point, respectively.

▶ Plot 2D + Rectangular

$x^3 - 5x + 1$

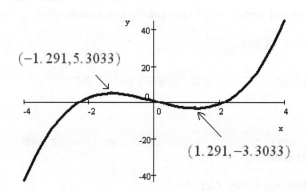

Curve Sketching

A default plot may well obscure some of the subtle, and even not so subtle, detail of a plot. You may need to adjust both the domain and the range to obtain a useful plot. For example, let us examine the graph of the function $f(x) = x^2 - 20x + 100$. In the default plot, a decreasing curve is visible, not giving much clue about the overall shape of the graph.

▶ Plot 2D + Rectangular

$x^2 - 20x + 100$

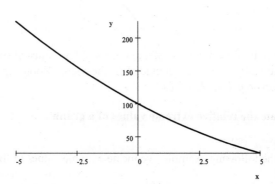

To see more detail, zoom and pan with the plotting tools, or turn off the **Default** on the **View** page of the **Plot Properties** dialog, then zoom out and experiment with different views such as the following.

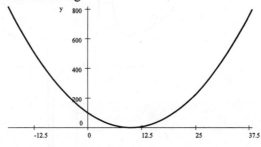

More striking, the first attempt at plotting the equation $7x^2 + 36xy - 50y^2 + 594x - 2363y - 26\,500 = 0$ will not create a plot, because there are no points on the graph in the default domain $-5 < x < 5$. Again, you can zoom out and experiment with different views and obtain the following.

▶ Plot 2D + Implicit

$7x^2 + 36xy - 50y^2 + 594x - 2363y - 26\,500 = 0$

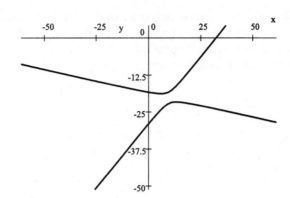

When feasible, the view of a graph should be adjusted so that the points where these extreme values occur are included in the view. Zooming and panning will often be necessary to accomplish this.

▶ **To locate the relative extreme values of a graph**

- Solve $f'(x) = 0$.

    In the following example, we locate extreme values of the function

$$f(x) = \frac{x^6 - 5x^3 + 10x^2 - 40x}{(x^2 - 4)^2}$$

The default plot of this expression gives a good view of the three extreme values.

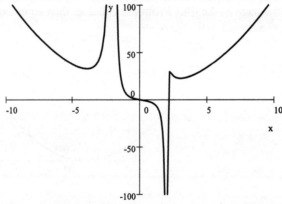

You can find the points where extreme values might occur with Solve.

▶ **Solve + Exact**

$$\frac{d}{dx}\left[\frac{x^6 - 5x^3 + 10x^2 - 40x}{(x^2 - 4)^2}\right] = 0, \text{ Solution is: } (\rho_1) \setminus \{-2, 2\} \text{ where } \rho_1 \text{ is a root of}$$
$$180X_{231}^2 - 80X_{231} - 20X_{231}^3 + 5X_{231}^4 - 24X_{231}^5 + 2X_{231}^7 + 160, X_{231}$$

You can find approximate real roots of this seventh-degree polynomial with Numeric from the Solve submenu. In the following, Digits Shown in Results is set at 5 on the General page of the Computation Setup dialog.

▶ Solve + Numeric

$180X_{231}^2 - 80X_{231} - 20X_{231}^3 + 5X_{231}^4 - 24X_{231}^5 + 2X_{231}^7 + 160 = 0$,

Solution is: $[X_{231} = 2.2359]$, $[X_{231} = -0.90484 + 1.6422i]$,

$[X_{231} = -0.90484 - 1.6422i]$, $[X_{231} = 3.0327]$, $[X_{231} = 0.21375 - 0.90433i]$,

$[X_{231} = 0.21375 + 0.90433i]$, $[X_{231} = -3.8864]$

The three real roots of f' give two local minimums: $f(-3.8864) = 32.812$ and $f(3.0327) = 22.553$, and one local maximum: $f(2.2359) = 29.656$.

You can gain additional insight into the graph of a rational function by rewriting it as a polynomial plus a fraction.

▶ Polynomials + Divide

$$\frac{x^6 - 5x^3 + 10x^2 - 40x}{\left(x^2 - 4\right)^2} = x^2 + \frac{1}{(x^2-4)^2}\left(58x^2 - 40x - 5x^3 - 128\right) + 8$$

Select and drag the expression $x^2 + 8$ to the view to see both curves in the same picture. Note how well the graph of $y = x^2 + 8$ matches the graph of $y = f(x)$ for large values of x.

▶ Plot 2D + Rectangular

$$\frac{x^6 - 5x^3 + 10x^2 - 40x}{\left(x^2 - 4\right)^2} \text{ and } x^2 + 8$$

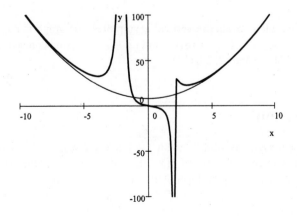

▶ **To determine concavity of a graph**

- Find intervals where the second derivative is positive or negative.

To locate the intervals where the graph of $f(x) = x^4 + 3x^3 - x^2 - 3x$ is concave upward, evaluate $f''(x)$ to obtain $f''(x) = 12x^2 + 18x - 2$, and solve the inequality $12x^2 + 18x - 2 > 0$.

▶ **Solve + Exact**

$12x^2 + 18x - 2 > 0$, Solution is: $\left(-\infty, -\frac{1}{12}\sqrt{105} - \frac{3}{4}\right) \cup \left(\frac{1}{12}\sqrt{105} - \frac{3}{4}, \infty\right)$

To solve more complicated inequalities or systems of inequalities, you can set expressions equal to zero and test for sign changes.

Example   You can answer the question of where the graph of $f(x) = \frac{x^6 - 5x^3 + 10x^2 - 40x}{(x^2 - 4)^2}$ is concave upward by investigating the sign of the second derivative. Apply **Evaluate** and **Factor** to find the second derivative:

$$f''(x) = 2(x+2)^{-4}(x-2)^{-4}$$
$$\cdot (320x^2 - 1200x - 400x^3 + 270x^4 - 5x^5 - 16x^6 + x^8 + 160)$$

Since the denominator is always nonnegative, it is sufficient to investigate the sign of the numerator. Apply **Solve + Numeric** to the equation

$$0 = 320x^2 - 1200x - 400x^3 + 270x^4 - 5x^5 - 16x^6 + x^8 + 160$$

to find the real solutions

$$[x = 0.13759], [x = 2.3414]$$

Compute the value at any point to the left, between, and to the right of these solutions, using **Evaluate Numerically**:

$$f''(0) = 1.25$$
$$f''(1) = -21.481$$
$$f''(2.4) = 40.964$$

Taking into account the vertical asymptotes, the graph is concave upward on the intervals $(-\infty, -2)$, $(-2, 0.1376)$, and $(2.3414, \infty)$, and concave downward on the intervals $(0.13759, 2)$ and $(2, 2.3414)$.

Indefinite Integration

An *antiderivative* of a function $f(x)$ is any function $g(x)$ whose derivative is $f(x)$. If $g(x)$ is an antiderivative of $f(x)$, then $g(x) + C$ is another antiderivative. In fact, every antiderivative is of the form $g(x) + C$ for some constant C.

The *indefinite integral* of $f(x)$ is the family of all antiderivatives of $f(x)$ and is denoted $\int f(x)\, dx$.

▶ **To evaluate an indefinite integral**

1. Place the insertion point anywhere in the expression.

2. Choose **Evaluate**, press CTRL + E, or click .

▶ Evaluate

$$\int \left(2x^2 + 3x + 5\right)\,dx = \tfrac{2}{3}x^3 + \tfrac{3}{2}x^2 + 5x$$

The system does not automatically return the constant of integration—the "arbitrary constant"—so you must remain alert and add the constant when needed. Simply enter $+\,C$ to change from

$$\int \left(2x^2 + 3x + 5\right)\,dx = \frac{2}{3}x^3 + \frac{3}{2}x^2 + 5x$$

to

$$\int \left(2x^2 + 3x + 5\right)\,dx = \frac{2}{3}x^3 + \frac{3}{2}x^2 + 5x + C$$

Such constants are needed, for example, if you have a formula for acceleration and you wish to find an expression for velocity.

Tip   It is common, although hardly necessary, to add a thin space between $f(x)$ and dx in an integral $\int f(x)\,dx$. The **Thin Space** (found under **Insert + Spacing + Horizontal Space** or by clicking ⬛) is for readability only, and in no way affects the way in which an integral is interpreted by the underlying compute engine.

You can evaluate indefinite integrals of piecewise-defined functions. You can define a function from a piecewise expression, or work directly with the picewise expression, as shown in the following examples. Turn on **Helper Lines** to see the null brackets on the right.

▶ Definitions + New Definition

$$f(x) = \begin{cases} x & \text{if } x < 0 \\ 3x^2 & \text{if } x \geq 0 \end{cases}$$

▶ Evaluate

$$\int f(x)\,dx = \begin{cases} \frac{1}{2}x^2 & \text{if } x \leq 0 \\ x^3 & \text{if } 0 < x \end{cases}$$

▶ Evaluate

$$\int \begin{cases} x & \text{if } x < 0 \\ 3x^2 & \text{if } x \geq 0 \end{cases} dx = \begin{cases} \frac{1}{2}x^2 & \text{if } x \leq 0 \\ x^3 & \text{if } 0 < x \end{cases}$$

Interpreting an Expression

The computer algebra system interprets many expressions that might be considered ambiguous. You can check the interpretation without evaluating an expression, when you are in doubt about whether your expression is being interpreted as you intended.

▶ **To interpret an expression without evaluation**

- Place the insertion point in the expression and choose Compute + Interpret or press CTRL + ?.

▶ **Compute + Interpret or** CTRL + ?

$$xy/z = x\tfrac{y}{z} \qquad\qquad \sin x/y = \sin \tfrac{x}{y}$$

$$\int ax^3 = \int ax^3 \, d? \qquad\qquad \int x^3 a = \int x^3 a \, d?$$

Even though the interpretation of the integral expressions do not indicate the variable of integration, it shows that the expressions are interpreted as indefinite integrals. If such an expression is evaluated, a choice will be made, generally based on the alphabetical order of the characters.

▶ **Evaluate**

$$\int ax^3 = \tfrac{1}{4}ax^4 \qquad\qquad \int x^3 a = \tfrac{1}{4}ax^4$$

If f is not defined as a function, then it is treated as a variable or constant.

▶ **Evaluate**

$$\int f = \tfrac{1}{2}f^2 \qquad\qquad \int f(x)dx = \tfrac{1}{2}fx^2$$

In these expressions, f behaves the same as any other variable, and $f(x)$ is interpreted as simply the product of f and x.

Sequences of Operations

Sometimes the answers that are returned are not in the form you need. You can take advantage of commands such as Simplify, Combine, Expand, Rewrite, and Factor on the Compute menu to rewrite these answers.

▶ **Evaluate, Factor**

$$\int e^{ax} \cos bx \, dx = \tfrac{a(\cos bx)e^{ax}+b(\sin bx)e^{ax}}{a^2+b^2} = \left(a^2 + b^2\right)^{-1}(a\cos bx + b\sin bx)\left(e^{ax}\right)$$

The integral $\int 2^x \cos bx \, dx$ looks just as straightforward as the one just computed, but requires several extra steps to get the answer into an equivalent final form. (See the exercises at the end of this chapter.) Simplification of algebraic expressions is more subtle than it might at first appear. Sometimes you just have to experiment with the menu items Simplify, Combine, Expand, Rewrite, and Factor to get expressions into a manageable form. This interaction with the system is a natural part of the machine-human interface.

Methods of Integration

Even though you can evaluate many integrals directly, several standard techniques of integration—such as integration by parts, change of variables by substitution, and partial fractions—are also available.

Integration by Parts

The *integration by parts* formula states that

$$\int u \, dv = uv - \int v \, du$$

This formula comes from the product formula for differentials
$$d(uv) = udv + vdu$$

and the linearity of integration, which implies that

$$\int d(uv) = \int udv + \int vdu$$

and the fundamental theorem of calculus, which allows you to replace $\int d(uv)$ by uv in the formula for integration by parts.

▶ **To use integration by parts**

1. Leave the insertion point in the integral.

2. From the Calculus submenu, choose Integrate by Parts.

3. Enter in the dialog box an appropriate expression for the Part to be Differentiated.

4. Choose OK.

For the integral $\int x \ln x \, dx$, for example, choosing $\ln x$ for the Part to be Differentiated gives the following result.

▶ Calculus + Integrate by Parts (Part to be Differentiated: $\ln x$)

$$\int x \ln x \, dx = \tfrac{1}{2}x^2 \ln x - \int \tfrac{1}{2}x \, dx$$

Since $\int \frac{1}{2}x\,dx$ can easily be integrated, this solves the problem of integrating $x \ln x\,dx$. Note that in this example, $u = \ln x$ and $dv = x\,dx$, so that $du = \frac{1}{x}\,dx$ and $v = \frac{1}{2}x^2$.

Change of Variables

It follows from the chain rule that if $u = g(x)$, then $du = g'(x)\,dx$. This yields the *change of variables formula* for integration:

$$\int f(g(x))\,g'(x)\,dx = \int f(u)\,du$$

▶ **To perform a change of variables**

1. Enter the integral $\int x \sin x^2\,dx$.

2. From the Calculus submenu, choose Change Variable.

3. Enter in the dialog box an appropriate substitution $\varphi(u) = g(x)$.

4. Choose OK.

For the integral $\int x \sin x^2\,dx$, the substitution $u = x^2$ gives the following:

▶ **Calculus + Change Variable (Substitution: $u = x^2$)**

$$\int x \sin x^2\,dx = \int \frac{1}{2} \sin u\,du$$

This replaces the problem of integrating $x \sin x^2\,dx$ by two much easier problems: first integrating $\frac{1}{2} \sin u\,du$ and then replacing u by x^2 in the result. Note that $u = g(x) = x^2$, $f(u) = \sin u$, and $du = 2x\,dx$.

For the integral $\int x^5 \sqrt{x^3 + 1}\,dx$, the substitution $u = x^3 + 1$ is useful.

▶ **Calculus + Change Variable (Substitution: $u = x^3 + 1$)**

$$\int x^5 \sqrt{x^3 + 1}\,dx = \int \frac{1}{3}\sqrt{u}\,(u - 1)\,du$$

▶ **Evaluate**

$$\int \frac{1}{3}\sqrt{u}\,(u - 1)\,du = \frac{2}{15}u^{\frac{5}{2}} - \frac{2}{9}u^{\frac{3}{2}}$$

Then perform an in-place replacement with $u = x^3 + 1$:

$$\frac{2}{15}u^{\frac{5}{2}} - \frac{2}{9}u^{\frac{3}{2}} = \frac{2}{15}\left(x^3 + 1\right)^{\frac{5}{2}} - \frac{2}{9}\left(x^3 + 1\right)^{\frac{3}{2}}$$

Partial Fractions

The *method of partial fractions* is based on the fact that a factorable rational function can be written as a sum of simpler fractions. Notice how evaluation of the following integral gives the answer as a sum of terms.

▶ Evaluate

$$\int \frac{3x^2 + 2x + 4}{(x-1)^2(x^2+1)^2}\, dx = \tfrac{1}{8}\pi - \tfrac{1}{4}\arctan x - \tfrac{5}{2}\ln(x-1)$$

$$+\tfrac{5}{4}\ln(x^2+1) - \frac{2}{x - x^2 + x^3 - 1} + \frac{x}{4x - 4x^2 + 4x^3 - 4}$$

$$-11\frac{x^2}{4x - 4x^2 + 4x^3 - 4}.$$

To gain an appreciation for how this calculation might be done internally, consider the method of partial fractions.

▶ **To use the method of partial fractions**

- Replace a rational function by its partial fractions expansion before carrying out its integration.

Example   Here is how you use this method on the integral $\displaystyle\int \frac{3x^2 + 2x + 4}{(x-1)^2(x^2+1)^2}\, dx.$

1. Enter the rational expression $\dfrac{3x^2 + 2x + 4}{(x-1)^2(x^2+1)^2}.$

2. With the insertion point in this expression, choose **Partial Fractions** from the **Calculus** or **Polynomials** submenu.

$$\frac{3x^2 + 2x + 4}{(x-1)^2(x^2+1)^2} = \frac{9}{4(x-1)^2} - \frac{5}{2(x-1)} + \frac{\tfrac{1}{2}x - 1}{(x^2+1)^2} + \frac{\tfrac{5}{2}x + \tfrac{1}{4}}{x^2+1}$$

3. Select the sum of rational expression with the mouse and click $\boxed{\text{(□)}}$. Enter an integral sign on the left and dx on the right.

This gives the following integral expression:

$$\int \left(\frac{9}{4(x-1)^2} - \frac{5}{2(x-1)} + \frac{1}{4}\frac{1+10x}{x^2+1} + \frac{1}{2}\frac{x-2}{(x^2+1)^2} \right) dx$$

4. Write the preceding integral as a sum of four integrals.

$$\int \frac{9}{4(x-1)^2}\, dx - \int \frac{5}{2(x-1)}\, dx + \frac{1}{4}\int \frac{1+10x}{x^2+1}\, dx + \frac{1}{2}\int \frac{x-2}{(x^2+1)^2}\, dx$$

5. Evaluate each of these integrals.

$$\int \frac{9}{4\,(x-1)^2}\,dx = -\frac{9}{4x-4}$$

$$-\int \frac{5}{2\,(x-1)}\,dx = -\frac{5}{2}\ln(x-1)$$

$$\frac{1}{4}\int \frac{1+10x}{x^2+1}\,dx = \frac{1}{4}\arctan x - \frac{1}{8}\pi + \frac{5}{4}\ln(x^2+1)$$

$$\frac{1}{2}\int \frac{-2+x}{(x^2+1)^2}\,dx = \frac{1}{4}\pi - \frac{1}{2}\arctan x - \frac{1}{2}\frac{x}{x^2+1} - \frac{1}{2\,(2x^2+2)}$$

6. The original integral is the sum of the expressions above on the right,

$$\int \frac{3x^2+2x+4}{(x-1)^2(x^2+1)^2}\,dx = -\frac{9}{4\,(x-1)} - \frac{5}{2}\ln(x-1)$$
$$+\frac{1}{4}\arctan x - \frac{1}{8}\pi + \frac{5}{4}\ln(x^2+1)$$
$$+\frac{1}{4}\pi - \frac{1}{2}\arctan x - \frac{1}{2}\frac{x}{x^2+1} - \frac{1}{2\,(2x^2+2)}$$

which simplifies to the answer previously computed directly with **Evaluate**.

Definite Integrals

The *definite integral* $\int_a^b f(x)\,dx$ of a function $f(x)$ defined on the interval $[a,b]$ is given by

$$\int_a^b f(x)\,dx = \lim_{\|P\|\to 0} \sum_{i=1}^n f(\bar{x}_i)\,\Delta x_i$$

where $\bar{x}_i$ is a point in the ith subinterval of the partition

$$P : a = x_0 < x_1 < x_2 < \cdots < x_n = b$$

of the interval $[a,b]$ and $\|P\| = \max\{\Delta x_i\}$. The sum

$$\sum_{i=1}^n f(\bar{x}_i)\,\Delta x_i$$

is called a *Riemann sum*. The function f is *integrable* on $[a,b]$ if the preceding limit exists.

If f is integrable on $[a,b]$, then

$$\int_a^b f(x)\,dx = \lim_{n\to\infty} \frac{b-a}{n} \sum_{i=1}^n f\left(a + i\frac{b-a}{n}\right)$$

In particular, if f is continuous on $[a,b]$, then f is integrable on $[a,b]$.

For positive-valued functions f, the sum

$$\frac{b-a}{n} \sum_{i=1}^n f\left(a + i\frac{b-a}{n}\right)$$

can be interpreted as the sum of areas of rectangles of base $\frac{b-a}{n}$ with height determined by the value of the function f at right endpoints of subintervals. For example, assume $a = -1$, $b = 1$, $n = 10$, and $f(x) = \frac{1}{x^2+1}$. Then

$$\frac{1-(-1)}{10} \sum_{i=1}^{n} f\left(-1 + i\frac{1-(-1)}{10}\right)$$

represents the sum of the areas of the 10 rectangles in the following figure.

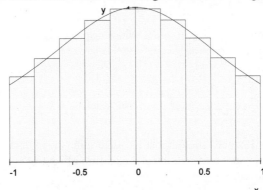

(See page 266 for a discussion of Riemann sums using left and right boxes.)

Entering and Evaluating Definite Integrals

▶ **To enter a definite integral**

1. Click $\int$, press CTRL + I, or choose Insert + Operator and choose the integral sign.

2. Click N_x , press CTRL + DOWN ARROW, or choose Insert + Subscript, and enter the lower limit.

3. Press TAB and enter the upper limit of integration. (Limits of integration work the same as any other subscripts or superscripts.)

4. Press the SPACEBAR or the RIGHT ARROW to move out of the superscript, and type the rest of the expression.

▶ **To evaluate a definite integral**

- Leave the insertion point in the expression and choose Evaluate or Evaluate Numerically.

▶ **Evaluate, Evaluate Numerically**

$$\int_0^1 x^2 \sqrt{x^3 + 1}\, dx = \tfrac{4}{9}\sqrt{2} - \tfrac{2}{9} = 0.4063171388$$

▶ **Evaluate Numerically**

$$\int_0^1 x^2 \sqrt{x^3 + 1}\, dx = 0.4063171388$$

Integrals involving absolute values or piecewise-defined functions can be treated as any other function.

▶ **Evaluate**

$$\int_{-2}^2 \left| x^2 - 1 \right|\, dx = 4$$

To understand this computation, determine the intervals for which $x^2 - 1$ is positive or negative, and write the integral as a sum of several integrals with the absolute value sign removed.

▶ **Solve + Exact**

$$x^2 - 1 > 0, \text{ Solution is: } (1, \infty) \cup (-\infty, -1)$$

It follows that $\left| x^2 - 1 \right| = x^2 - 1$ for $x < -1$ or $1 < x$, and $\left| x^2 - 1 \right| = -\left(x^2 - 1 \right) = 1 - x^2$ for $-1 < x < 1$. Thus, you can write the integral as the sum of the following three integrals:

$$\int_{-2}^2 \left| x^2 - 1 \right|\, dx = \int_{-2}^{-1} \left(x^2 - 1 \right)\, dx + \int_{-1}^1 \left(1 - x^2 \right)\, dx + \int_1^2 \left(x^2 - 1 \right)\, dx$$

▶ **Evaluate**

$$\int_{-2}^{-1} \left(x^2 - 1 \right)\, dx + \int_{-1}^1 \left(1 - x^2 \right)\, dx + \int_1^2 \left(x^2 - 1 \right)\, dx = 4$$

You can find the definite integral of a piecewise function either by integrating the expression directly or by defining a piecewise function $f(x)$.

▶ **Definitions + New Definition**

$$f(x) = \begin{cases} x^2 & \text{if } x < 0 \\ x & \text{if } x \geq 0 \end{cases}$$

▶ **Evaluate**

$$\int_{-2}^3 f(x)\, dx = \tfrac{43}{6} \qquad\qquad \int_{-2}^3 \left(\begin{cases} x^2 & \text{if } x < 0 \\ x & \text{if } x \geq 0 \end{cases} \right) dx = \tfrac{43}{6}$$

To understand this computation, write the integral as a sum of integrals involving ordinary functions. This yields

$$\int_{-2}^{3} f(x)\, dx = \int_{-2}^{0} x^2\, dx + \int_{0}^{3} x\, dx = \frac{8}{3} + \frac{9}{2} = \frac{43}{6}$$

Methods of Integration With Definite Integrals

Methods that were introduced for indefinite integration — integration by parts, change of variables, and partial fractions — can also be applied to definite integrals. See page 251 for general details about these methods.

▶ **To integrate by parts with a definite integral**

1. Place the insertion point in a definite integral.

2. From the Calculus submenu, choose Integrate by Parts.

3. Enter in the dialog box an appropriate expression for the Part to be Differentiated.

4. Choose OK.

▶ Calculus + Integration by Parts (Part to be Differentiated: $\ln x$)

$\int_{1}^{2} x \ln x\, dx = 2 \ln 2 - \int_{1}^{2} \frac{1}{2} x\, dx$

▶ **To use a change of variables with a definite integral**

1. Place the insertion point in a definite integral.

2. From the Calculus submenu, choose Change Variable.

3. Enter in the dialog box an appropriate substitution $\varphi(u) = g(x)$.

4. Choose OK.

▶ Calculus + Change Variable (Substitution: $u = x^3 + 1$)

$\int_{0}^{2} x^5 \sqrt{x^3 + 1}\, dx = \int_{1}^{9} \frac{1}{3} \sqrt{u}\,(u - 1)\, du$

This gives an integral that can be computed by elementary methods. Note that the limits have changed to match the new variable.

▶ **To use partial fractions with a definite integral**

• Replace a rational expression with its partial fractions expansion.

▶ **Calculus + Partial Fractions**

$$\frac{3x^2 + 2x + 4}{(x-1)^2} = \frac{8}{x-1} + \frac{9}{(x-1)^2} + 3$$

Thus

$$\int_3^7 \frac{3x^2 + 2x + 4}{(x-1)^2} dx = \int_3^7 3dx + \int_3^7 \frac{9}{(x-1)^2} dx + \int_3^7 \frac{8}{x-1} dx = 12 + 3 + 8\ln 3$$

▶ **Evaluate, Evaluate Numerically**

$$\int_3^7 3dx + \int_3^7 \frac{9}{(x-1)^2} dx + \int_3^7 \frac{8}{x-1} dx = 8\ln 6 - 8\ln 2 + 15 = 23.789$$

$$\int_3^7 \frac{3x^2 + 2x + 4}{(x-1)^2} dx = 8\ln 6 - 8\ln 2 + 15 = 23.789$$

Improper Integrals

If the proper integral $\int_a^b f(x)\, dx$ exists for every $b \geq a$, the limit

$$\int_a^\infty f(x)\, dx = \lim_{b \to \infty} \int_a^b f(x)\, dx$$

defines an *improper integral of the first kind*. The integral is said to converge if this limit exists and is finite.

▶ **To compute an improper integral of the first kind**

- Place the insertion point in the integral and choose **Evaluate** or **Evaluate Numerically**.

▶ **Evaluate**

$$\int_1^\infty x^{-2} dx = 1 \qquad \int_1^\infty x^{-1} dx = \infty \qquad \int_0^\infty e^{-3x} dx = \tfrac{1}{3}$$

$$\int_{-\infty}^0 e^{-3x} dx = \infty \qquad \int_{-\infty}^\infty e^{-x^2} dx = \sqrt{\pi}$$

▶ **Evaluate Numerically**

$$\int_{-\infty}^\infty e^{-x^2} dx = 1.7725$$

A definite integral for which the integrand has a discontinuity or a place where it is not defined within the interval of integration is an *improper integral of the second kind*. The discontinuity may occur either in the interior or at one or both endpoints of the interval of integration.

▶ **To evaluate an improper integral of the second kind**

- Place the insertion point in the integral and choose Evaluate or Evaluate Numerically.

▶ Evaluate

$$\int_0^1 \ln x\, dx = -1$$

If f has a discontinuity at a point c, and both $\int_a^c f(x)\, dx$ and $\int_c^b f(x)\, dx$ are convergent, then $\int_a^b f(x)\, dx = \int_a^c f(x)\, dx + \int_c^b f(x)\, dx$. If either diverges, then so does $\int_a^b f(x)\, dx$.

▶ Evaluate

$$\int_1^3 \frac{dx}{x-1} = \infty \qquad \int_{1/2}^3 \frac{dx}{x-1} = \text{undefined} \qquad \int_{-1}^1 \ln |x|\, dx = -2$$

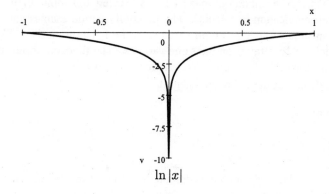

$\ln |x|$

Use special care when working with improper integrals and make certain that answers look reasonable. The following example illustrates a potential problem that occurs when the limits straddle a discontinuity. In this case, the system recognizes the discontinuity and does not attempt to evaluate the integral.

▶ Evaluate

$$\int_{-\pi}^{\pi} \frac{1+\sin x}{(x-\cos x)^2}\, dx = \int_{-\pi}^{\pi} \frac{\sin x+1}{(x-\cos x)^2}\, dx$$

▶ Evaluate Numerically

$$\int_{-\pi}^{\pi} \frac{1+\sin x}{(x-\cos x)^2}\, dx = (\text{numeric}) \int_{-\pi}^{\pi} \frac{\sin x+1}{(x-\cos x)^2}\, dx$$

However, the indefinite integral produces a solution.

▶ Evaluate

$$\int \frac{1+\sin x}{(x-\cos x)^2}\, dx = \frac{4\cos x - 4x}{4x^2 - 8x\cos x + 2\cos 2x + 2}$$

A naive approach to this problem, namely computing the indefinite integral and evaluating at the endpoints,

$$\left.\frac{4\cos x - 4x}{4x^2 - 8x\cos x + 2\cos 2x + 2}\right]_{-\pi}^{\pi} = \frac{-2\pi}{\pi^2 - 1}$$

gives an answer that is quite wrong. It is important to observe that the function $\frac{1+\sin x}{(x-\cos x)^2}$ is not defined when $x = \cos x$. This improper integral is examined further in the exercises at the end of this chapter.

Assumptions About Variables

The four functions assume, additionally, about, and unassume, were discussed in Chapter 5, beginning on page 122. We review this topic briefly to add an example of their application in calculus. The function assume enables you to place a restraint on a specific variable or on all variables. The function additionally allows you to place additional restraints on the same variable. The function about shows which restraints are active. The function unassume removes restraints.

Consider the following integral.

▶ Evaluate

$$\int_0^1 x^{2n-1}dx = \int_0^1 x^{2n-1}\, dx$$

This integral cannot be computed with no restraints because it converges for $n \geq 0$, but fails to converge for $n < 0$. You can evaluate this integral after applying the function assume to restrict possible values of n.

▶ Evaluate

$$\text{assume}(n, \text{positive})$$

▶ Evaluate

$$\int_0^1 x^{2n-1}dx = \frac{1}{2n}$$

The available assumptions on varuables include real, complex, integer, positive, negative, and nonzero. These assumptions can be made locally (for a specific variable) or globally. Additional information about making assumptions is available on page 122.

Definite Integrals from the Definition

You can use word processing and computing in place to fill in the steps for finding definite integrals from the definition.

Example   Define f by the equation $f(x) = x^3$. Calculate the integral $\int_1^4 f(x)dx$ as follows.

1. Enter the equation

$$\int_1^4 f(x)dx = \lim_{n \to \infty} \sum_{i=1}^{n} f\left(1 + i\frac{4-1}{n}\right)\frac{4-1}{n}$$

2. Select the term to the right of the summation sign.

3. Press and hold down the CTRL key and apply Evaluate, then apply Factor.

4. Select the series.

5. Press and hold down the CTRL key and apply Evaluate, then apply Expand, then attach parentheses.

6. With the insertion point in the expression, apply Evaluate.

These steps produce the following sequence of expressions.

$$
\begin{aligned}
\int_1^4 f(x)dx &= \lim_{n \to \infty} \sum_{i=1}^{n} f\left(1 + i\frac{4-1}{n}\right)\frac{4-1}{n} \\
&= \lim_{n \to \infty} \sum_{i=1}^{n} 3\frac{(n+3i)^3}{n^4} \\
&= \lim_{n \to \infty} \left(\frac{189}{2n} + \frac{135}{4n^2} + \frac{255}{4}\right) \\
&= \frac{255}{4}
\end{aligned}
$$

For comparison, you can compute this integral directly.

▶ Evaluate

$$\int_1^4 f(x)dx = \frac{255}{4}$$

Pictures of Riemann Sums

You can plot pictures of Riemann sums obtained from midpoints, left endpoints, or right endpoints of subintervals.

Middle Boxes

The Riemann sum determined by the midpoints is given by

$$\frac{b-a}{n} \sum_{i=0}^{n-1} f\left(a + \frac{b-a}{2n} + i\frac{b-a}{n}\right)$$

which is the sum of the areas of rectangles whose heights are determined by midpoints of subintervals.

▶ **To make a** Middle Boxes **plot**

1. Leave the insertion point inside the expression $x \sin x$.

2. From the Calculus submenu, choose Plot Approximate Integral. A middle-boxes plot will appear with default range settings.

3. Click the plot to select the frame, or double-click the plot to select the view.

4. Click the properties button $\boxed{\mathsf{Q}}$ and choose the Items Plotted page.

5. Choose Variables and Intervals and reset the Plot Interval as desired. Choose OK.

1. Reset the number of boxes as desired. Choose OK.

The following Middle Boxes plot uses $0 < x < 3$ and Number of Boxes is 10.

▶ Calculus + Plot Approximate Integral, Edit + Properties

$x \sin x$

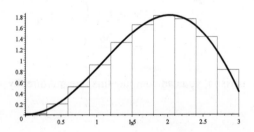

Applied to the expression $x \sin x$, with four rectangles and limits 0 and 3, the approximating Riemann sum is

$$\frac{3}{4} \sum_{k=0}^{3} \left(\frac{3}{8} + \frac{3}{4}k\right) \sin\left(\frac{3}{8} + \frac{3}{4}k\right) = 3.1784$$

Direct evaluation using Evaluate Numerically produces

$$\int_0^3 x \sin x \, dx = 3.1111$$

Left Boxes

The sum of the areas enclosed by rectangles is the *Riemann sum*

$$\frac{b-a}{n} \sum_{i=0}^{n-1} f\left(a + i\frac{b-a}{n}\right)$$

where the heights of the rectangles are determined by the function values at the left-hand endpoints of the subintervals.

▶ **To make a left-boxes plot**

1. Leave the insertion point inside the expression to be plotted.

2. From the Calculus submenu, choose Plot Approximate Integral. A middle-boxes plot will appear with default range settings.

3. Click the plot to select the frame, or double-click the plot to select the view.

4. Click the Properties button and choose the Items Plotted page.

5. Choose Variables and Intervals and reset the Plot Interval as desired. Choose OK.

1. Check Left Boxes. Reset the number of boxes as desired. Choose OK.

▶ Calculus + Plot Approximate Integral, Edit + Properties

$x \sin x$

Applied to the expression $x \sin x$, with four rectangles and $0 \leq x \leq 3$, the approximating Riemann sum is

$$\frac{3}{4} \sum_{k=0}^{3} \left(\frac{3}{4}k\right) \sin\left(\frac{3}{4}k\right) = 2.8186$$

Right Boxes

For right boxes , the sum of the areas enclosed by rectangles is the Riemann sum

$$\frac{b-a}{n}\sum_{i=1}^{n}f\left(a+i\frac{b-a}{n}\right)$$

where the heights of the rectangles are determined by the function values at the right-hand endpoints of the subintervals.

▶ **To make a right-boxes plot**

- Revise a middle-boxes plot or a left-boxes plot, this time choosing **Right Boxes**.

▶ Calculus + Plot Approximate Integral

$x\sin x$

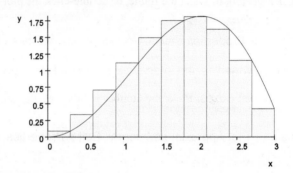

Applied to the expression $x\sin x$, with four rectangles and $0 \leq x \leq 3$, the approximating Riemann sum is

$$\frac{3}{4}\sum_{k=1}^{4}\left(\frac{3}{4}k\right)\sin\left(\frac{3}{4}k\right) = 3.1361$$

Left and Right Boxes

▶ **To make a plot that shows both left and right boxes**

- Revise a middle-boxes plot or a left-boxes plot, this time choosing **Left and Right Boxes**.

▶ Calculus + Plot Approximate Integral

$x\sin x$

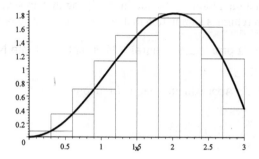

Approximation Methods

You can use the midpoint method, the trapezoidal rule, and Simpson's rule for approximating definite integrals. To apply each of these approximation methods, place the insertion point in a mathematical expression, from the **Calculus** submenu choose **Approximate Integral**, and then choose the appropriate method in the dialog box.

Midpoint Rule

In general, the *midpoint approximation* M_n for $\int_a^b f(x)dx$ with n subdivisions is given by

$$\int_a^b f(x)dx \approx M_n = \frac{b-a}{n} \sum_{i=0}^{n-1} f\left(a + \frac{b-a}{2n} + i\frac{b-a}{n}\right)$$

with an error bound of

$$\left| M_n - \int_a^b f(x)dx \right| \leq K\frac{(b-a)^3}{24n^2}$$

where K is any number such that $|f''(x)| \leq K$ for all $x \in [a,b]$.

▶ **To approximate $\int_a^b f(x)\,dx$ using the midpoint method**

1. Leave the insertion point in the expression $\int_a^b f(x)\,dx$.

2. From the **Calculus** submenu, choose **Approximate Integral**

3. In the dialog that appears, choose **Midpoint** and specify the number of **Subintervals**.

   or

1. Leave the insertion point in the expression $f(x)$.

2. From the **Calculus** submenu, choose **Approximate Integral**.

3. In the dialog, choose **Midpoint**, specify the number of **Subintervals**, and specify **Lower End of Range** and **Upper End of Range**.

To obtain the following output, in the dialog that appears, specify **10** Subintervals. The system returns a summation that you can evaluate numerically.

▶ Calculus + Approximate Integral + Midpoint, Evaluate Numerically

$$\int_0^\pi x \sin x \, dx \text{ Approximate integral (midpoint rule) is}$$

$$\tfrac{1}{10}\pi \sum_{i_3=0}^{9} \tfrac{1}{10}\pi \left(i_3 + \tfrac{1}{2}\right) \sin \tfrac{1}{10}\pi \left(i_3 + \tfrac{1}{2}\right) = 3.1545$$

For the following output, specify **10** Subintervals, enter **0** as Lower End of Range, and enter **3.14159** as Upper End of Range.

▶ Calculus + Approximate integral + Midpoint, Evaluate

$x \sin x$ Approximate integral (midpoint rule) is

$$0.314\,16 \sum_{i_4=0}^{9} (0.314\,16 i_4 + 0.157\,08) \sin (0.314\,16 i_4 + 0.157\,08) = 3.1545$$

Compare these results with direct computations of the integral.

▶ Evaluate, Evaluate Numerically

$$\int_0^\pi x \sin x \, dx = \pi \qquad \int_0^\pi x \sin x \, dx = 3.\,141\,6$$

Left and Right Boxes

In general, the *left endpoint approximation* L_n for $\int_a^b f(x)dx$ with n subdivisions is given by

$$\int_a^b f(x)dx \approx L_n = \frac{b-a}{n} \sum_{i=0}^{n-1} f\left(a + i\frac{b-a}{n}\right)$$

and the *right endpoint approximation* R_n for $\int_a^b f(x)dx$ with n subdivisions is given by

$$\int_a^b f(x)dx \approx R_n = \frac{b-a}{n} \sum_{i=1}^{n} f\left(a + i\frac{b-a}{n}\right)$$

▶ **To approximate $\int_a^b f(x)\, dx$ using left [right] boxes**

1. Leave the insertion point in the expression $\int_a^b f(x)\, dx$.

2. From the Calculus submenu, choose Approximate Integral

3. In the dialog that appears, choose Left [Right] Boxes and specify the number of Subintervals.

or

1. Leave the insertion point in the expression $f(x)$.

2. From the Calculus submenu, choose Approximate Integral.

3. In the dialog, choose Left [Right] Boxes, specify the number of Subintervals, and specify Lower End of Range and Upper End of Range.

For the following output, in the dialog that appears, specify **10 Subintervals**. The system returns a summation that you can evaluate numerically.

▶ Calculus + Approximate Integral + Left Boxes, Evaluate Numerically

$$\int_0^{\pi/2} x \sin x \, dx \text{ Approximate integral (left boxes) is}$$

$$\frac{1}{20}\pi \sum_{i_5=0}^{9} \frac{1}{20}\pi i_5 \sin \frac{1}{20}\pi i_5 = 0.878\,69$$

▶ Calculus + Approximate Integral + Right Boxes, Evaluate Numerically

$$\int_0^{\pi/2} x \sin x \, dx \text{ Approximate integral (right boxes) is}$$

$$\frac{1}{20}\pi \sum_{i_6=1}^{10} \frac{1}{20}\pi i_6 \sin \frac{1}{20}\pi i_6 = 1.125\,4$$

To obtain the following outputs, specify **10 Subintervals**, and enter **0** as Lower End of Range and **1.5708** as Upper End of Range.

▶ Calculus + Approximate Integral + Left Boxes, Evaluate Numerically

$x \sin x$ Approximate integral (left boxes) is

$$0.157\,08 \sum_{i_7=0}^{9} 0.157\,08 i_7 \sin 0.157\,08 i_7 = 0.878\,69$$

▶ Calculus + Approximate Integral + Right Boxes, Evaluate Numerically

$x \sin x$ Approximate integral (right boxes) is

$$0.157\,08 \sum_{i_8=1}^{10} 0.157\,08 i_8 \sin 0.157\,08 i_8 = 1.125\,4$$

Trapezoid Rule

The formula for the *trapezoid rule* approximation T_n is given by

$$\int_a^b f(x)dx \approx T_n = \frac{b-a}{2n}\left(f(a) + 2\sum_{i=1}^{n-1} f\left(a + i\frac{b-a}{n}\right) + f(b)\right)$$

with an error bound of

$$\left|T_n - \int_a^b f(x)dx\right| \leq K\frac{(b-a)^3}{12n^2}$$

where K is any number such that $|f''(x)| \leq K$ for all $x \in [a, b]$.

▶ **To approximate $\int_0^\pi x\sin x\,dx$ using the trapezoid rule**

1. Leave the insertion point in the expression $\int_0^\pi x\sin x\,dx$.

2. From the Calculus submenu, choose Approximate Integral.

3. In the dialog that appears, choose Trapezoid and specify the number of Subintervals.

   or

1. Leave the insertion point in the expression $f(x)$.

2. From the Calculus submenu, choose Approximate Integral.

3. In the dialog, choose Trapezoid, specify the number of Subintervals, and specify Lower End of Range and Upper End of Range.

To obtain the following output, specify **10 Subintervals**.

▶ Calculus + Approximate Integral + Trapezoid, Evaluate Numerically

$\int_0^\pi x\sin x\,dx$ Approximate integral (trapezoid rule) is

$$\tfrac{1}{10}\pi \sum_{i_9=1}^{9} \tfrac{1}{10}\pi i_9 \sin \tfrac{1}{10}\pi i_9 = 3.1157$$

To obtain the following output, in the dialog that appears, specify **10 Subintervals**, enter **0** as Lower End of Range, and enter **3.14159** as Upper End of Range.

▶ Calculus + Approximate Integral + Trapezoid, Evaluate Numerically

$x\sin x$ Approximate integral (trapezoid rule) is

$$0.31416 \sum_{i_{10}=1}^{9} 0.31416 i_{10} \sin 0.31416 i_{10} + 1.3095 \times 10^{-6} = 3.1157$$

Simpson's Rule

The *Simpson's rule* approximation S_n (n an even positive integer) is given for an arbitrary function f by

$$\int_a^b f(x)\, dx \approx S_n$$

$$= \frac{b-a}{3n}\left(f(a) + f(b) + 4\sum_{i=1}^{n/2} f\left(a + (2i-1)\frac{b-a}{n}\right)\right.$$

$$\left. +2\sum_{i=1}^{-1+n/2} f\left(a + 2i\frac{b-a}{n}\right)\right)$$

The error bound for Simpson's rule is given by

$$\left| S_n - \int_a^b f(x)\, dx \right| \leq K\frac{(b-a)^5}{180n^4}$$

where K is any number such that $\left| f^{(4)}(x) \right| \leq K$ for all $x \in [a, b]$. In particular, Simpson's rule is exact for integrals of polynomials of degree at most 3 (because the fourth derivative of such a polynomial is identically zero).

▶ **To approximate $\int_a^b f(x)\, dx$ using Simpson's rule**

1. Leave the insertion point in the expression $\int_a^b f(x)\, dx$.

2. From the Calculus submenu, choose Approximate Integral.

3. In the dialog that appears, choose Simpson and specify the number of Subintervals.

or

1. Leave the insertion point in the expression $f(x)$.

2. From the Calculus submenu, choose Approximate Integral.

3. In the dialog, choose Simpson, specify the number of Subintervals, and specify Lower End of Range and Upper End of Range.

For the following output, specify **10 Subintervals** in the dialog that appears.

▶ Calculus + Approximate Integral + Simpson, Evaluate Numerically

$\int_0^\pi x \sin x\, dx$ Approximate integral (Simpson's rule) is

$$\frac{1}{30}\pi\left(2\sum_{i_{11}=1}^{4}\frac{1}{5}\pi i_{11}\sin\frac{1}{5}\pi i_{11} + 4\sum_{i_{11}=1}^{5}\frac{1}{10}\pi(2i_{11}-1)\sin\frac{1}{10}\pi(2i_{11}-1)\right)$$

$$= 3.1418$$

For the following output, specify **10** Subintervals, and enter **0** as Lower End of Range and **3.14159** as Upper End of Range.

▶ Calculus + Approximate Integral + Simpson, Evaluate Numerically

$x \sin x$ Approximate integral (Simpson's rule) is

$$0.209\,44 \sum_{i_{12}=1}^{4} 0.628\,32 i_{12} \sin 0.628\,32 i_{12}$$

$$+0.418\,88 \sum_{i_{12}=1}^{5} (0.628\,32 i_{12} - 0.314\,16) \sin (0.628\,32 i_{12} - 0.314\,16)$$

$$+8.729\,9 \times 10^{-7} = 3.141\,8$$

Example   To find the number of subdivisions required to approximate $\int_0^1 e^{-x^2}\,dx$ using Simpson's rule with an error of at most 10^{-5}, you need to find an upper bound for the fourth derivative of e^{-x^2} on the interval $[0,1]$. One way you can do this is by plotting the fourth derivative on the interval $[0,1]$. Define $f(x) = e^{-x^2}$. Then evaluate the expression $f^{(4)}(x)$

$$f^{(4)}(x) = 12e^{-x^2} - 48x^2 e^{-x^2} + 16x^4 e^{-x^2}$$

and with the insertion point in this expression, choose **Plot 2D + Rectangular**.

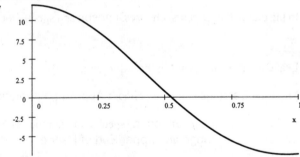

From the graph, you can see that $f^{(4)}(x)$ has a maximum value on this interval of $f^{(4)}(0) = 12$. Solve the inequality

$$12\frac{(1-0)^5}{180n^4} \leq 10^{-5}$$

to find the potential solutions

$$\left\{ n \leq -\frac{10}{3}\sqrt{3}\sqrt[4]{6} \right\}, \{n = 0\}, \left\{ \frac{10}{3}\sqrt{3}\sqrt[4]{6} \leq n \right\}$$

Since n must be an even positive integer, and also $n \geq \frac{10}{3}\sqrt{3}\sqrt[4]{6} = 9.036$, we take $n = 10$. Calculating,

$$S_{10} \approx \frac{1}{30} + \frac{1}{30}e^{-1} + \frac{2}{15}\sum_{i=1}^{5} e^{-\left(\frac{1}{5}i-\frac{1}{10}\right)^2} + \frac{1}{15}\sum_{i=1}^{4} e^{-\frac{1}{25}i^2} = 0.7468249483$$

Direct evaluation using **Evaluate Numerically** yields

$$\int_0^1 e^{-x^2}\, dx = 0.7468241328$$

and the approximation just computed is within the specified margin of error.

$$|0.7468241328 - 0.7468249483| = 8.155 \times 10^{-7} < 10^{-5}$$

Numerical Integration

Many integrals (such as $\int_0^1 e^{-x^2}\, dx$ and $\int_0^\pi \frac{\sin t}{t}\, dt$) cannot be evaluated exactly, but you can obtain numerical approximations by choosing **Evaluate Numerically**. See page 29 for information on changing settings that affect these approximations.

▶ **Evaluate Numerically**

$$\int_0^1 e^{-x^2}\, dx = 0.7468241328 \qquad \int_0^\pi \frac{\sin t}{t}\, dt = 1.851937052$$

Given a curve $y = f(x)$, the *arc length* between $x = a$ and $x = b$ is given by the integral

$$\int_a^b \sqrt{1 + (f'(x))^2}\, dx$$

For example, given $f(x) = x \sin x$, which has derivative $f'(x) = \sin x + x \cos x$, you can find the length of the arc between $x = 0$ and $x = \pi$ by applying **Evaluate Numerically**. Integrals associated with arc lengths of curves can almost never be evaluated exactly.

▶ **Evaluate, Evaluate Numerically**

$$\int_0^\pi \sqrt{1 + (f'(x))^2}\, dx = \int_0^\pi \sqrt{1 + (\sin x + x \cos x)^2}\, dx = 5.04040692$$

Curves in the plane or three-dimensional space can be represented parametrically. In the following we compute the arc length of the circular helix $(\cos\theta, \sin\theta, \theta)$ for $0 \le \theta \le 2\pi$ and then plot a view of this helix.

▶ **Definitions + New Definition**

$$x = \cos\theta \qquad\qquad y = \sin\theta \qquad\qquad z = \theta$$

▶ **Evaluate Numerically**

$$\int_0^{2\pi} \sqrt{\left(\frac{dx}{d\theta}\right)^2 + \left(\frac{dy}{d\theta}\right)^2 + \left(\frac{dz}{d\theta}\right)^2}\, d\theta = 8.885\,8$$

▶ **Plot 3D + Rectangular**

$(\cos\theta, \sin\theta, \theta)$

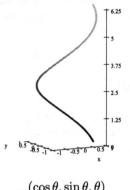

$(\cos\theta, \sin\theta, \theta)$

In polar coordinates arc length is given by the integral

$$\int_{\alpha}^{\beta} \sqrt{r^2 + \left(\frac{dr}{d\theta}\right)^2}\, d\theta$$

Following are the plot and arc length for the spiral $r = \theta$ with $0 \le \theta \le 6.2832$.

▶ **Plot 2D + Polar**

θ (Plot Interval $0 \le \theta \le 6.2832$)

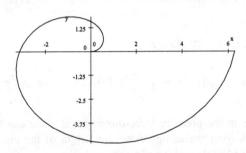

▶ **Definitions + New Definition**

$r = \theta$

▶ **Evaluate, Evaluate Numerically**

$\int_0^{2\pi} \sqrt{r^2 + \left(\frac{dr}{d\theta}\right)^2}\, d\theta = \pi\sqrt{4\pi^2 + 1} + \frac{1}{2}\ln\left(2\pi + \sqrt{4\pi^2 + 1}\right) = 21.256\,294\,15$

Visualizing Solids of Revolution

Problems of finding volumes and surface areas can be simplified by visualizing the solid.

Rectangular Coordinates

Assume the curve $y = 1 - x^2$ is rotated about the x-axis to form a solid. First, sketch the curve.

▶ Plot 2D + Rectangular

$1 - x^2$

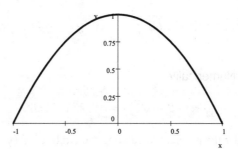

Then use a tube plot to visualize the surface.

▶ Plot 3D + Tube (Radius: $1 - x^2$)

$(0, x, 0)$

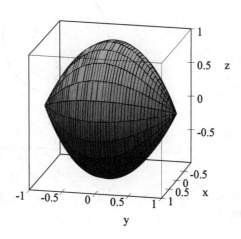

The volume is given by the integral

$$\int_{-1}^{1} \pi y^2 \, dx = \pi \int_{-1}^{1} \left(1 - x^2\right)^2 \, dx$$

▶ Evaluate

$$\pi \int_{-1}^{1} \left(1 - x^2\right)^2 \, dx = \frac{16}{15}\pi$$

The surface area is given by

$$\int 2\pi y \, ds = 2\pi \int_{-1}^{1} \left(1 - x^2\right) \sqrt{1 + \left(\frac{d}{dx}\left(1 - x^2\right)\right)^2} \, dx$$

▶ Evaluate

$$2\pi \int_{-1}^{1} \left(1 - x^2\right) \sqrt{1 + \left(\frac{d}{dx}\left(1 - x^2\right)\right)^2} \, dx = 2\pi \left(\frac{7}{16}\sqrt{5} - \frac{17}{32}\ln\left(-2 + \sqrt{5}\right)\right)$$

▶ Evaluate Numerically

$$2\pi \left(\frac{7}{16}\sqrt{5} - \frac{17}{32}\ln\left(-2 + \sqrt{5}\right)\right) = 10.\,965\,484\,66$$

Consider the problem of rotating the circle $x^2 + (y - 2)^2 = 1$ about the x-axis. We first sketch the circle.

▶ Plot 2D + Rectangular

$(\cos t, 2 + \sin t)$

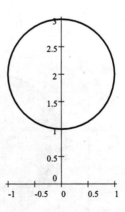

To rotate this circle about the x-axis, use a tube plot with spine $(2 \cos t, 0, 2 \sin t)$ and radius 1.

▶   Plot 3D + Tube (Radius: 1)

$$(2\cos t, 0, 2\sin t)$$

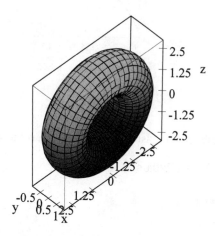

A differential of volume is equal to $(2\pi y)\,2x\,dy$ and hence the volume is equal to the integral

$$4\pi \int_1^3 y\sqrt{1-(y-2)^2}\,dy$$

▶   Evaluate

$$4\pi \int_1^3 y\sqrt{1-(y-2)^2}\,dy = 4\pi^2$$

The result $4\pi^2$ is intuitive because the volume is generated by rotating a circle of area π and the center of the circle travels a distance of 4π.

Parametric Equations

To find the volume generated by rotating the region bounded by the x-axis and one cycle of the curve $x = t + \sin t$, $y = 1 - \cos t$, we first draw the curve.

▶   Plot 2D + Rectangular

$$(t - \sin t, 1 - \cos t)$$

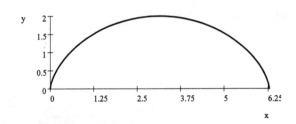

Use a tube plot to visualize the solid of revolution.

▶ Plot 3D + Tube (Radius: $1 - \cot t$)

$(0, t - \sin t, 0)$

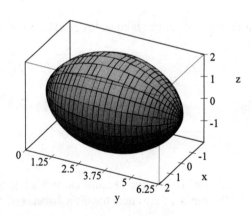

To compute the volume, note that a differential of volume is given by $\pi y^2 \, dx$ and hence the volume is

$$\int_0^{2\pi} \pi y^2 \, dx = \pi \int_0^{2\pi} (1 - \cos t)^2 (1 - \cos t) \, dt$$
$$= \pi \int_0^{2\pi} (1 - \cos t)^3 \, dt$$

▶ Evaluate

$\pi \int_0^{2\pi} (1 - \cos t)^3 \, dt = 5\pi^2$

Polar Coordinates

To find the volume of the solid generated by rotating $r = 1 - \cos\theta$ $(0 \le \theta \le \pi)$ about the x-axis, we note that $x = r\cos\theta = (1 - \cos\theta)\cos\theta$ and $y = r\sin\theta = (1 - \cos\theta)\sin\theta$.

▶ Plot 2D + Polar

$1 - \cos\theta$

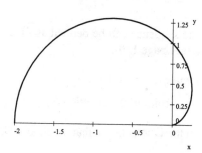

Use a tube plot to visualize the surface of revolution.

▶ Plot 3D + Tube (Radius: $(1 - \cos t)\sin t$)

$(0, (1 - \cos t)\cos t, 0)$

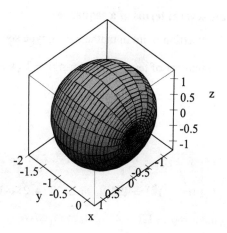

Sequences and Series

A sequence can be thought of as an infinite list, and a series as a sum of the terms of a sequence.

Sequences

A *sequence* $\{a_n\}_{n=1}^{\infty}$ is a function whose domain is the set of positive integers. Calculate limits of sequences by selecting an expression such as $\lim_{n\to\infty}\left(1+\frac{1}{n}\right)^n$ and choosing Evaluate, or by defining a_n, writing $\lim_{n\to\infty} a_n$, and choosing Evaluate.

▶ Evaluate

$$\lim_{n\to\infty}\left(1+\tfrac{1}{n}\right)^n = e$$

The terms of the sequence can be defined as function values, with the subscript as function argument (see page 110).

▶ **To define the sequence** $a_n = \left(1+\frac{1}{n}\right)^n$

1. With the insertion point in the equation $a_n = \left(1+\frac{1}{n}\right)^n$, click the New Definition button ![f(x)] on the Compute toolbar, or choose Definitions + New Definition.

2. In the Interpret Subscript dialog that appears, check A Function Argument and choose OK.

▶ Evaluate

$$\lim_{n\to\infty} a_n = e$$

▶ **To compute several terms of a sequence**

1. With the insertion point in mathematics, type seq. (It should turn gray.)

2. Enter the number of terms in the form $n = 1..4$ as a subscript, to obtain $\text{seq}_{n=1..4}$

3. Enter the general expression and evaluate.

▶ Evaluate

$$\text{seq}_{n=1..4}\left(\left(1+\tfrac{1}{n}\right)^n\right) = 2, \tfrac{9}{4}, \tfrac{64}{27}, \tfrac{625}{256}$$

$$\text{seq}_{n=1..4}\left(\left(1.0+\tfrac{1}{n}\right)^n\right) = 2.0, 2.\,25, 2.\,370\,4, 2.\,441\,4$$

$$\text{seq}_{x=1..5} \cos x = \cos 1, \cos 2, \cos 3, \cos 4, \cos 5$$

A sequence such as $\left\{\left(1+\frac{1}{n}\right)^n\right\}_{n=1}^{\infty}$ can be visualized graphically by plotting the expression $\left(1+\frac{1}{n}\right)^n$ at integer values of n.

▶ Plot + Rectangular

$$\left(1+\tfrac{1}{n}\right)^n$$

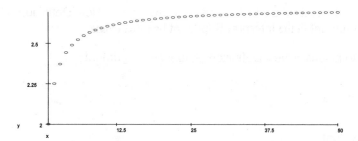

You can generate this figure by plotting $\left(1 + \frac{1}{n}\right)^{n}$, then revising the **Items Plotted** page so that the **Plot Style** is Point, the **Point Marker** is Circle, the **Plot Interval** is 1 to 50, and the **Sample Size** is 50.

This plot indicates that $\lim_{n \to \infty} \left(1 + \frac{1}{n}\right)^{n} \approx 2.7$. Indeed, **Evaluate** yields e and **Evaluate Numerically** produces $e = 2.718281828$.

Note   For further information on finite sequences, see page 33.

Series

The *partial sums* of the series $\sum_{k=1}^{\infty} a_k$ are the finite sums $s_n = \sum_{k=1}^{n} a_k$. These partial sums form a sequence $\{s_n\}$. If $\lim_{n \to \infty} s_n = s$ exists, then s is called the *sum* of the series $\sum_{k=1}^{\infty} a_k$. To sum a series, place the insertion point in the series and choose **Evaluate**. (See page 54 for details on entering the symbols $\sum_{k=1}^{\infty} a_k$.)

▶ **Evaluate**

$$\sum_{n=1}^{\infty} (0.99)^n = 99.0 \qquad \sum_{n=0}^{\infty} \frac{20^n}{n!} = e^{20}$$

$$\sum_{n=1}^{\infty} \frac{(-1)^n}{n} = -\ln 2 \qquad \sum_{n=1}^{\infty} \frac{1}{n^2} = \tfrac{1}{6}\pi^2$$

$$\sum_{n=1}^{\infty} \frac{1}{n^3} = \zeta(3) \qquad \sum_{n=1}^{\infty} \sin n\pi = \sum_{n=1}^{\infty} \sin n\pi$$

Occasionally, a result is obtained that may be obscure, such as the responses to $\sum_{n=1}^{\infty} \sin n\pi$ and $\sum_{n=1}^{\infty} \frac{1}{n^3}$. The lack of response to $\sum_{n=1}^{\infty} \sin n\pi$ is the computational engine version of "I give up" or "I do not have enough information." These series and the values of the zeta function $\zeta(\cdot)$ can be estimated numerically.

▶ **Evaluate Numerically**

$$\zeta(3) = 1.202056903 \qquad \sum_{n=1}^{\infty} \frac{1}{n^3} = 1.202056903$$

To sum a series in a form similar to $\sum_{n=1}^{\infty} a_n$, first enter an equation such as $a_n = \frac{n^2}{2^n}$. Then, from the Definitions submenu, choose New Definition. Check A function argument in the Interpret Subscript box that opens.

▶ Definitions + New Definition (A function argument)

$a_n = \frac{n^2}{2^n}$

▶ Evaluate

$\sum_{n=1}^{\infty} a_n = 6$

Ratio Test

A series $\sum_{n=1}^{\infty} a_n$ *converges absolutely* if $\sum_{n=1}^{\infty} |a_n|$ converges, in which case the series $\sum_{n=1}^{\infty} a_n$ also converges. The *ratio test* states that a series $\sum_{n=1}^{\infty} a_n$ converges absolutely (and therefore converges) if

$$\lim_{n \to \infty} \left| \frac{a_{n+1}}{a_n} \right| = L < 1$$

To verify the convergence of $\sum_{n=1}^{\infty} \frac{n^2}{2^n}$ using the ratio test, define $a_n = \frac{n^2}{2^n}$ and note the following.

▶ Evaluate

$\lim_{n \to \infty} \frac{a_{n+1}}{a_n} = \frac{1}{2}$

Thus, $L = \frac{1}{2}$, which is less than 1, so the series converges absolutely.

Root Test

The *root test* states that a series $\sum_{n=1}^{\infty} a_n$ converges absolutely (and therefore converges) if

$$\lim_{n \to \infty} \sqrt[n]{|a_n|} = L < 1$$

To verify convergence of $\sum_{n=1}^{\infty} \frac{n^2}{2^n}$ using the root test, define $a_n = \frac{n^2}{2^n}$ and note the following.

▶ Evaluate

$\lim_{n \to \infty} \sqrt[n]{|a_n|} = \frac{1}{2}$

Thus, $L = \frac{1}{2}$, which is less than 1, again showing that the series converges absolutely.

Integral Test

The *integral test* states that a series $\sum_{n=1}^{\infty} a_n$ converges absolutely if there exists a positive decreasing function f such that $f(n) = |a_n|$ for each positive integer n and

$$\int_1^{\infty} f(x)dx < \infty$$

To verify convergence of $\sum_{n=1}^{\infty} \frac{n^2}{2^n}$ using the integral test, define f by $f(x) = \frac{x^2}{2^x}$ and note the following.

▶ Evaluate, Evaluate Numerically

$$\int_1^{\infty} \frac{x^2}{2^x}dx = \frac{1}{2\ln 2} + \frac{1}{\ln^2 2} + \frac{1}{\ln^3 2} = 5.805497209$$

Thus, this integral is finite. (Although for $f(x) = \frac{x^2}{2^x}$, it is true that $f(1) < f(2) < f(3)$, you can verify that f is decreasing for $x > 3$. In fact,

$$f'(x) = 2\frac{x}{2^x} - \frac{x^2}{2^x}\ln 2$$

is positive only on the interval $0 < x < \frac{2}{\ln 2} = 2.885390082$, so f is decreasing on $3 < x < \infty$. Since convergence of a series depends on the tail end of the series only, it is sufficient that the sequence of terms be eventually decreasing.)

Maclaurin Series

The *Maclaurin series* of a function f is the series

$$\sum_{n=0}^{\infty} \frac{f^{(n)}(0)}{n!} x^n$$

where $f^{(n)}(0)$ indicates the nth derivative of f evaluated at 0. It is a power series expanded about $x = 0$.

▶ **To expand a function $f(x)$ in a Maclaurin series**

1. Place the insertion point in the expression $f(x)$.

2. Choose **Power Series**.

3. Specify the desired **Number of Terms**.

4. Specify **Expand in Powers of** x.

5. Choose **OK**.

With $f(x) = \frac{\sin x}{x}$ and 10 terms, the result is as follows.

▶ Power Series

$$\frac{\sin x}{x} = 1 - \frac{1}{6}x^2 + \frac{1}{120}x^4 - \frac{1}{5040}x^6 + \frac{1}{362880}x^8 + O\left(x^9\right)$$

The $O\left(x^9\right)$ term indicates that all the remaining terms in the series contain at least x^9 as a factor. (In fact, the truncation error is of order x^{10} in this case.) The odd powers of x have coefficients of 0.

Plot 2D provides an excellent visual comparison between a function and an approximating polynomial.

▶ Plot 2D + Rectangular

$$\frac{\sin x}{x} \quad \text{and} \quad 1 - \tfrac{1}{6}x^2 + \tfrac{1}{120}x^4$$

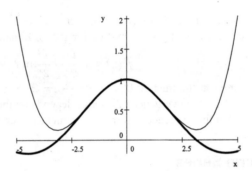

To determine which graph corresponds to which equation, evaluate one of the expressions where the graphs show some separation. For example, $\frac{\sin 4}{4} = -0.1892006238$, and hence the graph of $\frac{\sin x}{x}$ is the one that is negative at $x = 4$.

The following are additional examples of Maclaurin series expansions.

▶ Power Series (Expand in Powers of x)

$$e^x = 1 + x + \tfrac{1}{2}x^2 + \tfrac{1}{6}x^3 + \tfrac{1}{24}x^4 + \tfrac{1}{120}x^5 + \tfrac{1}{720}x^6 + O\left(x^7\right)$$

$$\sin x = x - \tfrac{1}{6}x^3 + \tfrac{1}{120}x^5 - \tfrac{1}{5040}x^7 + \tfrac{1}{362880}x^9 + O\left(x^{10}\right)$$

$$e^x \sin x = x + x^2 + \tfrac{1}{3}x^3 - \tfrac{1}{30}x^5 - \tfrac{1}{90}x^6 + O\left(x^7\right)$$

Remember that output can be copied and pasted (with ordinary word-processing tools) to create input for further calculations. In particular, select and delete the $+O\left(x^n\right)$ expression to convert the series into a polynomial. It is reassuring to note that, if the first few terms of the Maclaurin series for e^x are multiplied by the first few terms of the Maclaurin series for $\sin x$, then the result is the same as the first few terms of the Maclaurin series for $e^x \sin x$.

▶ Expand, Polynomials + Sort

$$\left(1 + x + \tfrac{1}{2}x^2 + \tfrac{1}{6}x^3 + \tfrac{1}{24}x^4 + \tfrac{1}{120}x^5\right)\left(x - \tfrac{1}{6}x^3 + \tfrac{1}{120}x^5\right)$$

$$= x + \tfrac{1}{3}x^3 - \tfrac{1}{30}x^5 + x^2 - \tfrac{1}{90}x^6 - \tfrac{1}{360}x^7 + \tfrac{1}{2880}x^9 + \tfrac{1}{14400}x^{10}$$

$$= \tfrac{1}{14400}x^{10} + \tfrac{1}{2880}x^9 - \tfrac{1}{360}x^7 - \tfrac{1}{90}x^6 - \tfrac{1}{30}x^5 + \tfrac{1}{3}x^3 + x^2 + x$$

Taylor Series

The Maclaurin series is a special case of the more general Taylor series. The *Taylor series* of f expanded about $x = a$ is given by

$$\sum_{n=0}^{\infty} \frac{f^{(n)}(a)}{n!}(x - a)^n$$

and hence is expanded in powers of $x - a$.

▶ **To expand a function $f(x)$ in a Taylor series**

1. Place the insertion point in the expression $f(x)$.

2. Choose **Power Series**.

3. Specify the desired **Number of Terms**.

4. Specify **Expand in Powers of** $x - a$.

5. Choose **OK**.

To find the Taylor series of $\ln x$ expanded about $x = 1$, choose **Power Series**. In the dialog box, select the desired number of terms and expand about the point $x - 1$.

▶ Power Series

$$\ln x = (x - 1) - \tfrac{1}{2}(x - 1)^2 + \tfrac{1}{3}(x - 1)^3 - \tfrac{1}{4}(x - 1)^4 + O\left((x - 1)^5\right)$$

A comparison between $\ln x$ and the polynomial $(x - 1) - \tfrac{1}{2}(x - 1)^2 + \tfrac{1}{3}(x - 1)^3 - \tfrac{1}{4}(x - 1)^4$ is illustrated graphically in the following figure. Note how closely the polynomial fits the graph of $\ln x$ in the neighborhood of the point $x = 1$.

▶ Plot 2D + Rectangular

$$\ln x \quad \text{and} \quad (x - 1) - \tfrac{1}{2}(x - 1)^2 + \tfrac{1}{3}(x - 1)^3 - \tfrac{1}{4}(x - 1)^4$$

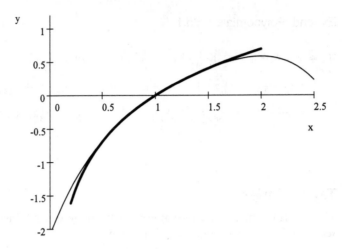

Multivariable Calculus

Multivariable calculus extends the fundamental ideas of differential and integral calculus to functions of several variables. The Compute menu commands that have been described for one-variable calculus easily adapt to functions of several variables. We look first at the general area of optimization which calls upon many of the ideas of differential calculus. Following that we will briefly consider Taylor polynomials in two variables and total differentials, and then describe the general approach for working with iterated integrals.

Optimization

Optimization of functions of several variables requires special techniques. The example immediately following demonstrates a direct approach, locating pairs where the partial derivatives are zero. Also see Lagrange multipliers (page 287) and Calculus + Find Extrema (page 286).

Extreme Values on a Surface

To find all candidates for the extreme values of a function such as $f(x, y) = x^3 - 3xy + y^3$, it is sufficient to locate all pairs (x, y) where both partial derivatives are zero. Since only real solutions are pertinent, it is useful to assume the variables represent real numbers.

▶ Evaluate

$\text{assume (real)} = \text{real}$

▶ Definitions + New Definition

$$f(x,y) = x^3 - 3xy + y^3$$

▶ Solve + Exact

$\frac{\partial}{\partial x} f(x,y) = 0$
$\frac{\partial}{\partial y} f(x,y) = 0$, Solution is: $[x = 1, y = 1], [x = 0, y = 0]$

Thus the only candidates for real extreme values are $(0,0)$ and $(1,1)$. You can identify the nature of these two points using the *second derivative test*:

$$\left[D_{xx}f(x,y)D_{yy}f(x,y) - (D_{xy}f(x,y))^2 \right]_{x=0,y=0} = -9 < 0$$

hence $(0,0)$ represents a saddle point; and

$$\left[D_{xx}f(x,y)D_{yy}f(x,y) - (D_{xy}f(x,y))^2 \right]_{x=1,y=1} = 27 > 0$$

$$\text{and} \qquad [D_{xx}f(x,y)]_{x=1,y=1} = 6 > 0$$

so the surface has a local minimum at $(1,1)$.

You can visualize the local minimum at $(1,1)$ by generating a plot of the surface. To
create the following plot, with the insertion point in the expression $x^3 - 3xy + y^3$, click
the Plot 3D Rectangular button 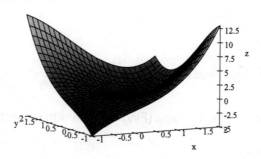 on the Compute toolbar. In the Items Plotted
page of the Plot Properties dialog, choose Patch and Mesh. Choose Variables and
Intervals and set the Plot Intervals to $-1 \le x \le 2$ and $-1 \le y \le 2$.. On the Axes
page, set Axes Type to Framed.

▶ Plot 3D + Rectangular

$$x^3 - 3xy + y^3$$

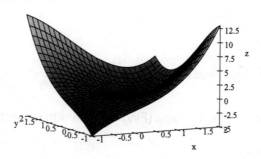

The level curve $x^3 - 3xy + y^3 = 0$ goes through the point $(0,0,0)$. For a better
view of this level curve, make a 2D plot of $x^3 - 3xy + y^3 = 0$ and add the level curve
$x^3 - 3xy + y^3 = -0.5$ to the plot.

▶ **Plot 2D + Implicit**

$$x^3 - 3xy + y^3 = 0$$

- Drag to the frame $x^3 - 3xy + y^3 = -0.5$

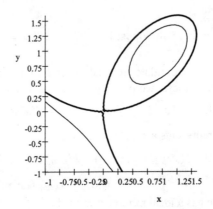

The thick curve is the level curve at 0 and the thin curves are components of the level curve at -0.5. This view gives an idea of where the z-values are positive and where they are negative. Note that the z-values on the surface $z = x^3 - 3xy + y^3$ are negative inside the loop in the first quadrant and in the lower left corner of the x-y-plane.

Extreme values of differentiable functions such as $x^3 - 3xy + y^3$ can also be found using Find Extrema from the Calculus submenu. In general, each application of Find Extrema reduces the number of variables by one and rephrases the problem in one less variable. Using this method with two or more variables requires multiple appropriate applications of the command.

Choose Find Extrema for the following examples. Use floating point coefficients for these problems to obtain numeric solutions. It is convenient to restrict the computations to real variables.

▶ Evaluate

assume $(\text{real}) = \text{real}$

▶ Calculus + Find Extrema (Find extrema with respect to which Variable(s): y)

$$x^3 - 3xy + y^3$$

Candidate(s) for extrema: $\left\{ x^3 - 2x^{\frac{3}{2}}, x^3 + 2x^{\frac{3}{2}} \right\}$, at $\{ [y = \sqrt{x}], [y = -\sqrt{x}] \}$

Note that for $y = \pm\sqrt{x}$, the expression $x^3 - 3xy + y^3$ simplifies to $x^3 \mp 2x^{\frac{3}{2}}$. To find the extreme values, apply the command again to the simplified expressions.

▶ Calculus + Find Extrema

$x^3 - 2x^{\frac{3}{2}}$ Candidate(s) for extrema: $\{-1, 0\}$, at $\{[x = 0], [x = 1]\}$

$x^3 + 2x^{\frac{3}{2}}$ Candidate(s) for extrema: $\{0\}$, at $\{[x = 0]\}$

The solution $y = \sqrt{x}$ yields the points $(0, 0)$ and $(1, 1)$, and $y = -\sqrt{x}$ yields the point $(0, 0)$. To determine the nature of these two critical points, use the second derivative test (page 285).

Lagrange Multipliers

You can use Lagrange multipliers to find constrained optima. To find extreme values of $f(x, y)$ subject to a constraint $g(x, y) = k$, it is sufficient to find all values of x, y, and λ such that

$$\nabla f(x, y) = \lambda \nabla g(x, y)$$

and $g(x, y) = k$ where ∇ is the gradient operator

$$\nabla f(x, y) = \left(\frac{\partial f}{\partial x}(x, y), \frac{\partial f}{\partial y}(x, y) \right)$$

The variable λ is called the *Lagrange multiplier*.

▶ **To find x and y whose sum is 5 and whose product is as large as possible**

1. Define $f(x, y) = xy$ and $g(x, y) = x + y$.

2. Solve the equation $\nabla f(x, y) = \lambda \nabla g(x, y)$ subject to $g(x, y) = 5$.

▶ Definitions + New Definition

$f(x, y) = xy$ $g(x, y) = x + y$

▶ Evaluate

$$\nabla f(x, y) = \begin{matrix} y \\ x \\ 0 \end{matrix} \qquad \nabla g(x, y) = \begin{matrix} 1 \\ 1 \\ 0 \end{matrix}$$

▶ Solve + Exact

$y = \lambda$
$x = \lambda$
$x + y = 5$, Solution is : $\left[x = \frac{5}{2}, y = \frac{5}{2}, \lambda = \frac{5}{2} \right]$

Optimization problems may require numerical solutions within given search intervals.

▶ Evaluate

$$\nabla (x + 2y) = \begin{pmatrix} 1 \\ 2 \\ 0 \end{pmatrix} \qquad \nabla (ye^x + xe^y) = \begin{pmatrix} e^y + ye^x \\ e^x + xe^y \\ 0 \end{pmatrix}$$

▶ Solve + Numeric

$$\begin{aligned} 1 &= \lambda \left(ye^x + e^y \right) \\ 2 &= \lambda \left(e^x + xe^y \right) \\ ye^x + xe^y &= 5 \\ x &\in (0, 10) \\ y &\in (0, 10) \\ \lambda &\in (-5, 5) \end{aligned} \quad , \text{Solution is} : [x = 1.666\,5, y = 0.450\,56, \lambda = 0.252\,90]$$

For $f(x, y) = x + 2y$ and $g(x, y) = ye^x + xe^y$, these numbers give

$$\begin{aligned} g(1.666\,5, 0.450\,56) &= 5.000\,1 \\ f(1.666\,5, 0.450\,56) &= 2.567\,6 \end{aligned}$$

The point $(1.666\,5, 0.450\,56)$ gives a possible extreme value for $f(x, y)$ satisfying the constraint $g(x, y) = 5$.

Taylor Polynomials in Two Variables

Let z be a function of two variables. The *second-degree Taylor polynomial* of z at (a, b) is given by

$$\begin{aligned} T_2(x, y) &= z(a, b) + D_x z(a, b)(x - a) + D_y z(a, b)(y - b) \\ &+ \frac{1}{2} D_{xx} z(a, b)(x - a)^2 + D_{xy} z(a, b)(x - a)(y - b) \\ &+ \frac{1}{2} D_{yy} z(a, b)(y - b)^2 \end{aligned}$$

▶ **To evaluate a partial derivative of a function z at (a, b)**

1. Evaluate the partial derivative at (x, y) using an expression such as $\frac{\partial}{\partial x} z(x, y)$, $D_x z(x, y)$, $\frac{\partial^2}{\partial x^2} z(x, y)$, or $D_{xy} z(x, y)$.

2. Evaluate at (a, b) using square brackets with the subscript $x = a, y = b$.

These steps can be combined into a single step.

▶ Evaluate

$$\left[\frac{\partial}{\partial x} \left(x^2 y \right) \right]_{x=1, y=2} = 4$$

To find the second-degree Taylor polynomial of $z = \frac{1}{1+x^2+y^2}$ at $(0, 0)$, first define the function $z(x, y)$, then compute the second degree Taylor polynomial as follows.

▶ Definitions + New Definition

$$z(x,y) = \frac{1}{1+x^2+y^2}$$

▶ Evaluate

$z(0,0) = 1$

$[D_y z(x,y)]_{x=0,y=0} = 0$

$[D_{xy} z(x,y)]_{x=0,y=0} = 0$

$[D_x z(x,y)]_{x=0,y=0} = 0$

$[D_{xx} z(x,y)]_{x=0,y=0} = -2$

$[D_{yy} z(x,y)]_{x=0,y=0} = -2$

These steps yield the second degree Taylor polynomial
$$T_2(x,y) = 1 - x^2 - y^2$$

The following plot has **Plot Intervals** $-0.5 \le x \le 0.5$ and $-0.5 \le y \le 0$, **Turn** 75 and **Tilt** 75. This cutaway plot shows how well the second-degree Taylor polynomial (the lower surface) matches the function z near $(0,0)$.

▶ Plot 3D + Rectangular

$$\frac{1}{1+x^2+y^2}$$

Drag to the plot: $1 - x^2 - y^2$

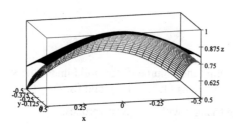

Total Differential

To compute the total differential of a function of two variables, define a function $u(x,y)$, represent each differential (du, dx, and dy) by a **Math Name** so that it will be treated as a variable. Then evaluate the expression

$$\mathrm{d}u = \frac{\partial}{\partial x} u(x,y)\, \mathrm{d}x + \frac{\partial}{\partial y} u(x,y)\, \mathrm{d}y$$

A similar procedure produces the total differential of a function of three variables.

▶ Definitions + New Definition

$$u\left(x,y\right) = x^3y^2$$

▶ **To create the grayed function names** du, dx, **and** dy

1. Choose Insert + Math Name.

2. Type the function name in the Name box and choose OK.

▶ Evaluate

$$du = \tfrac{\partial}{\partial x}u(x,y)\,dx + \tfrac{\partial}{\partial y}u(x,y)\,dy = 3x^2y^2\,dx + 2x^3y\,dy$$

Iterated Integrals

You can enter and evaluate iterated integrals. If $a \leq b$, $f(x) \leq g(x)$ for all $x \in [a,b]$, and $k(x,y) \geq 0$ for all $x \in [a,b]$ and all $y \in [f(x), g(x)]$, then the iterated integral

$$\int_a^b \int_{f(x)}^{g(x)} k(x,y)\,dy\,dx$$

can be interpreted as the volume of the solid bounded by the inequalities

$$a \leq x \leq b$$
$$f(x) \leq y \leq g(x)$$
$$0 \leq z \leq k(x,y)$$

Example Find the volume of the solid under the surface $z = 1 + xy$ and above the triangle with vertices $(1,1)$, $(4,1)$, and $(3,2)$

1. Plot the triangle with the given vertices.

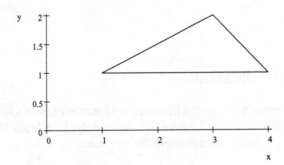

2. Find the equations of the bounding lines. $y = \frac{1}{2}x + \frac{1}{2}$
$$y = 5 - x$$
$$y = 1$$

3. Solve for x in terms of y. $x = 2y - 1$
$$x = 5 - y$$

4. Set up and evaluate an iterated integral

$$\int_1^2 \int_{2y-1}^{5-y} (1 + xy) \, dx \, dy = \frac{55}{8}$$

Here is a plot that shows the top of the surface $z = 1 + xy$ with the bounding triangle.

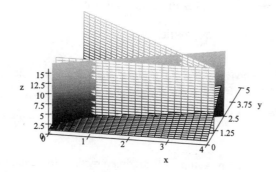

To make this plot, start with $[x, y, 1 + xy]$ and choose Plot 3D Rectangular. Add the items $[x, 1, 8y]$, $[x, \frac{1}{2}x + \frac{1}{2}, 8y]$, and $[x, 5 - x, 8y]$. In the Plot Properties dialog, change Surface Style to Hidden Line, set Turn to -80 and Tilt to 19, and set the View Intervals and all of the Plot Intervals to $0 < x < 4$, $0 < y < 2$, and $0 < z < 8$.

The Heaviside function provides another way to show the surface.

▶ **To enter the function name** Heaviside

- Click [sin cos] or choose Insert + Math Name, in the Name box, type **Heaviside**, and choose OK.

▶ Plot 3D + Rectangular

$(1 + xy) \, \text{Heaviside} \left(\frac{1}{2}x + \frac{1}{2} - y \right) \text{Heaviside} \left(5 - x - y \right) \text{Heaviside} \left(y - 1 \right)$

- In the Plot Properties dialog, change Surface Style to Hidden Line, set Turn to -80 and Tilt to 19, and set the Plot Intervals to $0 < x < 4$ and $0 < y < 2$. Change the Axes Type to Boxed.

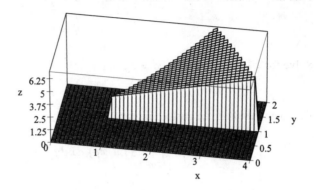

Here are two examples of iterated integrals.

▶ **Evaluate, Evaluate Numerically**

$$\int_0^1 \int_0^x x^2 \cos y \, dy \, dx \quad = \cos 1 + 2 \sin 1 - 2 \quad = 0.22324$$

$$\int_0^3 \int_0^{x/3} e^{x^2} \, dy \, dx \quad = \tfrac{1}{6} e^9 - \tfrac{1}{6} \quad = 1350.\,3$$

Following is an example illustrating a method for reversing the order of integration.

Example   Attempting to evaluate the double integral

$$\int_0^1 \int_{\sqrt{y}}^1 \sqrt{x^3 + 1} \, dx \, dy$$

exactly leads to frustration. However, you can reverse the order of integration by looking carefully at the region of integration in the plane.

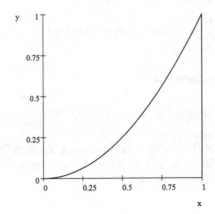

This region is bounded above by $y = x^2$ and below by $y = 0$. The new integral is

$$\int_0^1 \int_0^{x^2} \sqrt{x^3 + 1}\, dy\, dx$$

This double integral can be evaluated directly. You can gain some insight by iterated integration. The inner integral is just

$$\int_0^{x^2} \sqrt{x^3 + 1}\, dy = x^2 \sqrt{x^3 + 1}$$

You can integrate the resulting outer integral $\int_0^1 x^2 \sqrt{x^3 + 1}\, dx$ by applying **Calculus + Change Variable**, say with $u = x^3 + 1$. Then choosing **Evaluate and Evaluate Numerically**, yields

$$
\begin{aligned}
\int_0^1 \sqrt{(x^3 + 1)}x^2\, dx &= \int_1^2 \frac{1}{3}\sqrt{u}\, du \\
&= \frac{4}{9}\sqrt{2} - \frac{2}{9} \\
&= 0.4063171388
\end{aligned}
$$

For double and triple *indefinite* integrals you can use either repeated integral signs or the double and triple integrals available in the **Operators** dialog. Analogous to single indefinite integrals, for which you must add an arbitrary constant to the result of computing an indefinite integral, for a double integral $\iint f(x, y)\, dx\, dy$ you must add an arbitrary function of the form $\varphi(x) + \psi(y)$. For a triple integral $\iiint f(x, y, z)\, dx\, dy\, dz$ you must add an arbitrary function of the form $\varphi(x, y) + \psi(y, z) + \lambda(x, z)$.

▶ **To enter and evaluate a double or triple integral**

1. Click $\boxed{\Sigma\!\!\int}$ or choose **Insert + Operator**.

2. Click the double or triple integral and choose **OK**.

3. Enter the function and the differentials. (The latter are necessary.)

4. With the insertion point in the integral, choose **Evaluate**.

5. For a double integral, add an arbitrary function of the form $\varphi(x) + \psi(y)$; for a triple integral, add $\varphi(x, y) + \psi(y, z) + \lambda(x, z)$.

▶ **Evaluate and add arbitrary function**

$\iint xy\, dx\, dy = \frac{1}{4}x^2 y^2 + \varphi(x) + \psi(y)$

$\iint x \sin x \cos y\, dx\, dy = (\sin x - x \cos x) \sin y + \varphi(x) + \psi(y)$

$\iiint xy^2 z\, dx\, dy\, dz = \frac{1}{12}x^2 y^3 z^2 + \varphi(x, y) + \psi(y, z) + \lambda(x, z)$

Exercises

1. Verify the formula $\frac{d}{dx}\left(x^8\right) = 8x^7$ by starting with the definition of derivative and choosing submenu items such as Expand and Simplify.

2. Use Newton's method on the function $f(x) = x^2 + 1$, starting with $x_0 = 0.5$. What conclusions can you draw?

3. Find the equation of one line that is tangent to the graph of
$$f(x) = x(x-1)(x-3)(x-6)$$
at two different points.

4. For $0 < k < 1$, the elliptic integral
$$E = \int_0^{\pi/2} \sqrt{1 - k\sin^2 t}\, dt$$
has no elementary solution. Use a series expansion of the integrand to estimate E.

5. Find all the solutions to $x^y = y^x$ for unequal positive integers x and y.

6. Blood flowing through an artery flows fastest at the center of the artery, and slowest near the walls of the artery where friction is a factor. In fact, the velocity is given by the formula $v(r) = \alpha(R^2 - r^2)$, where α is a constant, R is the radius of the artery, and r is the distance from the center.

   Set up an integral that gives the total blood flow through an artery. Show that if an artery is constricted to one-half of its original radius, the blood flow (assuming constant blood pressure) is reduced to $\frac{1}{16}$ of its original flow.

7. The mass of an object traveling at a velocity v with rest mass m_0 is given by
$$m = m_0\left(1 - \frac{v^2}{c^2}\right)^{-1/2}$$
where c is the speed of light. Use a Maclaurin series expansion to show the increase in mass at low velocities.

8. Evaluate $\int 2^x \cos bx\, dx$ and simplify the answer.

9. Evaluate $\displaystyle\int_{-\pi}^{\pi} \frac{1 + \sin x}{(x - \cos x)^2}\, dx$.

10. Evaluate $\lim_{h\to 0+} \int_0^\infty \sin(x^{1+h})\, dx$.

11. The Fundamental Theorem of Calculus says that if f is continuous on a closed interval $[a, b]$, then

    a. If g is defined by $g(x) = \int_a^x f(t)dt$ for $x \in [a, b]$, then $g'(x) = f(x)$, and
    b. If F is any antiderivative of f, then $\int_a^b f(x)dx = F(b) - F(a)$.

    Demonstrate that these two conditions hold for each of the three functions $f(x) = x^3$, $f(x) = xe^x$, and $f(x) = \sin^2 x \cos x$.

12. The **arithmetic-geometric** mean of two positive numbers $a > b$ was defined by Gauss as follows. Let $a_0 = a$ and $b_0 = b$. Given a_n and b_n, let a_{n+1} be the arithmetic mean of a_n and b_n, and b_{n+1} the geometric mean of a_n and b_n:

$$a_{n+1} = \frac{a_n + b_n}{2} \quad \text{and} \quad b_{n+1} = \sqrt{a_n b_n}$$

Using mathematical induction, you can show that $a_n > a_{n+1} > b_{n+1} > b_n$ and deduce that both series $\{a_n\}$ and $\{b_n\}$ are convergent, and, in fact, that $\lim_{n \to \infty} a_n = \lim_{n \to \infty} b_n$.

Compute the arithmetic-geometric mean of the numbers 2 and 1 to five decimal places.

13. Two numbers x and y are chosen at random in the unit interval $[0, 1]$. What is the average distance between two such numbers?

Solutions

1. By definition,

$$
\begin{aligned}
\frac{d}{dx}\left(x^8\right) &= \lim_{h \to 0} \frac{(x+h)^8 - x^8}{h} \\
&= \lim_{h \to 0} \frac{8x^7 h + 28x^6 h^2 + \cdots + 28x^2 h^6 + 8x h^7 + h^8}{h} \\
&= \lim_{h \to 0} \left(8x^7 + 28x^6 h + \cdots + 28x^2 h^5 + 8x h^6 + h^7\right) \\
&= 8x^7
\end{aligned}
$$

2. Defining g by $g(x) = x - f(x)/f'(x)$, from the Calculus submenu, choose Iterate to obtain

$$
\begin{bmatrix}
0.5 \\
-0.75 \\
0.29167 \\
-1.5684 \\
-0.4654 \\
0.84164
\end{bmatrix}
$$

If this result seems to be headed nowhere, it is doing so for good reason. The function f is always positive, so it has no zeroes. Newton's method is searching for something that does not exist.

3. It is sufficient to find three numbers a, b, and m that satisfy $f'(a) = m$, $f'(b) = m$, and $\frac{f(b)-f(a)}{b-a} = m$. Put these three equations inside a 3×1 matrix and, from the Solve submenu, choose Exact to get several solutions, including the real solutions

$$
\left[a = \frac{5}{2} - \frac{1}{2}\sqrt{21}, b = \frac{5}{2} - \frac{1}{2}\sqrt{21}, m = -8 \right]
$$

$$
\left[a = \frac{5}{2} - \frac{1}{2}\sqrt{21}, b = \frac{1}{2}\sqrt{21} + \frac{5}{2}, m = -8 \right]
$$

$$\left[a = \frac{1}{2}\sqrt{21} + \frac{5}{2}, b = \frac{5}{2} - \frac{1}{2}\sqrt{21}, m = -8\right]$$

$$\left[a = \frac{1}{2}\sqrt{21} + \frac{5}{2}, b = \frac{1}{2}\sqrt{21} + \frac{5}{2}, m = -8\right]$$

$$\left[a = \frac{5}{2}, b = \frac{5}{2}, m = -8\right]$$

Three of the solutions are not allowed, because the problem requires $a \neq b$. The two remaining solutions have the roles of a and b reversed. Assuming $a < b$, that leaves the solution $\left[a = \frac{5}{2} - \frac{1}{2}\sqrt{21}, b = \frac{1}{2}\sqrt{21} + \frac{5}{2}, m = -8\right]$. Evaluating and expanding,

$$f(a) = \left(\frac{5}{2} - \frac{1}{2}\sqrt{21}\right)\left(\frac{3}{2} - \frac{1}{2}\sqrt{21}\right)\left(-\frac{1}{2} - \frac{1}{2}\sqrt{21}\right)\left(-\frac{7}{2} - \frac{1}{2}\sqrt{21}\right)$$

$$= -21 + 4\sqrt{21}$$

so that

$$y = f(a) + m(x - a)$$

$$= -21 + 4\sqrt{21} - 8\left(x - \frac{5}{2} + \frac{1}{2}\sqrt{21}\right)$$

$$= -1 - 8x$$

Plot the two curves $x\,(x - 1)\,(x - 3)\,(x - 6)$ and $-1 - 8x$, just for visual verification. Use a viewing window with domain interval $-1 \leq x \leq 6.5$ to generate the following picture.

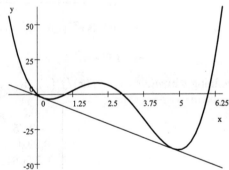

4. The series is given by $\sqrt{1 - k\sin^2 t} = 1 + \left(-\frac{1}{2}k\right)t^2 + \left(\frac{1}{6}k - \frac{1}{8}k^2\right)t^4 + O\left(t^5\right)$.
Thus, an estimate for E is given by

$$E \approx \int_0^{\pi/2}\left[1 + \left(-\frac{1}{2}k\right)t^2 + \left(\frac{1}{6}k - \frac{1}{8}k^2\right)t^4\right]dt$$

$$= \frac{1}{2}\pi - \frac{1}{48}\pi^3 k + \frac{1}{160}\pi^5\left(\frac{1}{6}k - \frac{1}{8}k^2\right)$$

As a check, $k = 1$ yields 1.0045 compared with the exact value

$$\int_0^{\pi/2}\sqrt{1 - \sin^2 t}\,dt = 1$$

and $k = 0$ yields $\frac{1}{2}\pi$, which agrees precisely with

$$\int_0^{\pi/2} dt = \frac{1}{2}\pi$$

5. Compute natural logs on both sides and separate variables to get $\frac{\ln x}{x} = \frac{\ln y}{y}$. Plot the graph of $\frac{\ln x}{x}$ on the interval $1 \le x \le 10$ and locate the extreme values of $\frac{\ln x}{x}$ by solving $\frac{d}{dx}\left(\frac{\ln x}{x}\right) = 0$. Note that 2 is the only integer between 1 and e, and verify

that $2^4 = 4^2$ is true.

6. The flow is given by the integral

$$\int_0^R \alpha(R^2 - r^2)2\pi r\, dr = \frac{1}{2}\alpha\pi R^4$$

If R is reduced by one-half, then R^4 is reduced to $\frac{1}{16}$ of the original amount.

7. The series expansion is given by

$$m_0\left(1 - \frac{v^2}{c^2}\right)^{-1/2} = m_0 + \frac{1}{2c^2}v^2 m_0 + \frac{3}{8c^4}v^4 m_0 + O\left(v^5\right)$$

If $\frac{v}{c}$ is small, then the model

$$m \approx m_0 + \frac{1}{2}\frac{m_0}{c^2}v^2$$

is useful for estimating the increased mass.

8. The following sequence requires Evaluate, Simplify, Combine Trig Functions, Factor, and Simplify.

$$\begin{aligned}
\int 2^x \cos bx\, dx &= \frac{1}{2b^2 + 2\ln^2 2}\left(b\left(\sin bx\right)2^{x+1} + \left(\cos bx \ln 2\right)2^{x+1}\right) \\
&= \frac{1}{2}\left(b^2 + \ln^2 2\right)^{-1}\left(b\sin bx + \cos bx \ln 2\right)\left(2^{x+1}\right) \\
&= 2^x\frac{b\sin bx + \cos bx \ln 2}{b^2 + \ln^2 2}
\end{aligned}$$

9. The integral

$$\int_{-\pi}^{\pi} \frac{1 + \sin x}{(x - \cos x)^2}\, dx$$

is improper, because $x - \cos x = 0$ has a root (≈ 0.73909) between $-\pi$ and π. Evaluate Numerically gives

$$\int_{-\pi}^{.739} \frac{1 + \sin x}{(x - \cos x)^2}\, dx = 7018.2$$

$$\int_{.7392}^{\pi} \frac{1 + \sin x}{(x - \cos x)^2}\, dx = 5201.4$$

Change Digits Shown in Results in the Settings dialog box to 10. Solving $\cos x = x$ numerically gives $x = 0.7390851332$. Using this as a limit, Evaluate Numerically gives

$$\int_{-\pi}^{0.73908} \frac{1+\sin x}{(x-\cos x)^2} dx = 116400.405\,5$$

$$\int_{0.73909}^{\pi} \frac{1+\sin x}{(x-\cos x)^2} dx = 122772.682\,2$$

providing some evidence that both integrals diverge.

10. You obtain the result

$$\lim_{h\to 0^+} \int_0^\infty \sin(x^{1+h})\,dx = 1$$

This result is reasonable, because the integral $f(h) = \int_0^\infty \sin(x^{1+h})\,dx$ can be viewed as a convergent alternating series for $h > 0$, and

$$g(y) = \int_0^y \sin x\,dx = 1 - \cos y$$

ranges in value between 0 and 2, with an average value of 1.

11. We need to show that for each of the three functions $f(x) = x^3$, $f(x) = xe^x$, and $f(x) = \sin^2 x \cos x$, (a) and (b) hold:

(a) If g is defined by $g(x) = \int_a^x f(t)dt$ for $x \in [a, b]$, then $g'(x) = f(x)$.

(b) If F is any antiderivative of f, then $\int_a^b f(x)dx = F(b) - F(a)$.

For $f(x) = x^3$,

$$g(x) = \int_a^x t^3 dt = \tfrac{1}{4}x^4 - \tfrac{1}{4}a^4$$

and

$$g'(x) = \frac{d}{dx}\left(\tfrac{1}{4}x^4 - \tfrac{1}{4}a^4\right) = x^3$$

The antiderivatives of f are of the form

$$F(x) = \int x^3 dx = \tfrac{1}{4}x^4 + C$$

for different constants C. Now

$$F(b) - F(x) = \left[\tfrac{1}{4}x^4 + C\right]_{x=a}^{x=b} = \tfrac{1}{4}b^4 - \tfrac{1}{4}a^4$$

which is the same as

$$\int_a^b x^3 dx = \tfrac{1}{4}b^4 - \tfrac{1}{4}a^4$$

For $f(x) = xe^x$,

$$g(x) = \int_a^x te^t dt = xe^x - e^x - ae^a + e^a$$

and

$$g'(x) = \frac{d}{dx}\left(xe^x - e^x - ae^a + e^a\right) = xe^x$$

The antiderivatives of f are of the form

$$F(x) = \int xe^x dx = xe^x - e^x + C$$

for different constants C. Now

$$F(b) - F(x) = [xe^x - e^x + C]_{x=a}^{x=b} = be^b - e^b - ae^a + e^a$$

which is the same as

$$\int_a^b xe^x dx = be^b - e^b - ae^a + e^a$$

For $f(x) = \sin^2 x \cos x$,

$$g(x) = \int_a^x \sin^2 t \cos t \, dt = \frac{1}{4}\sin x - \frac{1}{4}\sin a + \frac{1}{12}\sin 3a - \frac{1}{12}\sin 3x$$

and

$$g'(x) = \frac{d}{dx}\left(\frac{1}{4}\sin x - \frac{1}{4}\sin a + \frac{1}{12}\sin 3a - \frac{1}{12}\sin 3x\right)$$

$$= \frac{1}{4}\cos x - \frac{1}{4}\cos 3x$$

To check to see if this is the same as $f(x)$, apply **Combine + Trig Functions** to the expression $\sin^2 x \cos x$ to see that indeed

$$\sin^2 x \cos x = \frac{1}{4}\cos x - \frac{1}{4}\cos 3x$$

The antiderivatives of f are of the form

$$F(x) = \int \sin^2 x \cos x \, dx = \frac{1}{4}\sin x - \frac{1}{12}\sin 3x + C$$

for different constants C. Now

$$F(b) - F(x) = \left[\frac{1}{4}\sin x - \frac{1}{12}\sin 3x\right]_{x=a}^{x=b}$$

$$= \frac{1}{4}\sin b - \frac{1}{4}\sin a + \frac{1}{12}\sin 3a - \frac{1}{12}\sin 3b$$

while

$$\int_a^b \sin^2 x \cos x \, dx = \frac{1}{4}\sin b - \frac{1}{4}\sin a + \frac{1}{12}\sin 3a - \frac{1}{12}\sin 3b$$

12. Since the arithmetic-geometric mean lies between a_n and b_n for all n, we know the arithmetic-geometric mean to five decimal places when these two numbers agree to that many places.

- $a_1 = \dfrac{2+1}{2} = \dfrac{3}{2} = 1.5$ and $b_1 = \sqrt{2*1} = \sqrt{2} = 1.41421$

- $a_2 = \dfrac{\frac{3}{2} + \sqrt{2}}{2} = \frac{3}{4} + \frac{1}{2}\sqrt{2} = 1.45711$ and $b_2 = \sqrt{\frac{3}{2}\sqrt{2}} = 1.45648$

- $a_3 = \dfrac{\frac{\frac{3}{2}+\sqrt{2}}{2} + \sqrt{\frac{3}{2}\sqrt{2}}}{2} = \frac{3}{8} + \frac{1}{4}\sqrt{2} + \frac{1}{4}\sqrt{6}\sqrt[4]{2} = 1.45679$

  and $b_3 = \sqrt{\left(\frac{3}{4} + \frac{1}{2}\sqrt{2}\right)\sqrt{\frac{3}{2}\sqrt{2}}} = 1.45679$

13. An average value can be determined by evaluating an integral. The average distance between x and y is given by $\int_0^1 \int_0^1 |x - y| \, dy \, dx = \frac{1}{3}$. This can be verified using the following steps:

$$\int_0^1 \int_0^1 |x - y| \, dy \, dx \quad = \quad \int_0^1 \int_0^x |x - y| \, dy \, dx + \int_0^1 \int_x^1 |x - y| \, dy \, dx$$

$$= \quad \int_0^1 \int_0^x (x - y) \, dy \, dx + \int_0^1 \int_x^1 (y - x) \, dy \, dx$$

$$= \quad \frac{1}{6} + \frac{1}{6} = \frac{1}{3}$$

8 Matrix Algebra

Matrices are used throughout mathematics and in related fields such as physics, engineering, economics, and statistics. The algebra of matrices provides a model for the study of vector spaces and linear transformations.

Introduction

A rectangular array of mathematical expressions is called a *matrix*. A matrix with m rows and n columns is called an $m \times n$ matrix. Matrices are sometimes referred to simply as *arrays,* and an $m \times 1$ or $1 \times n$ array is also called a *vector*. Several methods for creating matrices are described in the ensuing sections.

Entries in matrices can be real or complex numbers, or mathematical expressions with real or complex coefficients. Most of the choices from the Matrices submenu operate on both real and complex matrices. The QR and SVD factorizations discussed later in this chapter assume real matrices.

Matrix entries are identified by their row and column number. The matrix can be considered as a function on pairs of positive integers. If the matrix is given a name, this feature can be used to retrieve the entries, with the arguments entered as subscripts.

▶ Definitions + New Definition

$$A = \begin{bmatrix} -85 & -55 & -37 \\ -35 & 97 & 50 \\ 79 & 56 & 49 \end{bmatrix}$$

▶ Evaluate

$A_{2,3} = 50$

$A_{3,3} = 49$

Note that the subscripted row and column numbers are separated by a comma.

Changing the Appearance of Matrices

You can make choices in the View menu that affect the appearance of matrices on the screen. Helper Lines and Input Boxes can be *shown* or *hidden*. The default is to show them to make it easier to handle entries on the screen. However, matrix helper lines and

301

input boxes do not appear when you preview or print the document. For this reason, you will usually want the matrix enclosed in brackets, either built-in or added manually. These two options provide the same screen appearance and the same mathematical properties, but the Typeset + Preview or Print appearance differs—when you typeset, the "built-in" brackets fit more tightly around the matrix entries than added brackets.

If you have a matrix without built-in delimiters, you will generally want to add brackets around it. The result of an operation on matrices generally appears with the same brackets as the original matrices.

▶ **To enclose a matrix in round or square brackets**

1. Select the matrix using the mouse or place the insertion point to the left (or right) of the matrix and press SHIFT + RIGHT (or LEFT) ARROW.

2. Click (□) or [□] .

or

1. Select the matrix with the insertion point, or click and drag.

2. Click 🔍 or choose Edit + Properties, and add Built-in Delimiters of the desired shape.

The choice of round, square, or curly brackets does not affect the mathematical properties of the matrix. Vertical straight brackets are interpreted as mathematical operations (*determinant* for single lines, *norm* for double lines) so they should be avoided for general use.

You can move about in a matrix with the arrow keys, by pressing TAB or BACKTAB, and by using the mouse. Pressing SPACEBAR moves your insertion point through the mathematics or out of the matrix.

Creating Matrices

You can create a matrix in four basically different ways: use the Matrix dialog; use a keyboard shortcut; or from the Matrices submenu, choose Fill Matrix or Random Matrix.

Each method involves different choices, as described in the following paragraphs.

▶ **To create a matrix with the Matrix dialog**

1. Click ▦ or choose Insert + Matrix.

2. Select the number of rows and columns.

3. Under Built-in Delimiters, choose None or select a type of built-in delimiter.

4. Choose OK.

5. Type entries in the input boxes.

The entries can be any valid mathematical expression. Both real and complex numbers are legitimate entries, as well as algebraic expressions. The built-in delimiters have the same appearance as expanding brackets on the screen, but they require less horizontal space when typeset.

▶ **To create a matrix with a keyboard shortcut**

- Press CTRL + S + M
  or

- Press CTRL + S + SHIFT + M

The first choice produces a matrix with the same attributes as your most recently created matrix. The second choice produces a 2×2 matrix.

You can create an $m \times n$ matrix that includes entries, using **Random Matrix** from the **Matrices** submenu. You can specify the range for random (integer) entries as well as specify some special types.

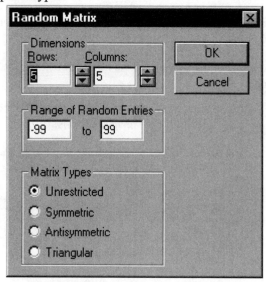

▶ **To create a matrix with Random Matrix**

1. From the **Matrices** submenu, choose **Random Matrix**.

2. Set the row and column numbers.

3. Set the range for random integers.

4. Check one of the **Matrix Types**.

5. Choose **OK**.

A matrix of the desired type appears inside the brackets.

▶ Matrices + Random Matrix

   (3×3, Range of Random Entries: -10 to 10)

$$\begin{bmatrix} -4 & 7 & 8 \\ 10 & -6 & -8 \\ -5 & 7 & 6 \end{bmatrix} \quad \begin{bmatrix} -6 & 0 & -10 \\ 0 & 5 & 1 \\ -10 & 1 & 1 \end{bmatrix} \quad \begin{bmatrix} 0 & -4 & 7 \\ 4 & 0 & 8 \\ -7 & -8 & 0 \end{bmatrix} \quad \begin{bmatrix} -4 & -3 & -10 \\ 0 & -8 & 5 \\ 0 & 0 & -1 \end{bmatrix}$$

   Unrestricted   Symmetric   Antisymmetric   Triangular

You can create an $m \times n$ matrix that includes entries, using Fill Matrix from the Matrices submenu.

▶ **To create a matrix with** Fill Matrix

1. Enter expanding brackets in the shape you prefer, and leave the insertion point inside the brackets.

2. From the Matrices submenu, choose Fill Matrix and set the Dimensions for Rows and Columns.

3. Choose one of the items from the Fill With box, respond to prompts if they appear, and choose OK.

The matrix appears inside the brackets, filled with the entries you chose. These choices are discussed in the next few paragraphs.

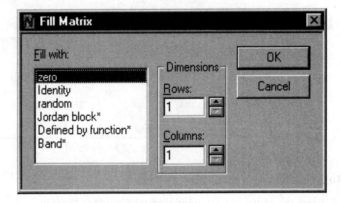

Zero

With the Zero option in Fill Matrix, you can create an $m \times n$ matrix whose entries are all zeroes, for any positive integers m and n.

▶ Matrices + Fill Matrix + Zero

$$\begin{pmatrix} 0 & 0 \\ 0 & 0 \end{pmatrix}$$

Identity

With the Identity option in Fill Matrix, you can create an $n \times n$ identity matrix for any positive integer n. You can also specify a nonsquare matrix and choose Identity.

▶ Matrices + Fill Matrix + Identity

$$
\begin{bmatrix} 1 & 0 & 0 \\ 0 & 1 & 0 \\ 0 & 0 & 1 \end{bmatrix}
\qquad
\begin{bmatrix} 1 & 0 & 0 & 0 \\ 0 & 1 & 0 & 0 \\ 0 & 0 & 1 & 0 \end{bmatrix}
\qquad
\begin{bmatrix} 1 & 0 & 0 \\ 0 & 1 & 0 \\ 0 & 0 & 1 \\ 0 & 0 & 0 \end{bmatrix}
$$

This operation produces a (square) identity matrix as large as possible and fills in remaining rows or columns, if any, with zeroes.

Random

With this option, you get a matrix filled with random integers between -999 and 999.

▶ Matrices + Fill Matrix + Random

$$
\begin{bmatrix}
-266 & -8 & 765 & 448 & -348 & 470 \\
-608 & -686 & 702 & -61 & -49 & -433 \\
966 & 902 & -942 & 712 & 761 & -892 \\
-564 & -826 & 251 & -414 & -44 & -214 \\
235 & -781 & 421 & -340 & 881 & 444
\end{bmatrix}
$$

Jordan Block

A *Jordan block* is a square matrix with the same expression along the main diagonal, ones on the superdiagonal, and zeroes elsewhere. The dialog box asks for the number of rows and columns and for an item for the diagonal. Jordan forms that are built from Jordan blocks are discussed in more detail later in this chapter.

▶ Matrices + Fill Matrix + Jordan block

- 2×2 with Item for Diagonal λ, and 5×5 with Item for Diagonal x

$$
\begin{pmatrix} \lambda & 1 \\ 0 & \lambda \end{pmatrix}
\qquad
\begin{pmatrix}
x & 1 & 0 & 0 & 0 \\
0 & x & 1 & 0 & 0 \\
0 & 0 & x & 1 & 0 \\
0 & 0 & 0 & x & 1 \\
0 & 0 & 0 & 0 & x
\end{pmatrix}
$$

Defined by Function

To use the Defined by Function option, first define a function $f(i, j)$ of two variables. Then use Defined by Function to create the $m \times n$ matrix with (i, j) entry equal to $f(i, j)$ for $1 \le i \le m$ and $1 \le j \le n$.

Example   Hilbert matrices

1. Define $f(i,j) = \dfrac{1}{i+j-1}$.

2. From the Fill Matrix dialog, choose **Defined by Function**.

3. Type f in the box for the function name.

4. Set rows and columns to 2 or 3.

5. Choose **OK**.

$$\begin{bmatrix} 1 & \frac{1}{2} \\ \frac{1}{2} & \frac{1}{3} \end{bmatrix} \qquad \begin{bmatrix} 1 & \frac{1}{2} & \frac{1}{3} \\ \frac{1}{2} & \frac{1}{3} & \frac{1}{4} \\ \frac{1}{3} & \frac{1}{4} & \frac{1}{5} \end{bmatrix}$$

Example   Vandermonde matrix

1. Define the function $g(i,j) = x_i^{j-1}$.

2. From the Fill Matrix dialog, choose **Defined by function**.

3. Enter g for the function name.

4. Set rows and columns to 4.

$$\begin{bmatrix} 1 & x_1 & x_1^2 & x_1^3 \\ 1 & x_2 & x_2^2 & x_2^3 \\ 1 & x_3 & x_3^2 & x_3^3 \\ 1 & x_4 & x_4^2 & x_4^3 \end{bmatrix}$$

Example   An "arbitrary" 3×3 matrix

1. Define the function $a(i,j) = a_{i,j}$.

2. From the Fill Matrix dialog, choose **Defined by function**.

3. Enter a for the function name.

4. Set rows and columns to 3.

$$\begin{bmatrix} a_{1,1} & a_{1,2} & a_{1,3} \\ a_{2,1} & a_{2,2} & a_{2,3} \\ a_{3,1} & a_{3,2} & a_{3,3} \end{bmatrix}$$

Note the comma between subscripts. Without the comma in the definition of the function $a(i,j)$, the subscript ij would be interpreted as a product!

You can use the following trick to create a general matrix up to 9×9 with no commas in the subscripts.

Example   Another form for an "arbitrary" 3×3 matrix

1. Define the function $a(i, j) = a_{10i+j}$.

2. From the Fill Matrix dialog, choose Defined by function.

3. Enter a for the function name.

4. Set rows and columns to 3.
$$\begin{bmatrix} a_{11} & a_{12} & a_{13} \\ a_{21} & a_{22} & a_{23} \\ a_{31} & a_{32} & a_{33} \end{bmatrix}$$

Example   Constant matrices

1. From the Fill Matrix dialog, choose Defined by function.

2. Enter 5 for the function name.
$$\begin{bmatrix} 5 & 5 \\ 5 & 5 \end{bmatrix}$$

Band

The Band option requires that you enter a list such as "a, b, c" with an odd number of entries. This option creates a matrix with a "band" of entries around the main diagonal, up to the width of your list, and with zeroes elsewhere. The first entry will be the item in the middle of your list.

Example   Band matrices

1. From the Fill Matrix dialog, choose Band.

2. Type a in the Enter Band list box.

3. Set rows to 2.
$$\begin{bmatrix} a & 0 \\ 0 & a \end{bmatrix}$$

1. From the Fill Matrix dialog, choose Band.

2. Type a, b, c in the Enter Band List box.

3. Set rows to 2 or 5, and columns to 2, 5, or 8.

$$\begin{bmatrix} b & c \\ a & b \end{bmatrix} \quad \begin{bmatrix} b & c & 0 & 0 & 0 \\ a & b & c & 0 & 0 \\ 0 & a & b & c & 0 \\ 0 & 0 & a & b & c \\ 0 & 0 & 0 & a & b \end{bmatrix} \quad \begin{bmatrix} b & c & 0 & 0 & 0 & 0 & 0 & 0 \\ a & b & c & 0 & 0 & 0 & 0 & 0 \\ 0 & a & b & c & 0 & 0 & 0 & 0 \\ 0 & 0 & a & b & c & 0 & 0 & 0 \\ 0 & 0 & 0 & a & b & c & 0 & 0 \end{bmatrix}$$

Choosing Band and entering the lists "0," "1," and "0, λ, 1" generates a zero matrix, an identity matrix, and a Jordan block, respectively.

$$\begin{bmatrix} 0 & 0 & 0 \\ 0 & 0 & 0 \\ 0 & 0 & 0 \end{bmatrix} \quad \begin{bmatrix} 1 & 0 & 0 \\ 0 & 1 & 0 \\ 0 & 0 & 1 \end{bmatrix} \quad \begin{bmatrix} \lambda & 1 & 0 \\ 0 & \lambda & 1 \\ 0 & 0 & \lambda \end{bmatrix}$$

Revising Matrices

You can add or delete rows, columns, or a full block of rows or columns from a matrix. The alignment of rows and columns can be reset. Entries in a rectangular block can be deleted or replaced.

Adding Rows and Columns

▶ **To add rows or columns to a matrix**

1. Select the matrix by placing the insertion point in a cell of the matrix or by placing the insertion point at the right of the matrix (but not outside of any brackets).

2. From the Edit menu, choose Insert Rows or Insert Columns.

   or

   Press and release the right mouse button and, from the pop-up menu, choose Insert Rows or Insert Columns.

3. Make appropriate choices from the dialog that appears and choose OK.

Deleting Rows and Columns

▶ **To delete a block of rows or columns**

1. Select a block of rows or columns with the mouse or with SHIFT + ARROW.

2. Press DEL.

   You can also use the procedure described above to delete entries from a rectangular block that does not include a complete row or column of a matrix.

   The choices Insert Row(s) and Insert Column(s) appear on the Edit menu only when a matrix is selected. If they do not appear, reposition the insertion point or select the matrix with click and drag, being careful to select only the inside of the matrix—that is, not including the exterior Helper Lines.

▶ **To lengthen a vector represented as an** $n \times 1$ **or** $1 \times n$ **matrix**

1. Place the insertion point in the last input box.

1. Press ENTER.

▶ **To shorten a vector represented as an** $n \times 1$ **or** $1 \times n$ **matrix**

1. Place the insertion point in the last input box.

1. Press BACKSPACE.

You can start with a display box, or the input boxes that appear with the fraction, radical, or bracket buttons, and make similar changes.

Changing Alignment

▶ **To change the alignment of entries**

1. Select the matrix using the mouse (or starting with the insertion point at the left of the matrix, press SHIFT + RIGHT ARROW).

2. Click the Properties button ![Q] on the Standard toolbar; or choose Edit + Properties; or press and release the right mouse button and choose Properties.

3. Make appropriate choices from the dialog box that appears.

Replacing a Rectangular Block

You can replace a rectangular block in an existing matrix with Copy and Paste or with Fill Matrix.

▶ **To replace a rectangular block with** Copy and Paste

1. Copy a rectangular matrix to the clipboard with Edit + Copy.

2. Select a rectangular portion of the same dimensions in any matrix, with the mouse or SHIFT + ARROW.

3. Choose Edit + Paste.

▶ **To change a matrix with** Fill Matrix

1. Select a rectangular portion of the matrix with the mouse or SHIFT + ARROW..

2. From the Matrices submenu, choose Fill Matrix.

3. Choose one of the items from the dialog.

4. Choose OK.

The selected region of the matrix is filled with the entries that you chose.

Example   To change the lower-right 2×2 corner of the matrix to the zero matrix, select the lower-right 2×2 corner of the matrix using the mouse; from the **Matrices** submenu, choose **Fill Matrix**; choose **Zero**; and choose **OK**.

$$\begin{bmatrix} 1 & 2 & 3 \\ 5 & 5 & 4 \\ 7 & 8 & 9 \end{bmatrix} \qquad \begin{bmatrix} 1 & 2 & 3 \\ 5 & 0 & 0 \\ 7 & 0 & 0 \end{bmatrix}$$

The lower-right corner is *replaced* by the 2×2 zero matrix. No new matrix is created.

You can delete a block of entries in a matrix by selecting a rectangular portion of the matrix with the mouse, and pressing DEL.

Example   To delete the entries in the lower-right 2×2 corner of the matrix, select the lower-right 2×2 corner of the matrix using the mouse and press DEL.

$$\begin{bmatrix} 1 & 2 & 3 \\ 5 & 5 & 4 \\ 7 & 8 & 9 \end{bmatrix} \qquad \begin{bmatrix} 1 & 2 & 3 \\ 5 & \square & \square \\ 7 & \square & \square \end{bmatrix}$$

Concatenating and Stacking Matrices

You can merge two matrices into one if they have the same number of rows or columns.

▶ **To concatenate two matrices with the same number of rows**

1. Place two matrices adjacent to each other.

2. Leave the insertion point in one of the matrices.

3. From the **Matrices** submenu, choose **Concatenate**.

► Concatenate

$$\begin{pmatrix} 1 & 2 \\ 3 & 4 \end{pmatrix} \begin{pmatrix} 5 & 6 \\ 7 & 8 \end{pmatrix}, \text{concatenate:} \begin{pmatrix} 1 & 2 & 5 & 6 \\ 3 & 4 & 7 & 8 \end{pmatrix}$$

$$\begin{pmatrix} x+1 & 2 \\ 3y & 4t+2 \end{pmatrix} \begin{pmatrix} 5+w \\ \sqrt{7}z \end{pmatrix}, \text{concatenate:} \begin{pmatrix} x+1 & 2 & 5+w \\ 3y & 4t+2 & \sqrt{7}z \end{pmatrix}$$

► **To stack two matrices with the same number of columns**

1. Place two matrices adjacent to each other.

2. Leave the insertion point in one of the matrices.

3. From the Matrices submenu, choose Stack.

► Stack

$$\begin{pmatrix} 1 & 2 \\ 3 & 4 \end{pmatrix} \begin{pmatrix} 5 & 6 \end{pmatrix}, \text{stack:} \begin{pmatrix} 1 & 2 \\ 3 & 4 \\ 5 & 6 \end{pmatrix}$$

► Stack

$$\begin{pmatrix} x+1 & 2 \\ 3y & 4t+2 \end{pmatrix} \begin{pmatrix} w+5 & z\sqrt{7} \end{pmatrix}, \text{stack:} \begin{pmatrix} x+1 & 2 \\ 3y & 4t+2 \\ w+5 & z\sqrt{7} \end{pmatrix}$$

Reshaping Lists and Matrices

A list of expressions entered in mathematics and separated by commas can be turned into a matrix whose entries, reading left to right and top to bottom, are the entries of the list in the given order.

► **To make a matrix from a list**

1. Place the insertion point within the list.

2. From the Matrices submenu, choose Reshape.

3. Specify the number of columns.

The number of rows depends on the length of the list. Extra input boxes at the end are left blank.

▶ Reshape

$$45, 21, 8, 19, 0, 5, 15, 6 \text{ to 3 columns:} \begin{bmatrix} 45 & 21 & 8 \\ 19 & 0 & 5 \\ 15 & 6 & \Box \end{bmatrix}$$

A matrix filled with data can be reshaped, with the new matrix corresponding to the same list as the original data.

▶ **To reshape a matrix**

1. Place the insertion point in the matrix.

2. From the Matrices submenu, choose Reshape.

3. Specify the new number of columns.

▶ Reshape

$$\begin{bmatrix} -85 & -55 & -37 & -35 \\ 97 & 50 & 79 & 56 \end{bmatrix} \text{ to 3 columns:} \begin{bmatrix} -85 & -55 & -37 \\ -35 & 97 & 50 \\ 79 & 56 & \Box \end{bmatrix}$$

See page 411 for further examples.

Standard Operations

You can perform standard operations on matrices, such as addition, subtraction, and multiplication, by evaluating expressions entered in natural notation.

Matrix Addition and Scalar Multiplication

You add two matrices of the same dimension by adding corresponding entries. The numbers or other expressions used as matrix entries are called *scalars*. You multiply a scalar with a matrix by multiplying every entry of the matrix by the scalar. You can do matrix addition and multiplication and other operations with scalars and matrices by choosing Evaluate. Place the insertion point anywhere inside the expression.

▶ Evaluate

$$\begin{bmatrix} 1 & 2 \\ 4 & 3 \end{bmatrix} + \begin{bmatrix} 5 & 6 \\ 8 & 7 \end{bmatrix} = \begin{bmatrix} 6 & 8 \\ 12 & 10 \end{bmatrix}$$

Note that the sum appears with the same brackets as the original matrices.

▶ Evaluate

$$\begin{pmatrix} a_{11} & a_{12} \\ a_{21} & a_{22} \end{pmatrix} + \begin{pmatrix} b_{11} & b_{12} \\ b_{21} & b_{22} \end{pmatrix} = \begin{pmatrix} a_{11} + b_{11} & a_{12} + b_{12} \\ a_{21} + b_{21} & a_{22} + b_{22} \end{pmatrix}$$

$$a \begin{bmatrix} 1 & 2 \\ 4 & 3 \end{bmatrix} = \begin{bmatrix} a & 2a \\ 4a & 3a \end{bmatrix}$$

$$a \begin{bmatrix} 1 & 2 \\ 4 & 3 \end{bmatrix} - b \begin{bmatrix} 5 & 6 \\ 8 & 7 \end{bmatrix} = \begin{bmatrix} a - 5b & 2a - 6b \\ 4a - 8b & 3a - 7b \end{bmatrix}$$

Inner Products and Matrix Multiplication

The product of a $1 \times n$ matrix with an $n \times 1$ matrix (the product of two vectors) produces a scalar called the *inner product* or *dot product* of the two vectors. The *matrix product* of an $m \times k$ matrix with a $k \times n$ matrix is an $m \times n$ matrix obtained by taking inner products of rows and columns, the ijth entry of the product AB being the inner product of the ith row of A with the jth column of B.

▶ Evaluate

$$\begin{pmatrix} a & b \end{pmatrix} \begin{pmatrix} c \\ d \end{pmatrix} = ac + bd \qquad \begin{pmatrix} 1 & 2 \\ 4 & 3 \end{pmatrix} \begin{pmatrix} 5 & 6 \\ 8 & 7 \end{pmatrix} = \begin{pmatrix} 21 & 20 \\ 44 & 45 \end{pmatrix}$$

▶ Evaluate

$$\begin{pmatrix} a & b \\ u & v \end{pmatrix} \begin{pmatrix} c \\ d \end{pmatrix} = \begin{pmatrix} ac + bd \\ uc + vd \end{pmatrix} \qquad \begin{bmatrix} 5 & 6 \\ 8 & 7 \end{bmatrix}^3 = \begin{bmatrix} 941 & 942 \\ 1256 & 1255 \end{bmatrix}$$

To put an exponent on a matrix, place the insertion point immediately to the right of the matrix, click $\boxed{\text{N}^\text{x}}$ or choose Insert + Superscript, and type the exponent in the input box.

Rows and Columns

You can find the vector that is the nth row or column of a matrix A with the functions $\text{row}(A, n)$ and $\text{col}(A, n)$. These function names automatically gray when typed in mathematics mode if Automatic Substitution is enabled. Otherwise, you can create them with the Insert + Math Name dialog.

► Evaluate

$$\mathrm{row}\left(\left[\begin{array}{cc} 1 & 2 \\ 4 & 3 \end{array}\right],2\right)=\left[\begin{array}{cc} 4 & 3 \end{array}\right] \qquad \mathrm{col}\left(\left[\begin{array}{cc} 1 & 2 \\ 4 & 3 \end{array}\right],2\right)=\left[\begin{array}{c} 2 \\ 3 \end{array}\right]$$

Identity and Inverse Matrices

The $n \times n$ *identity* matrix I has ones down the main diagonal (upper-left corner to lower-right corner) and zeroes elsewhere. The 3×3 identity matrix, for example, is

$$I=\left[\begin{array}{ccc} 1 & 0 & 0 \\ 0 & 1 & 0 \\ 0 & 0 & 1 \end{array}\right]$$

The *inverse* of an $n \times n$ matrix A is an $n \times n$ matrix B satisfying $AB = I$. To find the inverse of an invertible matrix A, place the insertion point in the matrix and choose Matrices + Inverse; or enter A with "-1" as a superscript and apply Evaluate.

► Matrices + Inverse

$$\left(\begin{array}{cc} 5 & 6 \\ 8 & 7 \end{array}\right), \text{inverse:} \left(\begin{array}{cc} -\frac{7}{13} & \frac{6}{13} \\ \frac{8}{13} & -\frac{5}{13} \end{array}\right)$$

► Evaluate

$$\left(\begin{array}{cc} 5 & 6 \\ 8 & 7 \end{array}\right)^{-1} = \left(\begin{array}{cc} -\frac{7}{13} & \frac{6}{13} \\ \frac{8}{13} & -\frac{5}{13} \end{array}\right)$$

To check that this matrix satisfies the defining property, evaluate the product.

$$\left(\begin{array}{cc} 5 & 6 \\ 8 & 7 \end{array}\right)\left(\begin{array}{cc} -\frac{7}{13} & \frac{6}{13} \\ \frac{8}{13} & -\frac{5}{13} \end{array}\right) = \left(\begin{array}{cc} 1 & 0 \\ 0 & 1 \end{array}\right)$$

The operation Evaluate Numerically gives you a numerical approximation of the inverse. The accuracy of this numerical approximation depends on properties of the matrix, as well as on the settings for Digits Used in Computation and Digits Shown in Result (see page 29).

► Evaluate Numerically

$$\left(\begin{array}{cc} 5 & 6 \\ 8 & 7 \end{array}\right)^{-1} = \left(\begin{array}{cc} -0.538\,46 & 0.461\,54 \\ 0.615\,38 & -0.384\,62 \end{array}\right)$$

Checking the product of a matrix with its inverse gives you an idea of the degree of accuracy of the approximation.

▶ Evaluate

$$\begin{pmatrix} 5 & 6 \\ 8 & 7 \end{pmatrix} \begin{pmatrix} -0.53846 & 0.46154 \\ 0.61538 & -0.38462 \end{pmatrix} = \begin{pmatrix} 0.999\,98 & -0.000\,02 \\ -0.000\,02 & 0.999\,98 \end{pmatrix}$$

Since $\left(A^n\right)^{-1} = \left(A^{-1}\right)^n$, you can compute negative powers of invertible matrices.

▶ Evaluate

$$\begin{pmatrix} 5 & 6 \\ 8 & 7 \end{pmatrix}^{-3} = \begin{pmatrix} -\frac{1255}{2197} & \frac{942}{2197} \\ \frac{1256}{2197} & -\frac{941}{2197} \end{pmatrix}$$

The $m \times n$ matrix with every entry equal to zero is the *identity for addition*; that is, for any $m \times n$ matrix A,

$$A + 0 = 0 + A = A$$

and the *additive inverse* of a matrix A is the matrix $(-1)\, A$.

▶ Evaluate

$$\begin{bmatrix} a_{1,1} & a_{1,2} & a_{1,3} \\ a_{2,1} & a_{2,2} & a_{2,3} \\ a_{3,1} & a_{3,2} & a_{3,3} \\ a_{4,1} & a_{4,2} & a_{4,3} \end{bmatrix} + \begin{bmatrix} -a_{1,1} & -a_{1,2} & -a_{1,3} \\ -a_{2,1} & -a_{2,2} & -a_{2,3} \\ -a_{3,1} & -a_{3,2} & -a_{3,3} \\ -a_{4,1} & -a_{4,2} & -a_{4,3} \end{bmatrix} = \begin{bmatrix} 0 & 0 & 0 \\ 0 & 0 & 0 \\ 0 & 0 & 0 \\ 0 & 0 & 0 \end{bmatrix}$$

Polynomials with Matrix Values

You can apply a polynomial function of one variable to a matrix, as in the following example.

Example   A polynomial expression, such as $x^2 - 5x - 2$, can be evaluated at a matrix.

1. Leave the insertion point in the expression $x = \begin{bmatrix} 1 & 2 \\ 4 & 3 \end{bmatrix}$, and from the Definitions submenu, choose **New Definition**.

2. Apply **Evaluate** to the polynomial.

▶ Evaluate

$$x^2 - 5x - 2 = \begin{bmatrix} 2 & -2 \\ -4 & 0 \end{bmatrix} \qquad x^2 - 5x - 2x^0 = \begin{bmatrix} 2 & -2 \\ -4 & 0 \end{bmatrix}$$

You can also define the function $f(x) = x^2 - 5x - 2x^0$ and apply **Evaluate**.

▶ Define + New Definition

$$f(x) = x^2 - 5x - 2x^0$$

▶ Evaluate

$$f\left(\begin{bmatrix} 1 & 2 \\ 4 & 3 \end{bmatrix}\right) = \begin{bmatrix} 2 & -2 \\ -4 & 0 \end{bmatrix}$$

The expression $-5 \begin{bmatrix} 1 & 2 \\ 4 & 3 \end{bmatrix} - 2$ is not, strictly speaking, a proper expression. However, when evaluated, the final 2 is interpreted in this context as $\begin{bmatrix} 2 & 0 \\ 0 & 2 \end{bmatrix}$, or twice the 2×2 identity matrix.

Operations on Matrix Entries

To operate on one entry of a matrix, select the entry, press and hold the CTRL key, and choose a command. The program will perform the operation in place, leaving the rest of the matrix unchanged. Because you are in a word-processing environment, you can edit individual entries (just click in the input box and then edit) and apply other word-processing features to entries, such as copy and paste or click and drag.

Many of the commands on the **Compute** menu operate directly on the entries when applied to a matrix, as can be seen from the following examples.

▶ Factor

$$\begin{bmatrix} 5 & 6 \\ 8 & 7 \end{bmatrix} = \begin{bmatrix} 5 & 2 \times 3 \\ 2^3 & 7 \end{bmatrix}$$

▶ Evaluate

$$\begin{bmatrix} \frac{d}{dx}\sin x & \int 6x^2\,dx \\ \frac{d^2}{dx^2}\ln x & x + 3x \end{bmatrix} = \begin{bmatrix} \cos x & 2x^3 \\ -\frac{1}{x^2} & 4x \end{bmatrix}$$

▶ **Evaluate Numerically**

$$\begin{bmatrix} \sin^2 \pi & e \\ \ln 5 & x + 3x \end{bmatrix} = \begin{bmatrix} 0.0 & 2.718\,3 \\ 1.609\,4 & 4.0x \end{bmatrix}$$

▶ **Combine + Trig Functions**

$$\begin{bmatrix} \sin^2 x + \cos^2 x & 6x^2 \\ 4 \sin 4x \cos 4x & \sin x \cos y + \sin y \cos x \end{bmatrix}$$
$$= \begin{bmatrix} 1 & 6x^2 \\ 2 \sin 8x & \sin (x + y) \end{bmatrix}$$

▶ **Evaluate**

$$\frac{d}{dx} \begin{bmatrix} x + 1 & 2x^3 - 3 \\ \sin 4x & 3 \sec x \end{bmatrix} = \begin{bmatrix} 1 & 6x^2 \\ 4 \cos 4x & \frac{3}{\cos^2 x} \sin x \end{bmatrix}$$

Row Operations and Echelon Forms

One of the elementary applications of matrix arrays is storing and manipulating coefficients of systems of linear equations. The various steps that you carry out in applying the technique of elimination to a system of linear equations

$$\begin{aligned} a_{11}x_1 + a_{12}x_2 + \ldots + a_{1n}x_n &= b_1 \\ a_{21}x_1 + a_{22}x_2 + \ldots + a_{2n}x_n &= b_2 \\ &\vdots \qquad \vdots \\ a_{m1}x_1 + a_{m2}x_2 + \ldots + a_{mn}x_n &= b_m \end{aligned}$$

can be applied equally well to the matrix of coefficients and scalars

$$\begin{bmatrix} a_{11} & a_{12} & \cdots & a_{1n} & b_1 \\ a_{21} & a_{22} & \cdots & a_{2n} & b_2 \\ \vdots & \vdots & \cdots & \vdots & \vdots \\ a_{m1} & a_{m2} & \cdots & a_{mn} & b_m \end{bmatrix}$$

For this and numerous other reasons, you perform *elementary row operations* on matrices. The goal of elementary row operations is to put the matrix in a special form, such as a *row echelon form,* in which the number of leading zeroes increases as the row number increases. The system provides several choices for obtaining a row echelon form, one of which gives the *reduced row echelon form* satisfying the following conditions:

- The number of leading zeroes increases as the row number increases.

- The first nonzero entry in each nonzero row is equal to 1.

- Each column that contains the leading nonzero entry for any row contains only zeroes above and below that entry.

Gaussian Elimination and Row Echelon Form

The three row echelon forms that can be obtained from the Matrices submenu are illustrated in the following examples.

▶ Matrices + Fraction-free Gaussian Elimination

$\begin{bmatrix} a & b \\ c & d \end{bmatrix}$, fraction-free Gaussian elimination: $\begin{bmatrix} a & b \\ 0 & ad - bc \end{bmatrix}$

$\begin{bmatrix} 8 & 2 & 3 \\ 2 & -5 & 8 \end{bmatrix}$, fraction-free Gaussian elimination: $\begin{bmatrix} 8 & 2 & 3 \\ 0 & -44 & 58 \end{bmatrix}$

▶ Matrices + Gaussian Elimination

$\begin{bmatrix} a & b \\ c & d \end{bmatrix}$, Gaussian elimination: $\begin{bmatrix} a & b \\ 0 & \frac{1}{a}(ad - bc) \end{bmatrix}$

$\begin{bmatrix} 8 & 2 & 3 \\ 2 & -5 & 8 \end{bmatrix}$, Gaussian elimination: $\begin{bmatrix} 8 & 2 & 3 \\ 0 & -\frac{11}{2} & \frac{29}{4} \end{bmatrix}$

▶ Matrices + Reduced Row Echelon Form

$\begin{bmatrix} a & b \\ c & d \end{bmatrix}$, Reduced row echelon form: $\begin{bmatrix} 1 & 0 \\ 0 & 1 \end{bmatrix}$

$\begin{bmatrix} 8 & 2 & 3 \\ 2 & -5 & 8 \end{bmatrix}$, Reduced row echelon form: $\begin{bmatrix} 1 & 0 & \frac{31}{44} \\ 0 & 1 & -\frac{29}{22} \end{bmatrix}$

Elementary Row Operations

You can perform elementary row operations by multiplying on the left by appropriate *elementary matrices*—the matrices obtained from an identity matrix by applying an elementary row operation. The technique is illustrated in the following examples.

To create an elementary matrix, choose Fill Matrix from the Matrices submenu; create an identity matrix of the appropriate dimension by making choices in the Fill Matrix dialog box, and perform an elementary row operation by editing the identity matrix. Choose Evaluate to get the following products.

▶ **Add λ times row 3 to row 1**

$$\begin{bmatrix} 1 & 0 & \lambda \\ 0 & 1 & 0 \\ 0 & 0 & 1 \end{bmatrix} \begin{bmatrix} -5 & -2 & -1 \\ 3 & -6 & 2 \\ 1 & 4 & 1 \end{bmatrix} = \begin{bmatrix} \lambda - 5 & 4\lambda - 2 & \lambda - 1 \\ 3 & -6 & 2 \\ 1 & 4 & 1 \end{bmatrix}$$

▶ **Interchange rows 2 and 3**

$$\begin{bmatrix} 1 & 0 & 0 \\ 0 & 0 & 1 \\ 0 & 1 & 0 \end{bmatrix} \begin{bmatrix} -50 & -12 & -18 \\ 31 & -26 & -62 \\ 1 & -47 & -91 \end{bmatrix} = \begin{bmatrix} -50 & -12 & -18 \\ 1 & -47 & -91 \\ 31 & -26 & -62 \end{bmatrix}$$

▶ **Multiply row 2 by λ**

$$\begin{bmatrix} 1 & 0 & 0 \\ 0 & \lambda & 0 \\ 0 & 0 & 1 \end{bmatrix} \begin{bmatrix} 80 & -2 & -18 \\ 33 & -26 & 82 \\ 14 & -47 & -91 \end{bmatrix} = \begin{bmatrix} 80 & -2 & -18 \\ 33\lambda & -26\lambda & 82\lambda \\ 14 & -47 & -91 \end{bmatrix}$$

You can perform other row or column operations that are available in the MuPAD library, as in the following example.

▶ **To access the MuPAD function swapRow and to name it S**

1. From the **Definitions** submenu, choose **Define MuPAD Name**.

2. Respond to the dialog box as follows:
   - **MuPAD Name: linalg::swapRow(x,i,j)**
   - **Scientific Notebook (WorkPlace) Name:** $S(x, i, j)$
   - In the area titled **The MuPAD Name is a Procedure,** check **That is built in to MuPAD or is automatically loaded.**

3. Check **OK**.

This procedure defines a function $S(x, i, j)$ that interchanges the rows i and j of a matrix x.

Example   Define

$$x = \begin{bmatrix} -85 & -55 & -37 & -35 \\ 97 & 50 & 79 & 56 \\ 49 & 63 & 57 & -59 \end{bmatrix}$$

and evaluate $S(x, 1, 2)$ to get

$$S(x, 1, 2) = \begin{bmatrix} 97 & 50 & 79 & 56 \\ -85 & -55 & -37 & -35 \\ 49 & 63 & 57 & -59 \end{bmatrix}$$

Equations

Elementary methods for solving systems of equations are discussed on page 74. The algebra of matrices provides you with additional tools for solving systems of linear equations, both directly and by translating into matrix equations.

Systems of Linear Equations

You identify a system of equations by entering the equations in an $n \times 1$ matrix, with one equation to a row. When you have the same number of unknowns as equations, put the insertion point anywhere in the system, and from the Solve submenu, choose Exact. The variables are found automatically without having to be specified, as in the following example.

▶ Solve + Exact

$$x + y - 2z = 1$$
$$2x - 4y + z = 0 \quad , \text{Solution is} : \left\{ x = \tfrac{17}{8}, y = \tfrac{11}{8}, z = \tfrac{5}{4} \right\}$$
$$2y - 3z = -1$$

To solve a system of equations with two equations and three unknowns, you must specify Variables to Solve for in a dialog box. Put the insertion point anywhere in the matrix and, from the Solve submenu, choose Exact. A dialog box opens asking you to specify the variables. Enter the variable names, separated by commas.

▶ Solve + Exact

Variable(s) to Solve for : x, y

$$\begin{array}{l} 2x - y = 1 \\ x + 3z = 4 \end{array} , \text{Solution is} : \{ y = -6z + 7, x = -3z + 4 \}$$

Variable(s) to Solve for : x, z

$$\begin{array}{l} 2x - y = 1 \\ x + 3z = 4 \end{array} , \text{Solution is} : \left\{ x = \tfrac{1}{2} + \tfrac{1}{2}y, z = \tfrac{7}{6} - \tfrac{1}{6}y \right\}$$

Matrix Equations

The system of equations

$$\begin{array}{rcl} a_{11}x_1 + a_{12}x_2 + \ldots + a_{1n}x_n & = & b_1 \\ a_{21}x_1 + a_{22}x_2 + \ldots + a_{2n}x_n & = & b_2 \\ & \vdots & \\ a_{m1}x_1 + a_{m2}x_2 + \ldots + a_{mn}x_n & = & b_m \end{array}$$

is the same as the following matrix equation:

$$\begin{bmatrix} a_{11} & a_{12} & \cdots & a_{1n} \\ a_{21} & a_{22} & \cdots & a_{2n} \\ \vdots & \vdots & \cdots & \vdots \\ a_{m1} & a_{m2} & \cdots & a_{mn} \end{bmatrix} \begin{bmatrix} x_1 \\ x_2 \\ \vdots \\ x_n \end{bmatrix} = \begin{bmatrix} b_1 \\ b_2 \\ \vdots \\ b_m \end{bmatrix}$$

You can solve these systems using **Exact** on the **Solve** submenu. There are advantages to solving systems of equations in this way, and often you can best deal with systems of linear equations by solving the matrix version of the system.

Example   Multiply the coefficient matrix $\begin{bmatrix} 1 & 1 & -2 \\ 2 & -4 & 1 \\ 0 & 2 & -3 \end{bmatrix}$ by the vector $\begin{bmatrix} x \\ y \\ z \end{bmatrix}$ to display the system of equations

$$\begin{aligned} x + y - 2z &= 1 \\ 2x - 4y + z &= 0 \\ 2y - 3z &= -1 \end{aligned}$$

in matrix form.

$$\begin{bmatrix} 1 & 1 & -2 \\ 2 & -4 & 1 \\ 0 & 2 & -3 \end{bmatrix} \begin{bmatrix} x \\ y \\ z \end{bmatrix} = \begin{bmatrix} x + y - 2z \\ 2x - 4y + z \\ 2y - 3z \end{bmatrix} = \begin{bmatrix} 1 \\ 0 \\ -1 \end{bmatrix}$$

▶ **Solve + Exact**

$$\begin{bmatrix} 1 & 1 & -2 \\ 2 & -4 & 1 \\ 0 & 2 & -3 \end{bmatrix} \begin{bmatrix} x \\ y \\ z \end{bmatrix} = \begin{bmatrix} 1 \\ 0 \\ -1 \end{bmatrix}, \text{Solution is :} \begin{bmatrix} \frac{17}{8} \\ \frac{11}{8} \\ \frac{5}{4} \end{bmatrix}$$

$$\begin{bmatrix} 2 & -1 & 0 \\ 1 & 0 & 3 \end{bmatrix} \begin{bmatrix} x \\ y \\ z \end{bmatrix} = \begin{bmatrix} 1 \\ 4 \end{bmatrix}, \text{Solution is :} \begin{bmatrix} -3\hat{t}_3 + 4 \\ -6\hat{t}_3 + 7 \\ \hat{t}_3 \end{bmatrix}$$

In the first case, you can also solve the equation by multiplying both the left and right sides of the equation by the inverse of the coefficient matrix, and evaluating the product.

▶ **Evaluate**

$$\begin{bmatrix} x \\ y \\ z \end{bmatrix} = \begin{bmatrix} 1 & 1 & -2 \\ 2 & -4 & 1 \\ 0 & 2 & -3 \end{bmatrix}^{-1} \begin{bmatrix} 1 \\ 0 \\ -1 \end{bmatrix} = \begin{bmatrix} \frac{17}{8} \\ \frac{11}{8} \\ \frac{5}{4} \end{bmatrix}$$

You can convert a system of linear equations to a matrix, and a matrix to a system of equations, using **Rewrite + Equations as Matrix** and **Rewrite + Matrix as Equations**.

▶ **To convert a system of equations to a matrix**

1. Place the insertion point in a system of equations that has been created as a list or one-column matrix.

2. Choose Rewrite + Equations as Matrix

3. In the dialog that appears, enter the variables separated by commas.

4. Choose OK.

▶ Rewrite + Equations as Matrix

$\{x + 2y - 3, 3x - 5y = 0\}$ (Variable List: x, y),

$$\text{Corresponding matrix:} \begin{bmatrix} 1 & 2 & 3 \\ 3 & -5 & 0 \end{bmatrix}$$

$$\begin{bmatrix} x + y - 2z = 1 \\ 2x - 4y + z = 0 \\ 2y - 3z = -1 \end{bmatrix} \text{(Variable List: } x, y, z\text{)},$$

$$\text{Corresponding matrix:} \begin{bmatrix} 1 & 1 & -2 & 1 \\ 2 & -4 & 1 & 0 \\ 0 & 2 & -3 & -1 \end{bmatrix}$$

▶ **To change a matrix to a system of equations**

1. Place the insertion point in an $m \times n$ matrix.

2. Choose Rewrite + Matrix as Equations.

3. In the dialog that appears, enter the variables separated by commas.

4. Choose OK.

▶ Rewrite + Matrix as Equations

$$\begin{pmatrix} 1 & 1 & -1 \\ 2 & -3 & 1 \end{pmatrix} \text{(Variable List: } x, y\text{)},$$

Corresponding equations: $\{x + y = -1, 2x - 3y = 1\}$

$$\begin{bmatrix} 1 & 1 & -2 & 1 \\ 2 & -4 & 1 & 0 \\ 0 & 2 & -3 & -1 \end{bmatrix},$$

Corresponding equations: $\{x + y - 2z = 1, 2x - 4y + z = 0, 2y - 3z = -1\}$

The response is a list of equations. If you want these equations in a one-column matrix, use Matrices + Reshape, and specify 1 column.

▶ Matrices + Reshape

$$\{x + y = -1, 2x - 3y = 1\}, \left\{ \begin{array}{l} x + y = -1 \\ 2x - 3y = 1 \end{array} \right\}$$

Matrix Operators

A matrix operator is a function that operates on matrices. The Matrices menu contains a number of matrix operators.

Trace

The *trace* of an $n \times n$ matrix is the sum of the diagonal elements. This operation applies to square matrices only.

▶ **To compute the trace of a square matrix**

1. Place the insertion point in the matrix.

2. From the Matrices submenu, choose Trace.

▶ Trace

$$\left(\begin{array}{cc} a & b \\ c & d \end{array} \right), \text{trace: } a + d \qquad \qquad \left(\begin{array}{ccc} -85 & -55 & -37 \\ -35 & 97 & 50 \\ 79 & 56 & 49 \end{array} \right), \text{trace: } 61$$

Transpose and Hermitian Transpose

The *transpose* of an $m \times n$ matrix is the $n \times m$ matrix that you obtain from the first matrix by interchanging the rows and columns.

▶ **To compute the transpose of a matrix**

1. Place the insertion point in the matrix.

2. From the Matrices submenu, choose Transpose.

▶ Matrices + Transpose

$$\begin{pmatrix} a & b \\ c & d \end{pmatrix}, \text{transpose: } \begin{pmatrix} a & c \\ b & d \end{pmatrix}$$

You can also compute the transpose of a matrix or vector by using the superscript T.

▶ Evaluate

$$\begin{pmatrix} a & b & c \\ d & e & f \end{pmatrix}^T = \begin{pmatrix} a & d \\ b & e \\ c & f \end{pmatrix} \qquad \begin{pmatrix} a \\ b \end{pmatrix}^T \begin{pmatrix} c \\ d \end{pmatrix} = ac + bd$$

The last example demonstrates a common way to take the inner product of vectors.

The *Hermitian transpose* of a matrix is the transpose together with the replacement of each entry by its complex conjugate. It is also referred to as the *adjoint* or *Hermitian adjoint* or *conjugate transpose* of a matrix (not to be confused with the classical adjoint or adjugate, discussed elsewhere in this chapter.)

▶ **To compute the Hermitian transpose of a matrix**

1. Place the insertion point in the matrix.

2. From the **Matrices** submenu, choose **Hermitian Transpose**.

$$\begin{pmatrix} 2+i & -i \\ 4-i & 2+i \end{pmatrix}, \text{Hermitian transpose: } \begin{pmatrix} 2-i & 4+i \\ i & 2-i \end{pmatrix}$$

$$\begin{pmatrix} a+ib & c+id \\ e+if & g+ih \end{pmatrix}, \text{Hermitian transpose: } \begin{pmatrix} a-ib & e-if \\ c-id & g-ih \end{pmatrix}$$

You can also compute the Hermitian transpose of a matrix using the superscript H.

▶ Evaluate

$$\begin{pmatrix} i & 2+i \\ 4i & 3-2i \end{pmatrix}^H = \begin{pmatrix} -i & -4i \\ 2-i & 3+2i \end{pmatrix}$$

Determinant

The *determinant* of an $n \times n$ matrix (a_{ij}) is the sum and difference of certain products of the entries. Specifically,

$$\det(a_{ij}) = \sum_{\sigma} (-1)^{sgn(\sigma)} a_{1\sigma(1)} a_{2\sigma(2)} \cdots a_{n\sigma(n)}$$

where σ ranges over all the permutations of $\{1, 2, \ldots, n\}$ and $(-1)^{sgn(\sigma)} = \pm 1$, depending on whether σ is an even or odd permutation.

Note that this operation applies to square matrices only.

▶ **To compute the determinant of a square matrix**

1. Place the insertion point in the matrix.

2. From the Matrices submenu, choose Determinant.

▶ Matrices + Determinant

$$\begin{bmatrix} a & b \\ c & d \end{bmatrix}, \text{determinant: } ad - bc \qquad \begin{bmatrix} -85 & -55 & -37 \\ -35 & 97 & 50 \\ 79 & 56 & 49 \end{bmatrix}, \text{determinant: } -121529$$

$$\begin{bmatrix} a_{1,1} & a_{1,2} & a_{1,3} \\ a_{2,1} & a_{2,2} & a_{2,3} \\ a_{3,1} & a_{3,2} & a_{3,3} \end{bmatrix}, \text{determinant: } \begin{array}{l} a_{1,1}a_{2,2}a_{3,3} - a_{1,1}a_{2,3}a_{3,2} - a_{2,1}a_{1,2}a_{3,3} \\ +a_{2,1}a_{1,3}a_{3,2} + a_{3,1}a_{1,2}a_{2,3} - a_{3,1}a_{1,3}a_{2,2} \end{array}$$

You can compute the determinant with Evaluate by denoting it with vertical expanding brackets or by using the function det.

$$\begin{vmatrix} -35 & 50 \\ 79 & 49 \end{vmatrix} \quad \text{or} \quad \det \begin{bmatrix} -35 & 50 \\ 79 & 49 \end{bmatrix}$$

The vertical brackets are available from the Matrix Properties dialog as built-in delimiters or as expanding brackets list from the Brackets dialog under ▨ .

To obtain the function det, type the letters **det** in mathematics, and they will turn gray when the t is typed. This function can also be chosen from the Name list in the Math Name dialog under ▨ .

▶ Evaluate

$$\det \begin{bmatrix} a & b \\ c & d \end{bmatrix} = ad - bc \qquad\qquad \begin{vmatrix} a & b \\ c & d \end{vmatrix} = ad - bc$$

$$\det \begin{bmatrix} -8 & -3 & 5 \\ 7 & 4 & -1 \\ 14 & 0 & 6 \end{bmatrix} = -304 \qquad\qquad \begin{vmatrix} -85 & -55 & 82 \\ -35 & 97 & -17 \\ 42 & 33 & -65 \end{vmatrix} = 223\,857$$

Adjugate

The *adjugate* or *classical adjoint* of a matrix A is the transpose of the matrix of cofactors of A. The i, j *cofactor* A_{ij} of A is the scalar $(-1)^{i+j} \det A\,(i|j)$, where $A\,(i|j)$ denotes the matrix that you obtain from A by removing the ith row and jth column.

▶ **Matrices + Adjugate**

$$\begin{pmatrix} a & b \\ c & d \end{pmatrix}, \text{adjugate:} \begin{pmatrix} d & -b \\ -c & a \end{pmatrix}$$

▶ **Matrices + Adjugate**

$$\begin{pmatrix} a & b & c \\ d & e & f \\ g & h & j \end{pmatrix}, \text{adjugate:} \begin{pmatrix} ej - fh & -bj + ch & bf - ce \\ -dj + fg & aj - cg & -af + cd \\ dh - eg & -ah + bg & ae - bd \end{pmatrix}$$

$$\begin{bmatrix} 9 & 6 & 7 & -5 \\ 4 & -8 & -3 & 92 \\ -3 & -6 & 7 & 6 \\ 5 & -5 & 0 & -1 \end{bmatrix}, \text{adjugate:} \begin{bmatrix} 3384 & 469 & -3183 & 7130 \\ 3329 & 301 & -3200 & -8153 \\ 4068 & -261 & 6896 & -2976 \\ 275 & 840 & 85 & -1116 \end{bmatrix}$$

The product of a matrix with its adjugate is diagonal, with the entries on the diagonal equal to the determinant of the matrix.

$$\begin{pmatrix} a & b \\ c & d \end{pmatrix}\begin{pmatrix} d & -b \\ -c & a \end{pmatrix} = \begin{pmatrix} ad - bc & 0 \\ 0 & ad - bc \end{pmatrix}$$

This relationship yields a well-known formula for the inverse of an invertible matrix A:

$$A^{-1} = \frac{1}{\det A} \text{adjugate } A$$

▶ **Evaluate**

$$\begin{bmatrix} 9 & 6 & 7 & -5 \\ 4 & -8 & -3 & 92 \\ -3 & -6 & 7 & 6 \\ 5 & -5 & 0 & -1 \end{bmatrix}\begin{bmatrix} 3384 & 469 & -3183 & 7130 \\ 3329 & 301 & -3200 & -8153 \\ 4068 & -261 & 6896 & -2976 \\ 275 & 840 & 85 & -1116 \end{bmatrix}$$

$$= \begin{bmatrix} 77531 & 0 & 0 & 0 \\ 0 & 77531 & 0 & 0 \\ 0 & 0 & 77531 & 0 \\ 0 & 0 & 0 & 77531 \end{bmatrix}$$

$$\det \begin{bmatrix} 9 & 6 & 7 & -5 \\ 4 & -8 & -3 & 92 \\ -3 & -6 & 7 & 6 \\ 5 & -5 & 0 & -1 \end{bmatrix} = 77531$$

Permanent

The *permanent* of an $n \times n$ matrix (a_{ij}) is the sum of certain products of the entries. Specifically,

$$\text{permanent}(a_{ij}) = \sum_{\sigma} a_{1\sigma(1)} a_{2\sigma(2)} \cdots a_{n\sigma(n)}$$

where σ ranges over all the permutations of $\{1, 2, \ldots, n\}$. This operation applies to square matrices only.

▶ **To compute the permanent of a matrix**

1. Place the insertion point in the matrix.

2. From the Matrices submenu, choose Permanent.

▶ Matrices + Permanent

$$\begin{bmatrix} a & b \\ c & d \end{bmatrix}, \text{permanent: } ad + bc$$

$$\begin{bmatrix} a_{1,1} & a_{1,2} & a_{1,3} \\ a_{2,1} & a_{2,2} & a_{2,3} \\ a_{3,1} & a_{3,2} & a_{3,3} \end{bmatrix}, \text{permanent: } \begin{array}{l} a_{1,1}a_{2,2}a_{3,3} + a_{1,1}a_{2,3}a_{3,2} + a_{2,1}a_{1,2}a_{3,3} \\ + a_{2,1}a_{1,3}a_{3,2} + a_{3,1}a_{1,2}a_{2,3} + a_{3,1}a_{1,3}a_{2,2} \end{array}$$

Maximum and Minimum Matrix Entries

The functions max and min applied to a matrix with integer entries will return the entry with maximum or minimum value.

▶ Evaluate

$$\max \begin{bmatrix} -85 & -55 & -37 & -35 & 97 \\ 50 & 79 & 56 & 49 & 63 \\ 57 & -59 & 45 & -8 & -93 \end{bmatrix} = 97$$

$$\min \begin{bmatrix} 92 & 43 & -62 & 77 & 66 \\ 54 & -5 & 99 & -61 & -50 \\ -12 & -18 & 31 & -26 & -62 \\ 1 & -47 & -91 & -47 & -61 \end{bmatrix} = -91$$

Matrix Norms

Choosing **Norm** from the **Matrices** submenu gives the 2-norm of a vector or matrix. The 2-*norm*, or *Euclidean norm*, of a vector is the Euclidean length of the vector.

$$\left\| \begin{array}{c} a \\ b \end{array} \right\| = \sqrt{a^2 + b^2} \qquad \left\| \begin{array}{c} a \\ b \\ c \\ d \end{array} \right\| = \sqrt{a^2 + b^2 + c^2 + d^2}$$

The 2-*norm*, or *Euclidean norm*, of a matrix A is its largest singular value—the number defined by

$$\|A\| = \max_{x \neq 0} \frac{\|Ax\|}{\|x\|}$$

This can also be computed as $\max \left\{ \sqrt{|E_i|} \right\}$ where the E_i's range over the eigenvalues of the matrix AA^H.

▶ **Matrices + Norm**

$$\left(\begin{array}{cc} 2 & 3 \\ 5 & 7 \end{array} \right), \text{2-norm: } 9.326\,8 \qquad \left(\begin{array}{ccc} 2 & -1 & 0 \\ -1 & 2 & -1 \\ 0 & -1 & 2 \end{array} \right), \text{2-norm: } 3.414\,2$$

$$\left(\begin{array}{cc} 2+3i & 5 \\ 6 & -7+2i \end{array} \right), \text{2-norm: } 9.$$

The 2-norm of a matrix can also be obtained with double brackets.

▶ **To put norm symbols around a matrix**

1. Select the matrix by using the mouse or by pressing SHIFT + RIGHT ARROW.

2. Click ⬚ or choose **Insert + Brackets** and select the norm symbols, or type CTRL + SHIFT + VERTICAL LINE. Choose OK.

▶ **Evaluate**

$$\left\| \begin{array}{cc} 0.2 & 0.3 \\ 0.5 & 0.7 \end{array} \right\| = 0.932\,68 \qquad \left\| \begin{array}{cc} 5 & 7 \\ -13 & 6 \end{array} \right\| = 14.454$$

$$\left\| \begin{array}{cc} 2+3i & 5 \\ 6 & -7+2i \end{array} \right\| = 9.937\,8$$

The 1-*norm* of a matrix is the maximum among the sums of the absolute values of the terms in a column:

$$\|A\|_1 = \max_{1 \leq j \leq n} \left(\sum_{i=1}^{n} |a_{ij}| \right)$$

▶ Evaluate

$$\left\| \begin{matrix} a & b \\ c & d \end{matrix} \right\|_1 = \max\left(|a|+|c|\,,|b|+|d|\right) \qquad \left\| \begin{matrix} 0.2234 & 0.3158 \\ -0.5624 & 0.7111 \end{matrix} \right\|_1 = 1.0269$$

$$\left\| \begin{matrix} 5 & 7 \\ -13 & 6 \end{matrix} \right\|_1 = 18 \qquad \left\| \begin{matrix} 5+3i & 7 \\ -13 & 6-5i \end{matrix} \right\|_1 = \sqrt{34}+13$$

The ∞-*norm* of a matrix is the maximum among the sums of the absolute values of the terms in a row:

$$\|A\|_\infty = \max_{1 \le i \le n}\left(\sum_{j=1}^{n} |a_{ij}|\right)$$

▶ Evaluate

$$\left\| \begin{matrix} a & b \\ c & d \end{matrix} \right\|_\infty = \max\left(|a|+|b|\,,|c|+|d|\right) \qquad \left\| \begin{matrix} 0.2234 & 0.3158 \\ -0.5624 & 0.7111 \end{matrix} \right\|_\infty = 1.2735$$

$$\left\| \begin{matrix} 5 & 7 \\ -13 & 6 \end{matrix} \right\|_\infty = 19 \qquad \left\| \begin{matrix} 5+3i & 7 \\ -13 & 6-5i \end{matrix} \right\|_\infty = 13+\sqrt{61}$$

The *Hilbert-Schmidt norm* (or *Frobenius norm*) $\|A\|_F$ of a matrix A is the square root of the sums of the squares of the terms of the matrix A.

$$\|A\|_F = \left(\sum_{\substack{1 \le j \le n \\ 1 \le i \le n}} |a_{ij}|^2\right)^{\frac{1}{2}}$$

▶ Evaluate

$$\left\| \begin{matrix} 5+3i & 7 \\ -13 & 6-5i \end{matrix} \right\|_F = \sqrt{313} \qquad \left\| \begin{matrix} a & b \\ c & d \end{matrix} \right\|_F = \sqrt{\left(|a|^2+|b|^2+|c|^2+|d|^2\right)}$$

$$\left\| \begin{matrix} 5 & 7 \\ -13 & 6 \end{matrix} \right\|_F = 3\sqrt{31} \qquad \left\| \begin{matrix} 0.2234 & 0.3158 \\ -0.5624 & 0.7111 \end{matrix} \right\|_F = 0.98569$$

Spectral Radius

The spectral radius of a real symmetric matrix is the largest of the absolute values of the eigenvalues of the matrix.

▶ Matrices + Spectral Radius

$$\begin{bmatrix} 5 & -3 & 1 \\ -3 & 0 & 5 \\ 1.0 & 5 & 4 \end{bmatrix}, \text{ spectral radius: } 7.762\,7$$

$$\begin{bmatrix} 5 & -4 \\ -4 & 3.0 \end{bmatrix}, \text{ spectral radius: } 8.123\,1$$

▶ Matrices + Eigenvalues

$$\begin{bmatrix} 5 & -3 & 1 \\ -3 & 0 & 5 \\ 1.0 & 5 & 4 \end{bmatrix}, \text{ eigenvalues: } \{-4.380\,1, 5.617\,4, 7.762\,7\}$$

$$\begin{bmatrix} 5 & -4 \\ -4 & 3.0 \end{bmatrix}, \text{ eigenvalues: } \{-0.123\,11, 8.123\,1\}$$

Condition Number

The *condition number* of an invertible matrix A is the product of the 2-norm of A and the 2-norm of A^{-1}. This number measures the sensitivity of some solutions of linear equations $Ax = b$ to perturbations in the entries of A and b. The matrix with condition number 1 is "perfectly conditioned."

▶ Matrices + Condition Number

$$\begin{bmatrix} 0 & 1 \\ 1 & 0 \end{bmatrix}, \text{ condition number: } 1.0 \qquad \begin{bmatrix} 18 & 7 \\ 3 & -4 \end{bmatrix}, \text{ condition number: } 4.031\,5$$

$$\begin{bmatrix} 1 & \frac{1}{2} & \frac{1}{3} & \frac{1}{4} \\ \frac{1}{2} & \frac{1}{3} & \frac{1}{4} & \frac{1}{5} \\ \frac{1}{3} & \frac{1}{4} & \frac{1}{5} & \frac{1}{6} \\ \frac{1}{4} & \frac{1}{5} & \frac{1}{6} & \frac{1}{7} \end{bmatrix}, \text{ condition number: } 15514.0$$

$$\begin{bmatrix} 1 & 1 \\ 1 & 1.00001 \end{bmatrix}, \text{ condition number: } 4.0 \times 10^5$$

These final two matrices are extremely ill-conditioned. Small changes in some entries of A or b may result in large changes in the solution to linear equations of the form $Ax = b$ in these two cases.

Exponential Functions

A natural way to define e^M is to imitate the power series for e^x

$$e^x = 1 + x + \frac{1}{2}x^2 + \frac{1}{6}x^3 + \frac{1}{24}x^4 + \cdots$$

$$e^M = 1 + M + \frac{1}{2}M^2 + \frac{1}{6}M^3 + \frac{1}{24}M^4 + \cdots$$

and more generally,

$$e^{tM} = \sum_{k=0}^{\infty} \frac{(tM)^k}{k!}$$

To evaluate the expression e^M (or $\exp(M)$) for a matrix M, leave the insertion point in the expression e^M and choose **Evaluate**, as shown in the following examples. Define

$$A = \begin{bmatrix} 1 & 2 \\ 0 & 3 \end{bmatrix}, B = \begin{bmatrix} 1 & 2 \\ 0 & 1 \end{bmatrix}, C = \begin{bmatrix} 0 & 1 & 0 \\ 0 & 0 & 1 \\ 0 & 0 & 0 \end{bmatrix}, D = \begin{bmatrix} 1 & 3 & 0 \\ 0 & 1 & 0 \\ 0 & 0 & 1 \end{bmatrix}$$

▶ Evaluate

$$e^A = \begin{bmatrix} e & -e + e^3 \\ 0 & e^3 \end{bmatrix} \qquad e^{tA} = \begin{bmatrix} e^t & -e^t + e^{3t} \\ 0 & e^{3t} \end{bmatrix}$$

$$\exp(A) = \begin{pmatrix} e & -e + e^3 \\ 0 & e^3 \end{pmatrix} \qquad \exp(tA) = \begin{bmatrix} e^t & -e^t + e^{3t} \\ 0 & e^{3t} \end{bmatrix}$$

$$e^{A+B} = \begin{bmatrix} e^2 & -2e^2 + 2e^4 \\ 0 & e^4 \end{bmatrix} \qquad e^A e^B = \begin{bmatrix} e^{2\times1} & 2e^{2\times1} + e\left(-e + e^3\right) \\ 0 & ee^3 \end{bmatrix}$$

$$De^{tC}D^{-1} = \begin{bmatrix} 1 & t & 3t + \frac{1}{2}t^2 \\ 0 & 1 & t \\ 0 & 0 & 1 \end{bmatrix} \qquad e^{DtCD^{-1}} = \begin{bmatrix} 1 & t & t^2\left(\frac{3}{t} + \frac{1}{2}\right) \\ 0 & 1 & t\left(\frac{3}{t} + 1\right) - 3 \\ 0 & 0 & 1 \end{bmatrix}$$

Note that one of the properties of exponents that holds for real numbers fails for matrices. The equality $e^{A+B} = e^A e^B$ requires that $AB = BA$, and this property fails to hold for the matrices in the example. However, exponentiation preserves the property of similarity, as demonstrated by $De^{tC}D^{-1} = e^{DtCD^{-1}}$.

Polynomials and Vectors Associated With a Matrix

A square matrix has a characteristic and a minimum polynomial. The characteristic polynomial determines eigenvalues and eigenvectors of the matrix. Eigenvalues are an important feature of any dynamical system. One important application is to the solution of a system of ordinary differential equations.

Characteristic Polynomial and Minimum Polynomial

The *characteristic polynomial* of a square matrix A is the determinant of the characteristic matrix $xI - A$.

▶ Matrices + Characteristic Polynomial

$$\begin{pmatrix} 4 & 1 & 0 \\ 0 & 4 & 0 \\ 0 & 0 & 4 \end{pmatrix}, \text{characteristic polynomial: } X^3 - 12X^2 + 48X - 64$$

▶ Evaluate (, Factor)

$$X \begin{pmatrix} 1 & 0 & 0 \\ 0 & 1 & 0 \\ 0 & 0 & 1 \end{pmatrix} - \begin{pmatrix} 4 & 1 & 0 \\ 0 & 4 & 0 \\ 0 & 0 & 4 \end{pmatrix} = \begin{pmatrix} -4+X & -1 & 0 \\ 0 & -4+X & 0 \\ 0 & 0 & -4+X \end{pmatrix}$$

▶ Evaluate, Factor

$$\det \begin{pmatrix} -4+X & -1 & 0 \\ 0 & -4+X & 0 \\ 0 & 0 & -4+X \end{pmatrix} = 48X - 12X^2 + X^3 - 64 = (X-4)^3$$

The *minimum polynomial* of a square matrix A is the monic polynomial $p(x)$ of smallest degree such that $p(A) = 0$. By the Cayley–Hamilton theorem, $f(A) = 0$ if $f(x)$ is the characteristic polynomial of A. The minimum polynomial of A is a factor of the characteristic polynomial of A.

▶ Matrices + Minimum Polynomial, Factor

$$\begin{pmatrix} 4 & 1 & 0 \\ 0 & 4 & 0 \\ 0 & 0 & 4 \end{pmatrix}, \text{minimum polynomial: } X^2 - 8X + 16$$

Example   This example illustrates the Cayley-Hamilton theorem.

Define $p(X) = X^2 - 8X + 16X^0$ and $A = \begin{pmatrix} 4 & 1 & 0 \\ 0 & 4 & 0 \\ 0 & 0 & 4 \end{pmatrix}$.

Apply **Evaluate** to get the following.

$$p(A) = \begin{pmatrix} 0 & 0 & 0 \\ 0 & 0 & 0 \\ 0 & 0 & 0 \end{pmatrix}$$

The minimum and characteristic polynomial operations have to return a variable for the polynomial. In the preceding examples, they returned X. However, the variable used depends on the matrix entries and you do not need to avoid X in the matrix. You will be asked to supply a name for the polynomial variable.

▶ Matrices + Minimum Polynomial, Polynomial Variable λ

$\begin{pmatrix} 3X & x \\ 5 & y \end{pmatrix}$ (Polynomial Variableλ),

minimum polynomial: $\lambda^2 + (-3X - y)\lambda + (-5x + 3Xy)$

Eigenvalues and Eigenvectors

Given a matrix A, the matrix commands Eigenvectors and Eigenvalues on the Matrices submenu find scalars c and nonzero vectors v for which $Av = cv$. If there is a floating-point number in the matrix, the result is a numerical solution. Otherwise, the result is an exact symbolic solution or no solution. When a solution is not found, change at least one entry to floating point to obtain a numeric solution.

These scalars and vectors are sometimes called *characteristic* values and *characteristic* vectors. The eigenvalues, or characteristic values, are roots of the characteristic polynomial.

▶ Matrices + Eigenvalues

$\begin{pmatrix} \cos\alpha & -\sin\alpha \\ \sin\alpha & \cos\alpha \end{pmatrix}$, eigenvalues: $\{\cos\alpha - i\sin\alpha, \cos\alpha + i\sin\alpha\}$

This matrix has characteristic polynomial $X^2 - 2X\cos\alpha + 1$. Replacing X by the eigenvalue $\cos\alpha + i\sin\alpha$ and applying Simplify gives
$$(\cos\alpha + i\sin\alpha)^2 - 2(\cos\alpha + i\sin\alpha)\cos\alpha + 1 = 0$$
demonstrating that eigenvalues are roots of the characteristic polynomial. Note the different results obtained using integer versus floating-point entries.

▶ Matrices + Eigenvalues

$\begin{pmatrix} 1 & 2 \\ 3 & 4 \end{pmatrix}$, eigenvalues: $\{\frac{5}{2} - \frac{1}{2}\sqrt{33}, \frac{1}{2}\sqrt{33} + \frac{5}{2}\}$

$\begin{pmatrix} 1.0 & 2 \\ 3 & 4 \end{pmatrix}$, eigenvalues: $\{-0.372\,28, 5.\,372\,3\}$

When you choose Eigenvectors from the Matrices submenu, the system returns eigenvectors paired with the corresponding eigenvalues. The eigenvectors are grouped by eigenvalues, making the multiplicity for each eigenvalue apparent. Symbolic solutions will be returned in some cases. When a symbolic solution is not found, change at least one entry to floating point to obtain a numeric solution.

▶ **Matrices + Eigenvectors**

$$\begin{pmatrix} 49 & -69 & 99 \\ 23 & -81 & 20 \\ 48 & 1.0 & -87 \end{pmatrix}, \text{ eigenvectors: } \left\{ \begin{pmatrix} 0.937\,33 \\ 0.186\,22 \\ 0.294\,51 \end{pmatrix} \right\} \leftrightarrow 66.\,398,$$

$$\left\{ \begin{pmatrix} 0.159\,9 \\ 0.887\,94 \\ 0.431\,27 \end{pmatrix} \right\} \leftrightarrow -67.\,144, \left\{ \begin{pmatrix} 0.540\,43 \\ 0.113\,89 \\ -0.833\,64 \end{pmatrix} \right\} \leftrightarrow -118.\,25$$

$$\begin{pmatrix} 5 & -6 & -6 \\ -1 & 4 & 2 \\ 3 & -6 & -4 \end{pmatrix}, \text{ eigenvectors: } \left\{ \begin{pmatrix} 1 \\ -\frac{1}{3} \\ 1 \end{pmatrix} \right\} \leftrightarrow 1, \left\{ \begin{pmatrix} 2 \\ 0 \\ 1 \end{pmatrix}, \begin{pmatrix} 2 \\ 1 \\ 0 \end{pmatrix} \right\} \leftrightarrow 2$$

In the preceding example, 1 is an eigenvalue occurring with multiplicity 1, and 2 is an eigenvalue occurring with multiplicity 2. The defining property $Av = cv$ is illustrated in the following example:

▶ **Evaluate**

$$\begin{pmatrix} 5 & -6 & -6 \\ -1 & 4 & 2 \\ 3 & -6 & -4 \end{pmatrix} \begin{pmatrix} 2 \\ 1 \\ 0 \end{pmatrix} = \begin{pmatrix} 4 \\ 2 \\ 0 \end{pmatrix} = 2 \begin{pmatrix} 2 \\ 1 \\ 0 \end{pmatrix}$$

$$\begin{pmatrix} 5 & -6 & -6 \\ -1 & 4 & 2 \\ 3 & -6 & -4 \end{pmatrix} \begin{pmatrix} 2 \\ 0 \\ 1 \end{pmatrix} = \begin{pmatrix} 4 \\ 0 \\ 2 \end{pmatrix} = 2 \begin{pmatrix} 2 \\ 0 \\ 1 \end{pmatrix}$$

$$\begin{pmatrix} 5 & -6 & -6 \\ -1 & 4 & 2 \\ 3 & -6 & -4 \end{pmatrix} \begin{pmatrix} -3 \\ 1 \\ -3 \end{pmatrix} = \begin{pmatrix} -3 \\ 1 \\ -3 \end{pmatrix}$$

Positive Definite Matrices

A square matrix is called *Hermitian* if it is equal to its conjugate transpose. A Hermitian matrix with real entries is the same as a symmetric matrix. A Hermitian matrix A is *positive definite* if all the eigenvalues of A are positive. Otherwise, the computational engine MuPAD classifies A as *indefinite*.

An indefinite Hermitian matrix A is sometimes classified as *positive semidefinite* if all the eigenvalues of A are nonnegative; as *negative definite* if all the eigenvalues are negative; and as *negative semidefinite* if all the eigenvalues are nonpositive.

▶ Matrices + Definiteness Tests

$$\begin{bmatrix} 2 & -1 \\ -1 & 2 \end{bmatrix}$$ is positive definite $$\begin{bmatrix} 1 & -1 \\ -1 & 1 \end{bmatrix}$$ is indefinite

$$\begin{bmatrix} 2 & -i \\ i & 1 \end{bmatrix}$$ is positive definite $$\begin{bmatrix} -2 & i \\ -i & -2 \end{bmatrix}$$ is indefinite

▶ Matrices + Eigenvalues

$$\begin{bmatrix} 2 & -1 \\ -1 & 2 \end{bmatrix}$$, eigenvalues: $1, 3$ $$\begin{bmatrix} 1 & -1 \\ -1 & 1 \end{bmatrix}$$, eigenvalues: $0, 2$

$$\begin{bmatrix} 2 & -i \\ i & 1 \end{bmatrix}$$, eigenvalues: $\frac{3}{2} - \frac{1}{2}\sqrt{5}, \frac{1}{2}\sqrt{5} + \frac{3}{2}$ $$\begin{bmatrix} -2 & i \\ -i & -2 \end{bmatrix}$$, eigenvalues: $-3, -1$

Vector Spaces Associated With a Matrix

Four vector spaces are naturally associated with an $m \times n$ matrix A: the row space, the column space, and the left and right nullspaces. A *basis* for a vector space is a linearly independent set of vectors that spans the space. Commands on the Matrices submenu find bases for these vector spaces. These bases are not unique and different methods may compute different bases.

The Row Space

The *row space* is the vector space spanned by the row vectors of A. Any choice or row basis has the same number of vectors and spans the same vector space. However, there is no natural choice for the vectors that make up a row basis. You can find other bases for the row space by choosing Reduced Row Echelon Form from the Matrices submenu, or by applying Fraction-Free Gaussian Elimination and then taking the nonzero rows from the result.

▶ **To find a basis for the row space**

1. Leave the insertion point in the matrix.

2. From the Matrices submenu, choose Row Basis.

▶ **Matrices + Row Basis**

$$\begin{bmatrix} -85 & -55 & -37 & -35 \\ 97 & 50 & 79 & 56 \\ 49 & 63 & 57 & -59 \\ -36 & 8 & 20 & -94 \end{bmatrix}, \text{row basis:}$$

$$[[\begin{array}{cccc} -85 & -55 & -37 & -35 \end{array}], [\begin{array}{cccc} 97 & 50 & 79 & 56 \end{array}], [\begin{array}{cccc} 49 & 63 & 57 & -59 \end{array}]]$$

▶ **Matrices + Reduced Row Echelon Form**

$$\begin{bmatrix} -85 & -55 & -37 & -35 \\ 97 & 50 & 79 & 56 \\ 49 & 63 & 57 & -59 \\ -36 & 8 & 20 & -94 \end{bmatrix}, \text{row echelon form:} \quad \begin{bmatrix} 1 & 0 & 0 & \frac{133337}{68264} \\ 0 & 1 & 0 & -\frac{74049}{34132} \\ 0 & 0 & 1 & -\frac{3085}{9752} \\ 0 & 0 & 0 & 0 \end{bmatrix}$$

The nonzero rows in the preceding matrix give the following basis for the row space:

$$\begin{bmatrix} 1 & 0 & 0 & \frac{133\,337}{68\,264} \end{bmatrix}, \begin{bmatrix} 0 & 1 & 0 & -\frac{74\,049}{34\,132} \end{bmatrix}, \begin{bmatrix} 0 & 0 & 1 & -\frac{3085}{9752} \end{bmatrix}.$$

▶ **Matrices + Fraction-free Gaussian Elimination**

$$\begin{bmatrix} -85 & -55 & -37 & -35 \\ 97 & 50 & 79 & 56 \\ 49 & 63 & 57 & -59 \\ -36 & 8 & 20 & -94 \end{bmatrix}, \text{fraction-free Gaussian elimination:}$$

$$\begin{bmatrix} -85 & -55 & -37 & -35 \\ 0 & 1085 & -3126 & -1365 \\ 0 & 0 & 136\,528 & -43\,190 \\ 0 & 0 & 0 & 0 \end{bmatrix}$$

The nonzero rows in the preceding matrix give the following basis for the row space:

$$\begin{bmatrix} -85 & -55 & -37 & -35 \end{bmatrix}, \begin{bmatrix} 0 & 1085 & -3126 & -1365 \end{bmatrix},$$
$$\begin{bmatrix} 0 & 0 & 136528 & -43190 \end{bmatrix}.$$

▶ **Matrices + Gaussian Elimination**

$$\begin{bmatrix} -85 & -55 & -37 & -35 \\ 97 & 50 & 79 & 56 \\ 49 & 63 & 57 & -59 \\ -36 & 8 & 20 & -94 \end{bmatrix} \longmapsto \begin{bmatrix} -85 & -55 & -37 & -35 \\ 0 & -\frac{217}{17} & \frac{3126}{85} & \frac{273}{17} \\ 0 & 0 & \frac{19\,504}{155} & -\frac{1234}{31} \\ 0 & 0 & 0 & 0 \end{bmatrix}$$

The nonzero rows in the preceding matrix give the following basis for the row space:

$$\begin{bmatrix} -85 & -55 & -37 & -35 \end{bmatrix}, \begin{bmatrix} 0 & -\frac{217}{17} & \frac{3126}{85} & \frac{273}{17} \end{bmatrix}, \begin{bmatrix} 0 & 0 & \frac{19\,504}{155} & -\frac{1234}{31} \end{bmatrix}$$

The Column Space

The *column space* is the vector space spanned by the columns of A.

▶ **To find a basis for the column space**

1. Leave the insertion point in the matrix.

2. From the **Matrices** submenu, choose **Column Basis**.

▶ **Matrices + Column Basis**

$$\begin{bmatrix} -85 & -55 & -37 & -35 \\ 97 & 50 & 79 & 56 \\ 49 & 63 & 57 & -59 \\ -36 & 8 & 20 & -94 \end{bmatrix},$$

$$\text{column basis: } \left[\begin{bmatrix} -85 \\ 97 \\ 49 \\ -36 \end{bmatrix}, \begin{bmatrix} -55 \\ 50 \\ 63 \\ 8 \end{bmatrix}, \begin{bmatrix} -37 \\ 79 \\ 57 \\ 20 \end{bmatrix} \right]$$

You can also take the transpose of A and apply to the transpose the various other methods demonstrated in the previous section, because the column space of A is the same as the row space of A^T.

The Left and Right Nullspaces

The (right) *nullspace* is the vector space consisting of all $n \times 1$ vectors X satisfying $AX = 0$. You find a basis for the nullspace by choosing **Nullspace Basis** from the **Matrices** submenu.

▶ **Matrices + Nullspace Basis**

$$\begin{bmatrix} -85 & -55 & -37 & -35 \\ 97 & 50 & 79 & 56 \\ 49 & 63 & 57 & -59 \\ -36 & 8 & 20 & -94 \end{bmatrix}, \text{ nullspace basis: } \begin{bmatrix} -\frac{133337}{68264} \\ \frac{74049}{34132} \\ \frac{3085}{9752} \\ 1 \end{bmatrix}$$

The *left nullspace* is the vector space consisting of all $1 \times m$ vectors Y satisfying $YA = 0$. You find a basis for the left nullspace by first taking the transpose of A and then choosing **Nullspace Basis** from the **Matrices** submenu.

▶ Evaluate

$$\begin{bmatrix} -85 & -55 & -37 & -35 \\ 97 & 50 & 79 & 56 \\ 49 & 63 & 57 & -59 \\ -36 & 8 & 20 & -94 \end{bmatrix}^T = \begin{bmatrix} -85 & 97 & 49 & -36 \\ -55 & 50 & 63 & 8 \\ -37 & 79 & 57 & 20 \\ -35 & 56 & -59 & -94 \end{bmatrix}$$

▶ Matrices + Nullspace Basis

$$\begin{bmatrix} -85 & 97 & 49 & -36 \\ -55 & 50 & 63 & 8 \\ -37 & 79 & 57 & 20 \\ -35 & 56 & -59 & -94 \end{bmatrix}, \text{nullspace basis:} \begin{bmatrix} -1 \\ 0 \\ -1 \\ 1 \end{bmatrix}$$

To check that this vector is in the left nullspace, take the transpose of the vector and check the product.

▶ Evaluate

$$\begin{bmatrix} -1 \\ 0 \\ -1 \\ 1 \end{bmatrix}^T \begin{bmatrix} -85 & -55 & -37 & -35 \\ 97 & 50 & 79 & 56 \\ 49 & 63 & 57 & -59 \\ -36 & 8 & 20 & -94 \end{bmatrix} = \begin{bmatrix} 0 & 0 & 0 & 0 \end{bmatrix}$$

Orthogonal Matrices

An *orthogonal matrix* is a real matrix for which the inner product of any two different columns is zero and the inner product of any column with itself is one. The matrix is said to have *orthonormal* columns. Such a matrix necessarily has orthonormal rows as well.

▶ Matrices + Orthogonality Test

$$\begin{pmatrix} 0 & 0 & 1 \\ 1 & 0 & 0 \\ 0 & 1 & 0 \end{pmatrix}, \text{orthogonal?} \; true$$

$$\begin{pmatrix} 0 & 1 \\ 1 & 1 \end{pmatrix}, \text{orthogonal?} \; false$$

The QR Factorization and Orthonormal Bases

Any real matrix A with at least as many rows as columns can be factored as a product QR, where Q is an *orthogonal* matrix—that is, the columns of Q are *orthonormal* (the inner product of any two different columns is 0, and the inner product of any column with itself is 1) and R is upper-right triangular with the same rank as A. If the original matrix A is square, then so is R. If A is a square matrix with linearly independent columns, R is invertible.

▶ **To obtain the QR factorization**

1. Leave the insertion point in a matrix.

2. From the **Matrices** submenu, choose **QR Decomposition**.

▶ **Matrices + QR Decomposition**

$$\begin{pmatrix} 3 & 0 \\ 4 & 5 \end{pmatrix} = \begin{pmatrix} \frac{3}{5} & -\frac{4}{5} \\ \frac{4}{5} & \frac{3}{5} \end{pmatrix} \begin{pmatrix} 5 & 4 \\ 0 & 3 \end{pmatrix}$$

$$\begin{pmatrix} -4 & 2 \\ 1 & -1 \\ 0 & 2 \end{pmatrix} = \begin{pmatrix} -\frac{4}{17}\sqrt{17} & -\frac{1}{612}\sqrt{17}\sqrt{72} & \frac{1}{18}\sqrt{18} \\ \frac{1}{17}\sqrt{17} & -\frac{1}{153}\sqrt{17}\sqrt{72} & \frac{2}{9}\sqrt{18} \\ 0 & \frac{1}{36}\sqrt{17}\sqrt{72} & \frac{1}{18}\sqrt{18} \end{pmatrix} \begin{pmatrix} \sqrt{17} & -\frac{9}{17}\sqrt{17} \\ 0 & \frac{1}{17}\sqrt{17}\sqrt{72} \\ 0 & 0 \end{pmatrix}$$

When A is a square matrix with linearly independent columns, the two matrices Q and $A = QR$ have the same column spaces.

Example   The preceding product comes from the following linear combinations.

$$\begin{pmatrix} 3 \\ 4 \end{pmatrix} = 5 \begin{pmatrix} \frac{3}{5} \\ \frac{4}{5} \end{pmatrix} + 0 \begin{pmatrix} -\frac{4}{5} \\ \frac{3}{5} \end{pmatrix}$$

and

$$\begin{pmatrix} 0 \\ 5 \end{pmatrix} = 4 \begin{pmatrix} \frac{3}{5} \\ \frac{4}{5} \end{pmatrix} + 3 \begin{pmatrix} -\frac{4}{5} \\ \frac{3}{5} \end{pmatrix}$$

Observe that the columns of A are linear combinations of the columns of Q. Then, since both column spaces have dimension 2 and one contains the other, it follows that they must be the same space.

This conversion of the columns of A into the orthonormal columns of Q is referred to as the *Gram–Schmidt orthogonalization* process. In general, since R is upper-right triangular, the subspace spanned by the first k columns of the matrix $A = QR$ is the same as the subspace spanned by the first k columns of the matrix Q.

Rank and Dimension

The *rank* of a matrix is the dimension of the column space. It is the same as the dimension of the row space or the number of nonzero singular values.

▶ Matrices + Rank

$$
\begin{bmatrix} -8 & -5 & 7 & -2 \\ 7 & 5 & 9 & 5 \\ 1 & 0 & -16 & -3 \\ 8 & 5 & -7 & 2 \end{bmatrix}, \text{ rank: } 2
$$

▶ Matrices + Row Basis

$$
\begin{bmatrix} -8 & -5 & 7 & -2 \\ 7 & 5 & 9 & 5 \\ 1 & 0 & -16 & -3 \\ 8 & 5 & -7 & 2 \end{bmatrix}, \text{ row basis: } \begin{bmatrix} \begin{bmatrix} -8 & -5 & 7 & -2 \end{bmatrix}, \begin{bmatrix} 7 & 5 & 9 & 5 \end{bmatrix} \end{bmatrix}
$$

▶ Matrices + Column Basis

$$
\begin{bmatrix} -8 & -5 & 7 & -2 \\ 7 & 5 & 9 & 5 \\ 1 & 0 & -16 & -3 \\ 8 & 5 & -7 & 2 \end{bmatrix}, \text{ column basis: } \begin{bmatrix} \begin{bmatrix} -8 \\ 7 \\ 1 \\ 8 \end{bmatrix}, \begin{bmatrix} -5 \\ 5 \\ 0 \\ 5 \end{bmatrix} \end{bmatrix}
$$

Normal Forms of Matrices

Any equivalence relation on a set of matrices partitions the set of matrices into a collection of equivalence classes. A normal form, or canonical form, for a matrix is a choice of another matrix that displays certain invariants for that equivalence class, usually together with an algorithm for constructing the form from the given matrix.

Two such equivalence relations are similarity and equivalence. Two $n \times n$ matrices A and B are *similar* if there is an invertible $n \times n$ matrix C such that $B = C^{-1}AC$. Two $m \times n$ matrices A and B are *equivalent* if one can be obtained from the other by a sequence of elementary row and column operations. In other words, $B = QAP$ for some invertible matrices Q and P.

When the context is matrices over the integers, "invertible" should be interpreted as "unimodular;" that is, both the matrix and its inverse have integer entries—in particular, a unimodular matrix has determinant 1. When the context is matrices over the ring $F[x]$ for a field F, "invertible" means both the matrix and its inverse have entries in $F[x]$.

Smith Normal Form

Every matrix A over a principal ideal domain (PID) is equivalent to a diagonal matrix of the form
$$diag(1, \ldots, 1, p_1, p_2, \ldots, p_k, 0, \ldots, 0)$$
where for each i, p_i is a factor of p_{i+1}. This matrix, which is uniquely determined by A, is called the *Smith normal form* of A. The diagonal entries of the Smith normal form of a matrix A are the *invariant factors* of A. The *Smith normal form* of A can be obtained as a matrix $S = QAP$ where Q and P are invertible over the PID.

Integer Matrices

The *Smith normal form* of an integer matrix A is a matrix $S = QAP$ where Q and P are unimodular—nonsingular matrices with integer entries whose inverses also have integer entries. In particular, Q and P have determinant 1. You can find the Smith normal form of a square integer matrix.

▶ Matrices + Smith Normal Form

$$\begin{pmatrix} 2 & 9 & 5 \\ 3 & 4 & 3 \\ 4 & 1 & -1 \end{pmatrix}, \text{ Smith normal form: } \begin{pmatrix} 1 & 0 & 0 \\ 0 & 1 & 0 \\ 0 & 0 & 56 \end{pmatrix}$$

The following product illustrates the equivalence relation. The two new matrices that occur are unimodular.
$$\begin{pmatrix} 2 & 9 & 5 \\ 3 & 4 & 3 \\ 4 & 1 & -1 \end{pmatrix} = \begin{pmatrix} 2 & 9 & -5 \\ 3 & 4 & -3 \\ 4 & 1 & -2 \end{pmatrix} \begin{pmatrix} 1 & 0 & 0 & 0 \\ 0 & 1 & 0 & 0 \\ 0 & 0 & 56 & 0 \end{pmatrix} \begin{pmatrix} 1 & 0 & 21 \\ 0 & 1 & 27 \\ 0 & 0 & 1 \end{pmatrix}$$

Matrices over $F[x]$

It is a remarkable fact that two $n \times n$ matrices with entries in a field F are similar if and only if their characteristic matrices $xI - A$ and $xI - B$ are equivalent. These characteristic matrices are matrices over the principal ideal domain $F[x]$, and two square matrices with polynomial entries are equivalent if and only if they have the same *Smith normal form*. The entries can be any polynomials, except that floating point coefficients cannot be used.

▶ Matrices + Smith Normal Form

$$\begin{pmatrix} x^2 & -2i\left(x^3 + x^2\right) + 2x^2 \\ 0 & \sqrt{2}i\left(x^3 + x^2\right) \end{pmatrix}, \text{ Smith normal form: } \begin{pmatrix} x^2 & 0 \\ 0 & x^2 + x^3 \end{pmatrix}$$

The Smith normal form can be used to test whether two matrices are similar. The field in question can be the rationals or any finite field extension of the rationals. We illustrate this with an example.

Example   Take two similar matrices: $A = \begin{bmatrix} 1 & 2 \\ 3 & 4 \end{bmatrix}$ and

$$B = \begin{bmatrix} 1 & 9 \\ -3 & 4 \end{bmatrix}^{-1} \begin{bmatrix} 1 & 2 \\ 3 & 4 \end{bmatrix} \begin{bmatrix} 1 & 9 \\ -3 & 4 \end{bmatrix} = \begin{bmatrix} \frac{61}{31} & -\frac{319}{31} \\ -\frac{24}{31} & \frac{94}{31} \end{bmatrix}$$

These matrices have the following characteristic matrices:

$$xI - A = \begin{bmatrix} x & 0 \\ 0 & x \end{bmatrix} - \begin{bmatrix} 1 & 2 \\ 3 & 4 \end{bmatrix} = \begin{bmatrix} x-1 & -2 \\ -3 & x-4 \end{bmatrix}$$

$$xI - B = \begin{bmatrix} x & 0 \\ 0 & x \end{bmatrix} - \begin{bmatrix} \frac{61}{31} & -\frac{319}{31} \\ -\frac{24}{31} & \frac{94}{31} \end{bmatrix} = \begin{bmatrix} x-\frac{61}{31} & \frac{319}{31} \\ \frac{24}{31} & x-\frac{94}{31} \end{bmatrix}$$

with Smith normal forms both equal to

$$\begin{bmatrix} 1 & 0 \\ 0 & x^2 - 5x - 2 \end{bmatrix}$$

See page 344 for another example relating Smith normal forms and characteristic polynomials.

Hermite Normal Form

Given a matrix A with entries in a PID, the *Hermite normal form* of A is a row echelon matrix $H = QA$ where Q is invertible in the ring of matrices over the PID. The first nonzero entry in each row is from a prespecified set of nonassociates, and the entries above that first nonzero entry are from a prespecified set of representatives of the ring modulo that entry. If the PID is the ring of integers, the first nonzero entry in each row is a positive integer n_{ij}, and the entries above that first nonzero entry are often chosen from the set $\{0, 1, 2, ..., n_{ij} - 1\}$.

▶ Matrices + Hermite Normal Form

$\begin{pmatrix} 7 & 34 & 46 \\ 4 & 20 & 27 \end{pmatrix}$, Hermite normal form: $\begin{pmatrix} 1 & 2 & 3 \\ 0 & 4 & 5 \end{pmatrix}$

$\begin{bmatrix} 2 & 5 \\ -4 & 5 \end{bmatrix}$, Hermite normal form: $\begin{bmatrix} 2 & 5 \\ 0 & 15 \end{bmatrix}$

Companion Matrix and Rational Canonical Form

The *companion matrix* of a monic polynomial $a_0 + a_1 X + \cdots + a_{n-1}X^{n-1} + X^n$ of degree n is the $n \times n$ matrix with a subdiagonal of ones, final column

$$\begin{bmatrix} -a_0 & -a_1 & \cdots & -a_{n-1} \end{bmatrix}^T$$

and other entries zero.

▶ Polynomials + Companion Matrix

$$x^4 + 3x^2 - 2x + 1, \text{Companion matrix:} \begin{bmatrix} 0 & 0 & 0 & -1 \\ 1 & 0 & 0 & 2 \\ 0 & 1 & 0 & -3 \\ 0 & 0 & 1 & 0 \end{bmatrix}$$

$$x^3 + ax^2 + bx + c, \text{Companion matrix:} \begin{bmatrix} 0 & 0 & -c \\ 1 & 0 & -b \\ 0 & 1 & -a \end{bmatrix}$$

Note that the first of the following matrices is the companion matrix of its own characteristic and minimum polynomials.

▶ Matrices + Minimum Polynomial

$$\begin{pmatrix} 0 & 0 & 0 & 0 & -a \\ 1 & 0 & 0 & 0 & -b \\ 0 & 1 & 0 & 0 & -c \\ 0 & 0 & 1 & 0 & -d \\ 0 & 0 & 0 & 1 & -e \end{pmatrix}, \text{minimum polynomial: } X^5 + sX^4 + dX^3 + cX^2 + bX + a$$

A *rational canonical form*, sometimes called a *Frobenius form*, is a block diagonal matrix with each block the companion matrix of its own minimum and characteristic polynomials. Each of the minimum polynomials of these blocks is a factor of the characteristic polynomial of the original matrix. The polynomials that determine the blocks of the rational canonical form sequentially divide one another.

Choosing **Rational Canonical Form** from the **Matrices** submenu produces a factorization of a square matrix as PBP^{-1}, where B is in rational canonical form. The matrix B will have entries from the smallest subring of the complex numbers containing the entries of the original matrix. The invertible matrices will have entries from the smallest subfield of the complex numbers containing the entries of the original matrix. For example, if the matrix has integer entries, the rational canonical form will also, and the invertible matrices will have rational entries.

▶ Matrices + Rational Canonical Form

$$\begin{bmatrix} 1 & 2 & 3 \\ 4 & 5 & 6 \\ 7 & 8 & 9 \end{bmatrix} = \begin{bmatrix} 1 & 1 & 30 \\ 0 & 4 & 66 \\ 0 & 7 & 102 \end{bmatrix} \begin{bmatrix} 0 & 0 & 0 \\ 1 & 0 & 18 \\ 0 & 1 & 15 \end{bmatrix} \begin{bmatrix} 1 & -2 & 1 \\ 0 & -\dfrac{17}{9} & \dfrac{11}{9} \\ 0 & \dfrac{7}{54} & -\dfrac{2}{27} \end{bmatrix}$$

Notice that the rational canonical form in the preceding example is the companion matrix of its minimum polynomial $X^3 - 15X^2 - 18X$. Now look at the companion matrix of this same matrix.

▶ Evaluate

$$x \begin{bmatrix} 1 & 0 & 0 \\ 0 & 1 & 0 \\ 0 & 0 & 1 \end{bmatrix} - \begin{bmatrix} 1 & 2 & 3 \\ 4 & 5 & 6 \\ 7 & 8 & 9 \end{bmatrix} = \begin{bmatrix} x-1 & -2 & -3 \\ -4 & x-5 & -6 \\ -7 & -8 & x-9 \end{bmatrix}$$

▶ Matrices + Smith Normal Form

$$\begin{bmatrix} x-1 & -2 & -3 \\ -4 & x-5 & -6 \\ -7 & -8 & x-9 \end{bmatrix}, \text{ Smith normal form: } \begin{bmatrix} 1 & 0 & 0 \\ 0 & 1 & 0 \\ 0 & 0 & -18x - 15x^2 + x^3 \end{bmatrix}$$

Notice that the polynomial occurring in the preceding Smith normal form is the same polynomial as occurred earlier.

▶ Matrices + Rational Canonical Form

$$\begin{bmatrix} 5 & -6 & -6 \\ -1 & 4 & 2 \\ 3 & -6 & -4 \end{bmatrix} = \begin{bmatrix} 1 & 5 & 2 \\ 0 & -1 & 0 \\ 0 & 3 & 1 \end{bmatrix} \begin{bmatrix} 0 & -2 & 0 \\ 1 & 3 & 0 \\ 0 & 0 & 2 \end{bmatrix} \begin{bmatrix} 1 & -1 & -2 \\ 0 & -1 & 0 \\ 0 & 3 & 1 \end{bmatrix}$$

There are two blocks in the preceding rational canonical form:

1. The companion matrix $\begin{bmatrix} 0 & -2 \\ 1 & 3 \end{bmatrix}$ of $X^2 - 3X + 2 = (X-1)(X-2)$

2. The companion matrix $[2]$ of $X - 2$

▶ Matrices + Characteristic Polynomial, Factor

$$\begin{bmatrix} 5 & -6 & -6 \\ -1 & 4 & 2 \\ 3 & -6 & -4 \end{bmatrix}, \text{ characteristic polynomial:}$$

$$X^3 - 5X^2 + 8X - 4 = (X-1)(X-2)^2$$

▶ Matrices + Minimum Polynomial, Factor

$$\begin{bmatrix} 5 & -6 & -6 \\ -1 & 4 & 2 \\ 3 & -6 & -4 \end{bmatrix}, \text{ minimum polynomial: } X^2 - 3X + 2 = (X-1)(X-2)$$

The characteristic matrix $xI - A$ of the preceding matrix A is

$$x \begin{bmatrix} 1 & 0 & 0 \\ 0 & 1 & 0 \\ 0 & 0 & 1 \end{bmatrix} - \begin{bmatrix} 5 & -6 & -6 \\ -1 & 4 & 2 \\ 3 & -6 & -4 \end{bmatrix} = \begin{bmatrix} x-5 & 6 & 6 \\ 1 & x-4 & -2 \\ -3 & 6 & x+4 \end{bmatrix}$$

▶ Matrices + Smith Normal Form

$$\begin{bmatrix} x-5 & 6 & 6 \\ 1 & x-4 & -2 \\ -3 & 6 & x+4 \end{bmatrix}, \text{ Smith normal form: } \begin{bmatrix} 1 & 0 & 0 \\ 0 & x-2 & 0 \\ 0 & 0 & x^2-3x+2 \end{bmatrix}$$

▶ Factor

$$\begin{bmatrix} 1 & 0 & 0 \\ 0 & x-2 & 0 \\ 0 & 0 & x^2-3x+2 \end{bmatrix} = \begin{bmatrix} 1 & 0 & 0 \\ 0 & (x-2) & 0 \\ 0 & 0 & (x-1)(x-2) \end{bmatrix}$$

These two examples illustrate a relationship among the Smith normal form, the characteristic matrix, and the rational canonical form of a matrix.

Note   The Smith normal form of the characteristic matrix of A displays the factors of the characteristic polynomial of A that determine the rational canonical form of A.

Jordan Form

Choosing Jordan Form from the Matrices submenu produces a factorization of a square matrix as PJP^{-1}, where J is in Jordan form. This form is a block diagonal matrix with each block an elementary Jordan matrix. More specifically, the *Jordan form* of an $n \times n$ matrix A with k linearly independent eigenvectors is a matrix of the form

$$J(A) = \begin{bmatrix} J_{n_1}(\lambda_1) & 0 & \cdots & 0 \\ 0 & J_{n_2}(\lambda_2) & \cdots & 0 \\ \vdots & \vdots & \ddots & \vdots \\ 0 & 0 & \cdots & J_{n_k}(\lambda_k) \end{bmatrix}$$

where $n_1+n_2+\cdots+n_k = n$, and each diagonal block $J_{n_i}(\lambda_i)$ is an $n_i \times n_i$ *elementary Jordan matrix* of the form

$$J_{n_i}(\lambda_i) = \begin{bmatrix} \lambda_i & 1 & \cdots & 0 & 0 \\ 0 & \lambda_i & \cdots & 0 & 0 \\ \vdots & \vdots & \ddots & \vdots & \vdots \\ 0 & 0 & \cdots & \lambda_i & 1 \\ 0 & 0 & \cdots & 0 & \lambda_i \end{bmatrix}$$

The matrix $J(A)$ is similar to A and its form is as nearly diagonal as possible among all matrices of the form $P^{-1}AP$.

▶ Matrices + Jordan Form

$$\begin{bmatrix} 2 & -1 & 0 \\ -1 & 2 & -1 \\ 0 & -1 & 2 \end{bmatrix}$$

$$= \begin{bmatrix} 1 & 1 & 1 \\ 0 & -\sqrt{2} & \sqrt{2} \\ -1 & 1 & 1 \end{bmatrix} \begin{bmatrix} 2 & 0 & 0 \\ 0 & \sqrt{2}+2 & 0 \\ 0 & 0 & 2-\sqrt{2} \end{bmatrix} \begin{bmatrix} \frac{1}{2} & 0 & -\frac{1}{2} \\ \frac{1}{4} & -\frac{1}{4}\sqrt{2} & \frac{1}{4} \\ \frac{1}{4} & \frac{1}{4}\sqrt{2} & \frac{1}{4} \end{bmatrix}$$

Thus, the Jordan form of

$$A = \begin{pmatrix} 2 & -1 & 0 \\ -1 & 2 & -1 \\ 0 & -1 & 2 \end{pmatrix}$$

is

$$J\left(\begin{bmatrix} 2 & -1 & 0 \\ -1 & 2 & -1 \\ 0 & -1 & 2 \end{bmatrix} \right) = \begin{bmatrix} 2 & 0 & 0 \\ 0 & \sqrt{2}+2 & 0 \\ 0 & 0 & -\sqrt{2}+2 \end{bmatrix}$$

In this case, $J(A)$ is diagonal, so each $J_{n_i}(\lambda_i)$ is a 1×1 matrix. The matrix A has the characteristic and minimum polynomial

$$-4 + 10X - 6X^2 + X^3 = (X-2)\left(X - 2 - \sqrt{2}\right)\left(X - 2 + \sqrt{2}\right)$$

whose roots $\{2,\ 2+\sqrt{2},\ 2-\sqrt{2}\}$ are the diagonal entries of the Jordan form.

▶ Matrices + Jordan Form

$$\begin{bmatrix} 2 & 0 & 0 & 0 \\ 1 & 2 & 0 & 0 \\ 0 & 0 & 2 & 0 \\ 0 & 0 & -3 & 2 \end{bmatrix} = \begin{bmatrix} 1 & 0 & 0 & 0 \\ 0 & 1 & 0 & 0 \\ 0 & 0 & -\frac{1}{3} & 0 \\ 0 & 0 & 0 & 1 \end{bmatrix} \begin{bmatrix} 2 & 0 & 0 & 0 \\ 1 & 2 & 0 & 0 \\ 0 & 0 & 2 & 0 \\ 0 & 0 & 1 & 2 \end{bmatrix} \begin{bmatrix} 1 & 0 & 0 & 0 \\ 0 & 1 & 0 & 0 \\ 0 & 0 & -3 & 0 \\ 0 & 0 & 0 & 1 \end{bmatrix}$$

Thus, the Jordan form is

$$J\left(\begin{bmatrix} 2 & 0 & 0 & 0 \\ 1 & 2 & 0 & 0 \\ 0 & 0 & 2 & 0 \\ 0 & 0 & -3 & 2 \end{bmatrix} \right) = \begin{bmatrix} 2 & 0 & 0 & 0 \\ 1 & 2 & 0 & 0 \\ 0 & 0 & 2 & 0 \\ 0 & 0 & 1 & 2 \end{bmatrix}$$

In this case, $J_{n_1}(\lambda_1) = J_{n_2}(\lambda_2) = \begin{bmatrix} 2 & 0 \\ 1 & 2 \end{bmatrix}$, the companion matrix of the minimum polynomial of

$$A = \begin{bmatrix} 2 & 0 & 0 & 0 \\ 1 & 2 & 0 & 0 \\ 0 & 0 & 2 & 0 \\ 0 & 0 & -3 & 2 \end{bmatrix}$$

The matrix A has characteristic polynomial $(X-2)^4$ with roots $\{2, 2, 2, 2\}$, and minimum polynomial $X^2 - 4X + 4 = (X-2)^2$.

▶ Matrices + Jordan Form

$$\begin{bmatrix} 2 & 0 & 0 & 0 \\ 0 & 2 & 0 & 0 \\ 0 & 0 & 2 & 0 \\ 0 & 0 & 0 & 2 \end{bmatrix} = \begin{bmatrix} 1 & 0 & 0 & 0 \\ 0 & 1 & 0 & 0 \\ 0 & 0 & 1 & 0 \\ 0 & 0 & 0 & 1 \end{bmatrix} \begin{bmatrix} 2 & 0 & 0 & 0 \\ 0 & 2 & 0 & 0 \\ 0 & 0 & 2 & 0 \\ 0 & 0 & 0 & 2 \end{bmatrix} \begin{bmatrix} 1 & 0 & 0 & 0 \\ 0 & 1 & 0 & 0 \\ 0 & 0 & 1 & 0 \\ 0 & 0 & 0 & 1 \end{bmatrix}$$

The preceding matrix is already in Jordan form. It has minimum polynomial $X - 2$ and characteristic polynomial $(X - 2)^4$, the same characteristic polynomial as the previous one, but a different minimum polynomial and a different Jordan form.

▶ Matrices + Jordan Form

$$\left(\begin{bmatrix} 1 & 2 \\ -1 & -1 \end{bmatrix} \right) = \begin{bmatrix} 1 & 1 \\ -\frac{1}{2} - \frac{1}{2}i & -\frac{1}{2} + \frac{1}{2}i \end{bmatrix} \begin{bmatrix} -i & 0 \\ 0 & i \end{bmatrix} \begin{bmatrix} \frac{1}{2} + \frac{1}{2}i & i \\ \frac{1}{2} - \frac{1}{2}i & -i \end{bmatrix}$$

In this case, $J_{n_1}(\lambda_1) = [i]$ and $J_{n_2}(\lambda_2) = [-i]$ are 1×1 matrices. The matrix $\begin{bmatrix} 1 & 2 \\ -1 & -1 \end{bmatrix}$ has the characteristic and minimum polynomial $x^2 + 1 = (x + i)(x - i)$.

Matrix Decompositions

There are various ways to decompose a matrix into the product of simpler matrices of special types. These decompositions are frequently useful in numerical matrix calculations.

Singular Value Decomposition (SVD)

Any $m \times n$ real matrix A can be factored into a product $A = UDV$, with U and V real orthogonal $m \times m$ and $n \times n$ matrices, respectively, and D a diagonal matrix with positive numbers in the first rank-A entries on the main diagonal, and zeroes everywhere else. The entries on the main diagonal of D are called the *singular values* of A. This factorization $A = UDV$ is called a *singular value decomposition* of A.

▶ Matrices + Singular Values

$$\begin{bmatrix} 5 & -5 & -3 \\ -3 & 0 & 5 \\ 1.0 & 5 & 4 \end{bmatrix}, \text{ singular values: } [10.053, 4.6119, 3.5588]$$

$$\begin{bmatrix} 5 & -5 & -3 \\ -3 & 0 & 5 \end{bmatrix}, \text{ singular values: } [8.8882, 3.7417]$$

▶ Matrices + SVD

$$\begin{bmatrix} 5 & -5 & -3 \\ -3 & 0 & 5 \\ 1 & 5 & 4 \end{bmatrix} = \begin{bmatrix} 0.721\,52 & 0.191\,19 & 0.665\,47 \\ -0.455\,04 & -0.593\,48 & 0.663\,87 \\ -0.521\,87 & 0.781\,81 & 0.341\,21 \end{bmatrix} \cdot$$

$$\begin{bmatrix} 10.\,053 & 0 & 0 \\ 0 & 4.\,611\,9 & 0 \\ 0 & 0 & 3.\,558\,8 \end{bmatrix} \begin{bmatrix} 0.442\,73 & -0.618\,41 & -0.649\,27 \\ 0.762\,85 & 0.640\,32 & -8.\,970\,6 \times 10^{-2} \\ 0.471\,22 & -0.455\,58 & 0.755\,25 \end{bmatrix}$$

These two outer matrices fail the orthogonality test because they are numerical approximations only. You can check the inner products of the columns to see that they are approximately orthogonal.

▶ Matrices + Singular Values, Matrices + SVD

$$\begin{pmatrix} 1 & 2.0 \\ 3 & 4 \end{pmatrix}, \text{ singular values: } [5.\,465\,0, 0.365\,97]$$

$$\begin{pmatrix} 1 & 2 \\ 3 & 4 \end{pmatrix} = \begin{pmatrix} 0.404\,55 & -0.914\,51 \\ 0.914\,51 & 0.404\,55 \end{pmatrix} \begin{pmatrix} 5.\,465\,0 & 0 \\ 0 & 0.365\,97 \end{pmatrix} \begin{pmatrix} 0.576\,05 & 0.817\,42 \\ 0.817\,42 & -0.576\,05 \end{pmatrix}$$

PLU Decomposition

Any $m \times n$ real or complex matrix A can be factored into a product $A = PLU$, with L and U lower and upper triangular $m \times m$ and $m \times n$ matrices, respectively, with 1's on the main diagonal of L, and with P a permutation matrix. This factorization $A = PLU$ is called the *PLU decomposition* of A. The matrices P and L are invertible and the matrix U is a row echelon form of A.

▶ Matrices + PLU Decomposition

$$\begin{bmatrix} 1 & 2 & 3 \\ 2 & 4 & 6 \\ 3 & 2 & 1 \end{bmatrix} = \begin{bmatrix} 1 & 0 & 0 \\ 0 & 0 & 1 \\ 0 & 1 & 0 \end{bmatrix} \begin{bmatrix} 1 & 0 & 0 \\ 3 & 1 & 0 \\ 2 & 0 & 1 \end{bmatrix} \begin{bmatrix} 1 & 2 & 3 \\ 0 & -4 & -8 \\ 0 & 0 & 0 \end{bmatrix}$$

$$\begin{bmatrix} .532 & 1.95 \\ 1.5 & .0013 \end{bmatrix} = \begin{bmatrix} 1 & 0 \\ 0 & 1 \end{bmatrix} \begin{bmatrix} 1 & 0 \\ 2.\,819\,5 & 1 \end{bmatrix} \begin{bmatrix} 0.532 & 1.95 \\ 0 & -5.\,496\,8 \end{bmatrix}$$

$$\begin{pmatrix} 5i & \sqrt{2} \\ -7 & 2\pi/3 \end{pmatrix} = \begin{pmatrix} 1 & 0 \\ 0 & 1 \end{pmatrix} \begin{pmatrix} 1 & 0 \\ \frac{7}{5}i & 1 \end{pmatrix} \begin{pmatrix} 5i & \sqrt{2} \\ 0 & \frac{2}{3}\pi - \frac{7}{5}i\sqrt{2} \end{pmatrix}$$

Note that the upper triangular matrix in the first line of the preceding example is the same as that in the following example.

▶ Matrices + Fraction-Free Gaussian Elimination

$$\begin{bmatrix} 1 & 2 & 3 \\ 2 & 4 & 6 \\ 3 & 2 & 1 \end{bmatrix}, \text{ fraction-free Gaussian elimination: } \begin{bmatrix} 1 & 2 & 3 \\ 0 & -4 & -8 \\ 0 & 0 & 0 \end{bmatrix}$$

In general, the upper triangular matrix in the PLU decomposition is the echelon form of the original matrix obtained by Gaussian elimination.

QR Decomposition

A real $m \times n$ matrix A with $m \geq n$ can be factored as a product QR, where Q is an *orthogonal* $m \times m$ matrix (the columns of Q are *orthonormal*—that is, QQ^T is the $m \times m$ identity matrix) and R is upper-right triangular with the same rank as A. If the original matrix A is square, then so is R. If A has linearly independent columns, then R is invertible.

▶ **To obtain the QR factorization**

1. Leave the insertion point in a matrix.

2. From the Matrices submenu, choose **QR Decomposition**.

▶ Matrices + QR Decomposition

$$\begin{pmatrix} 3 & 0 \\ 4 & 5 \end{pmatrix} = \begin{pmatrix} \frac{3}{5} & -\frac{4}{5} \\ \frac{4}{5} & \frac{3}{5} \end{pmatrix} \begin{pmatrix} 5 & 4 \\ 0 & 3 \end{pmatrix}$$

$$\begin{bmatrix} 1 & 1 \\ 0 & -2 \\ 3 & -1 \end{bmatrix} = \begin{bmatrix} \frac{1}{10}\sqrt{10} & \frac{3}{70}\sqrt{5}\sqrt{28} & \frac{3}{14}\sqrt{14} \\ 0 & -\frac{1}{14}\sqrt{5}\sqrt{28} & \frac{1}{7}\sqrt{14} \\ \frac{3}{10}\sqrt{10} & -\frac{1}{70}\sqrt{5}\sqrt{28} & -\frac{1}{14}\sqrt{14} \end{bmatrix} \begin{bmatrix} \sqrt{10} & -\frac{1}{5}\sqrt{10} \\ 0 & \frac{1}{5}\sqrt{5}\sqrt{28} \\ 0 & 0 \end{bmatrix}$$

Cholesky Decomposition

For a real square matrix that happens to be *symmetric* ($A = A^T$) and *positive definite* (all eigenvalues are positive), there is a particularly efficient triangular decomposition, significantly faster than alternative methods for solving linear equations.

An $n \times n$ real symmetric positive-definite matrix A can be factored into a product $A = GG^T$, with G a real positive-definite lower triangular $n \times n$ matrix. This factorization $A = GG^T$ is called the *Cholesky decomposition* of A.

▶ Matrices + Cholesky Decomposition

$$\begin{bmatrix} 2 & -1 \\ -1 & 2 \end{bmatrix} = \begin{bmatrix} \sqrt{2} & 0 \\ -\frac{1}{2}\sqrt{2} & \frac{1}{2}\sqrt{2}\sqrt{3} \end{bmatrix} \begin{bmatrix} \sqrt{2} & -\frac{1}{2}\sqrt{2} \\ 0 & \frac{1}{2}\sqrt{2}\sqrt{3} \end{bmatrix}$$

$$\begin{bmatrix} 2.0 & -1.0 \\ -1.0 & 2.0 \end{bmatrix} = \begin{bmatrix} 1.4142 & 0 \\ -0.70711 & 1.2247 \end{bmatrix} \begin{bmatrix} 1.4142 & -0.70711 \\ 0 & 1.2247 \end{bmatrix}$$

$$\begin{pmatrix} 1 & \frac{1}{2} & \frac{1}{3} \\ \frac{1}{2} & \frac{1}{3} & \frac{1}{4} \\ \frac{1}{3} & \frac{1}{4} & \frac{1}{5} \end{pmatrix} = \begin{pmatrix} 1 & 0 & 0 \\ \frac{1}{2} & \frac{1}{6}\sqrt{3} & 0 \\ \frac{1}{3} & \frac{1}{6}\sqrt{3} & \frac{1}{30}\sqrt{5} \end{pmatrix} \begin{pmatrix} 1 & \frac{1}{2} & \frac{1}{3} \\ 0 & \frac{1}{6}\sqrt{3} & \frac{1}{6}\sqrt{3} \\ 0 & 0 & \frac{1}{30}\sqrt{5} \end{pmatrix}$$

Exercises

1. The vectors $u = \begin{bmatrix} 1 & 1 & 0 \end{bmatrix}$ and $v = \begin{bmatrix} 1 & 1 & 1 \end{bmatrix}$ span a plane in $\mathbb{R}^3$. Find the projection matrix P onto the plane, and find a nonzero vector b that is projected to zero.

2. For the following matrix, find the characteristic polynomial, minimum polynomial, eigenvalues, and eigenvectors. Discuss the relationships among these, and explain the multiplicity of the eigenvalue.
$$\begin{bmatrix} 2 & 0 & 0 & 0 \\ 1 & 2 & 0 & 0 \\ 0 & 0 & 2 & 0 \\ 0 & 0 & -3 & 2 \end{bmatrix}$$

3. Which of the following statements are correct for the matrix $A = \begin{bmatrix} 1 & 1 & 1 \\ 1 & 0 & 2 \end{bmatrix}$?

The set of all solutions $x = \begin{bmatrix} x_1 \\ x_2 \\ x_3 \end{bmatrix}$ of the equation $Ax = \begin{bmatrix} 0 \\ 0 \end{bmatrix}$ is the column space of A; the row space of A; a nullspace of A; a plane; a line; a point.

4. The matrices that rotate the xy-plane are $A(\theta) = \begin{bmatrix} \cos\theta & -\sin\theta \\ \sin\theta & \cos\theta \end{bmatrix}$. Verify that $A(\theta)A(\varphi) = A(\theta + \varphi)$ and $A(-\theta) = A(\theta)^{-1}$, using matrix products and trigonometric identities.

5. Show that
$$\det \begin{pmatrix} x & y & 1 \\ a & b & 1 \\ c & d & 1 \end{pmatrix} = 0$$

is the equation of the line through the two points (a, b) and (c, d).

6. Verify that the 4×4 Vandermonde matrix

$$
\begin{bmatrix}
1 & x_1 & x_1^2 & x_1^3 \\
1 & x_2 & x_2^2 & x_2^3 \\
1 & x_3 & x_3^2 & x_3^3 \\
1 & x_4 & x_4^2 & x_4^3
\end{bmatrix}
$$

(see page 306) has determinant $(x_2 - x_1)(x_3 - x_1)(x_3 - x_2)(x_1 - x_4)(x_2 - x_4)$ $(x_4 - x_3)$.

Solutions

1. The projection matrix P onto the plane in $\mathbb{R}^3$ spanned by the vectors $u = [1, 1, 0]$ and $v = [1, 1, 1]$ is the product $P = A\left(A^T A\right)^{-1} A^T$, where u and v are the columns of A.

$$
P = \begin{bmatrix} 1 & 1 \\ 1 & 1 \\ 0 & 1 \end{bmatrix} \left(\begin{bmatrix} 1 & 1 \\ 1 & 1 \\ 0 & 1 \end{bmatrix}^T \begin{bmatrix} 1 & 1 \\ 1 & 1 \\ 0 & 1 \end{bmatrix} \right)^{-1} \begin{bmatrix} 1 & 1 \\ 1 & 1 \\ 0 & 1 \end{bmatrix}^T = \begin{bmatrix} \frac{1}{2} & \frac{1}{2} & 0 \\ \frac{1}{2} & \frac{1}{2} & 0 \\ 0 & 0 & 1 \end{bmatrix}
$$

Note that Pw is a linear combination of u and v for any vector $w = (x, y, z)$ in $\mathbb{R}^3$, so P maps $\mathbb{R}^3$ onto the plane spanned by u and v.

$$
\begin{bmatrix} \frac{1}{2} & \frac{1}{2} & 0 \\ \frac{1}{2} & \frac{1}{2} & 0 \\ 0 & 0 & 1 \end{bmatrix} \begin{bmatrix} x \\ y \\ z \end{bmatrix} = \begin{bmatrix} \frac{1}{2}x + \frac{1}{2}y \\ \frac{1}{2}x + \frac{1}{2}y \\ z \end{bmatrix} = \left(\frac{x+y}{2} - z \right) \begin{bmatrix} 1 \\ 1 \\ 0 \end{bmatrix} + z \begin{bmatrix} 1 \\ 1 \\ 1 \end{bmatrix}
$$

To find a nonzero vector b that is projected to zero, leave the insertion point in the matrix P and, from the Matrices submenu, choose Nullspace Basis.

$$
\begin{bmatrix} \frac{1}{2} & \frac{1}{2} & 0 \\ \frac{1}{2} & \frac{1}{2} & 0 \\ 0 & 0 & 1 \end{bmatrix}, \text{ nullspace basis: } \left\{ \begin{bmatrix} -1 \\ 1 \\ 0 \end{bmatrix} \right\}
$$

2. The matrix $\begin{bmatrix} 2 & 0 & 0 & 0 \\ 1 & 2 & 0 & 0 \\ 0 & 0 & 2 & 0 \\ 0 & 0 & -3 & 2 \end{bmatrix}$ has characteristic polynomial $(X - 2)^4$, minimum polynomial $4 - 4X + X^2 = (X - 2)^2$, and eigenvalue and eigenvectors

$$
\left\{ 2, 4, \begin{bmatrix} 0 \\ 0 \\ 0 \\ 1 \end{bmatrix}, \begin{bmatrix} 0 \\ 1 \\ 0 \\ 0 \end{bmatrix} \right\}
$$

The minimal polynomial is a factor of the characteristic polynomial. The eigenvalue 2 occurs with multiplicity 4 as a root of the characteristic polynomial $(X - 2)^4$. The eigenvalue 2 has two linearly independent eigenvectors. Note that

$$\begin{bmatrix} 2 & 0 & 0 & 0 \\ 1 & 2 & 0 & 0 \\ 0 & 0 & 2 & 0 \\ 0 & 0 & -3 & 2 \end{bmatrix} \begin{bmatrix} 0 \\ 0 \\ 0 \\ 1 \end{bmatrix} = \begin{bmatrix} 0 \\ 0 \\ 0 \\ 2 \end{bmatrix} = 2 \begin{bmatrix} 0 \\ 0 \\ 0 \\ 1 \end{bmatrix}$$

$$\begin{bmatrix} 2 & 0 & 0 & 0 \\ 1 & 2 & 0 & 0 \\ 0 & 0 & 2 & 0 \\ 0 & 0 & -3 & 2 \end{bmatrix} \begin{bmatrix} 0 \\ 1 \\ 0 \\ 0 \end{bmatrix} = \begin{bmatrix} 0 \\ 2 \\ 0 \\ 0 \end{bmatrix} = 2 \begin{bmatrix} 0 \\ 1 \\ 0 \\ 0 \end{bmatrix}$$

3. The solutions of this equation are in $\mathbb{R}^3$, and the column space of A is a subset of $\mathbb{R}^2$, so these solutions cannot be the column space of A. They do form the nullspace of A by the definition of nullspace; consequently, this set is a subspace of $\mathbb{R}^3$. The product of A with the first row of A is $\begin{bmatrix} 1 & 1 & 1 \\ 1 & 0 & 2 \end{bmatrix} \begin{bmatrix} 1 \\ 1 \\ 1 \end{bmatrix} = \begin{bmatrix} 3 \\ 3 \end{bmatrix}$, which is not $\begin{bmatrix} 0 \\ 0 \end{bmatrix}$, so the solution set is not the row space of A.

To determine whether this subspace of $\mathbb{R}^3$ is a point, line, or plane, solve the system of equations: from the Solve submenu, choose Exact

$$\begin{bmatrix} 1 & 1 & 1 \\ 1 & 0 & 2 \end{bmatrix} \begin{bmatrix} x_1 \\ x_2 \\ x_3 \end{bmatrix} = \begin{bmatrix} 0 \\ 0 \end{bmatrix}, \text{Solution is:} \begin{bmatrix} -2t_1 \\ t_1 \\ t_1 \end{bmatrix}$$

The subspace is the line that passes through the origin and the point $\begin{bmatrix} -2 & 1 & 1 \end{bmatrix}$.

4. Apply the following operations in turn.
   - New Definition: $A(\theta) = \begin{bmatrix} \cos\theta & -\sin\theta \\ \sin\theta & \cos\theta \end{bmatrix}$
   - Evaluate, Evaluate:

   $$A(\theta) A(\varphi) = \begin{bmatrix} \cos\theta & -\sin\theta \\ \sin\theta & \cos\theta \end{bmatrix} \begin{bmatrix} \cos\varphi & -\sin\varphi \\ \sin\varphi & \cos\varphi \end{bmatrix}$$
   $$= \begin{bmatrix} \cos\theta\cos\varphi - \sin\theta\sin\varphi & -\cos\theta\sin\varphi - \sin\theta\cos\varphi \\ \sin\theta\cos\varphi + \cos\theta\sin\varphi & \cos\theta\cos\varphi - \sin\theta\sin\varphi \end{bmatrix}$$

   - Leave the insertion point in the matrix and, from the Combine submenu, choose Trig Functions.

   $$\begin{bmatrix} \cos\theta\cos\varphi - \sin\theta\sin\varphi & -\cos\theta\sin\varphi - \sin\theta\cos\varphi \\ \sin\theta\cos\varphi + \cos\theta\sin\varphi & \cos\theta\cos\varphi - \sin\theta\sin\varphi \end{bmatrix} =$$

   $$\begin{bmatrix} \cos(\theta+\varphi) & -\sin(\theta+\varphi) \\ \sin(\theta+\varphi) & \cos(\theta+\varphi) \end{bmatrix}$$

This result proves the first identity. For the second part, carry out the following steps.

- Evaluate
$$A\left(\theta\right)A\left(-\theta\right) = \begin{bmatrix} \cos\theta & -\sin\theta \\ \sin\theta & \cos\theta \end{bmatrix} \begin{bmatrix} \cos\theta & \sin\theta \\ -\sin\theta & \cos\theta \end{bmatrix}$$

- Evaluate
$$\begin{bmatrix} \cos\theta & -\sin\theta \\ \sin\theta & \cos\theta \end{bmatrix} \begin{bmatrix} \cos\theta & \sin\theta \\ -\sin\theta & \cos\theta \end{bmatrix} =$$

$$\begin{bmatrix} \cos^2\theta + \sin^2\theta & 0 \\ 0 & \cos^2\theta + \sin^2\theta \end{bmatrix}$$

- Leave the insertion point in the matrix and choose **Simplify** or, from the **Combine** submenu, choose **Trig Functions**.

$$\begin{bmatrix} \cos^2\theta + \sin^2\theta & 0 \\ 0 & \cos^2\theta + \sin^2\theta \end{bmatrix} = \begin{bmatrix} 1 & 0 \\ 0 & 1 \end{bmatrix}$$

5. Solving the equation

$$\det \begin{pmatrix} x & y & 1 \\ a & b & 1 \\ c & d & 1 \end{pmatrix} = 0$$

for y gives the solution $y = \dfrac{xd - xb + cb - ad}{c - a}$, which can be rewritten as $y = \dfrac{d - b}{c - a}x + \dfrac{cb - ad}{c - a}$. If $c \neq a$, this is the equation of the line through the two points (a, b) and (c, d). If $c = a$,

$$\det \begin{pmatrix} x & y & 1 \\ a & b & 1 \\ a & d & 1 \end{pmatrix} = xb - xd + ad - ab = 0$$

has the solution $x = a$, which is the equation of the line through the points (a, b) and (a, d).

6. Use **Evaluate**, then **Factor** to obtain

$$\begin{vmatrix} 1 & x_1 & x_1^2 & x_1^3 \\ 1 & x_2 & x_2^2 & x_2^3 \\ 1 & x_3 & x_3^2 & x_3^3 \\ 1 & x_4 & x_4^2 & x_4^3 \end{vmatrix} = (x_2 - x_1)(x_3 - x_1)(x_3 - x_2)(x_1 - x_4)(x_2 - x_4)(x_4 - x_3)$$

9 Vector Calculus

Vector calculus is the calculus of functions that assign vectors to points in space. It can be applied to problems such as finding the work done by a force field in moving an object along a curve or finding the rate of fluid flow across a surface. This chapter provides information about vectors and the use of vectors in calculus.

Vectors

The term *vector* is used to indicate a quantity that has both magnitude and direction. A vector is often represented by an arrow or a directed line segment. The length of the arrow represents the magnitude of the vector and the arrow points in its direction. Two directed line segments are considered equivalent in the sense that they have the same length and point in the same direction. In other words, a vector $\mathbf{v}$ can be thought of as a set of equivalent directed line segments. A *two-dimensional vector* is an ordered pair $\mathbf{a} = (a_1, a_2)$ of real numbers. A *three-dimensional vector* is an ordered triple $\mathbf{a} = (a_1, a_2, a_3)$ of real numbers. An *n-dimensional vector* is an ordered n-tuple $\mathbf{a} = (a_1, a_2, \ldots, a_n)$ of real numbers. The numbers $a_1, a_2, \ldots, a_n$ are called the *components* of $\mathbf{a}$.

Notation for Vectors

You can represent vectors in any one of the following ways.

- n-tuples within parentheses or brackets: $(2, -1, 0)$, (x_1, x_2, x_3), $[3, 2, 1]$, $[x_1, x_2, x_3]$
- $1 \times n$ matrices: $\begin{bmatrix} 1 & 2 & 3 \end{bmatrix}$, $\begin{bmatrix} 5 & -1 & 3 & 17 & -8 & 2 \end{bmatrix}$, $\begin{bmatrix} x_1 & x_2 & x_3 & x_4 \end{bmatrix}$
- $n \times 1$ matrices: $\begin{bmatrix} 1 \\ 0 \\ -1 \end{bmatrix}$, $\begin{bmatrix} 35 \\ -4 \end{bmatrix}$, $\begin{pmatrix} x_1 \\ x_2 \end{pmatrix}$

This flexibility allows you to use the output of previous work as input, without undue worry about the shape of that output. For purposes of clear exposition, you will find it preferable to use consistent notation for vectors.

▶ **To create a vector in matrix form**

1. Click ▦ or choose Insert + Matrix.

2. Set the number of rows (or columns) to 1 and the number of columns (or rows) to the dimension of the vector.

3. Enter the values for the components in the input boxes.

4. Select the vector with the mouse and click one of the expanding brackets buttons to enclose the vector in brackets.

Vector Sums and Scalar Multiplication

The sum of two vectors $[x_1, x_2, ..., x_n]$ and $[y_1, y_2, ..., y_n]$ is defined by
$$[x_1, x_2, ..., x_n] + [y_1, y_2, ..., y_n] = [x_1 + y_1, x_2 + y_2, ..., x_n + y_n]$$
The product of a scalar a and a vector $[x_1, x_2, ..., x_n]$ is defined by
$$a[x_1, x_2, ..., x_n] = [ax_1, ax_2, ..., ax_n]$$

▶ **To evaluate a vector sum**

1. Type the expression in mathematics mode.

2. Choose **Evaluate**.

▶ Evaluate

$$(x_1, x_2, x_3) + (y_1, y_2, y_3) = \begin{bmatrix} x_1 + y_1 & x_2 + y_2 & x_3 + y_3 \end{bmatrix}$$

$$(6, -1 + i) + (2 - 3i, 3) = \begin{bmatrix} 8 - 3i & 2 + i \end{bmatrix}$$

$$\begin{pmatrix} a \\ b \end{pmatrix} + \begin{pmatrix} c \\ d \end{pmatrix} = \begin{pmatrix} a + c \\ b + d \end{pmatrix} \qquad \begin{bmatrix} 1 \\ 2 \end{bmatrix} - \begin{bmatrix} -3 \\ 1 \end{bmatrix} = \begin{bmatrix} 4 \\ 1 \end{bmatrix}$$

▶ **To evaluate the product of a scalar with a vector**

1. Type the expression in mathematics mode.

2. Choose **Evaluate**.

▶ Evaluate

$$a \begin{bmatrix} x_1 & x_2 & x_3 \end{bmatrix} = \begin{bmatrix} ax_1 & ax_2 & ax_3 \end{bmatrix} \qquad 6 \begin{bmatrix} 2 & 3 & -5 \end{bmatrix} = \begin{bmatrix} 12 & 18 & -30 \end{bmatrix}$$

$$i\sqrt{3} \begin{bmatrix} 2 & -6i & 5 - 3i \end{bmatrix} = \begin{bmatrix} 2i\sqrt{3} & 6\sqrt{3} & (3 + 5i)\sqrt{3} \end{bmatrix}$$

Dot Product

To enter the dot used for the dot product, click the dot on the **Symbol Cache** toolbar, or select it from the **Binary Operations** panel under [±÷] . The *dot product* (or *inner* *product*) of two real vectors $(a_1, a_2, ..., a_n)$ and $(b_1, b_2, ..., b_n)$ is defined by

$$(a_1, a_2, ..., a_n) \cdot (b_1, b_2, ..., b_n) = a_1 b_1 + a_2 b_2 + \cdots + a_n b_n$$

and the standard inner product of two vectors with complex entries is defined by

$$(a_1, a_2, ..., a_n) \cdot (b_1, b_2, ..., b_n) = a_1 b_1^* + a_2 b_2^* + \cdots + a_n b_n^*$$

where $b^* = x - iy$ is the complex conjugate of $b = x + iy$. For real numbers b, it is clear that $b^* = b$, so these two definitions are consistent. The dot product can also be obtained by matrix multiplication:

$$(a_1, a_2, ..., a_n) \cdot (b_1, b_2, ..., b_n) = \begin{bmatrix} a_1 & a_2 & \cdots & a_n \end{bmatrix} \begin{bmatrix} b_1^* \\ b_2^* \\ \vdots \\ b_n^* \end{bmatrix}$$

$$= \begin{bmatrix} a_1 \\ a_2 \\ \vdots \\ a_n \end{bmatrix}^T \begin{bmatrix} b_1^* \\ b_2^* \\ \vdots \\ b_n^* \end{bmatrix}$$

To compute a dot product, write the expression and apply **Evaluate**.

▶ **Evaluate**

$$(1, 2, 3) \cdot (3, 2, 1) = 10 \qquad\qquad [3x, -1, 5] \cdot [1, 1, 1] = 3x + 4$$

$$(1 + 2i, -3i) \cdot (5, 1 - i) = 8 + 7i \qquad \begin{pmatrix} 1 + 2i & -3i \end{pmatrix} \begin{pmatrix} 5 \\ (1-i)^* \end{pmatrix} = 8 + 7i$$

The standard default on variables returns complex solutions. You can change this default with the function assume. (See page 122.)

▶ **Evaluate**

$$\text{assume}\,(\text{complex}) = \text{complex}$$

$$(u, v, w) \cdot (x, y, z) = ux^* + vy^* + wz^*$$

$$\text{assume}\,(\text{real}) = \text{real}$$

$$(u, v, w) \cdot (x, y, z) = ux + vy + wz$$

For the following examples of dot products with $n = 3$, define

$$a = \begin{bmatrix} 1 & 2 & 3 \end{bmatrix}, b = \begin{pmatrix} 1 \\ 0 \\ -1 \end{pmatrix}, c = [3, 2, 1], \text{ and } d = (2, -1, 0)$$

with **New Definition** from the **Definitions** submenu.

▶ **Evaluate**

$$a \cdot c = 10 \qquad\qquad a \cdot b = -2 \qquad\qquad c \cdot d = 4$$

Cross Product

The *cross product* of three-dimensional vectors $a = (a_1, a_2, a_3)$ and $b = (b_1, b_2, b_3)$ is defined by

$$a \times b = (a_2 b_3 - a_3 b_2, a_3 b_1 - a_1 b_3, a_1 b_2 - a_2 b_1)$$

To enter the cross used for the cross product, click the cross on the **Symbol Cache** toolbar, or select it from the **Binary Operations** panel under [±÷] .

For the following examples, use the vectors $a = \begin{bmatrix} 1 & 2 & 3 \end{bmatrix}$, $b = \begin{pmatrix} 1 \\ 0 \\ -1 \end{pmatrix}$,

$c = [3, 2, 1]$, and $d = (2, -1, 0)$, as defined previously.

▶ **Evaluate**

$$a \times b = \begin{pmatrix} -2 & 4 & -2 \end{pmatrix} \qquad a \times c = \begin{pmatrix} -4 & 8 & -4 \end{pmatrix} \qquad c \times d = \begin{pmatrix} 1 & 2 & -7 \end{pmatrix}$$

$$\begin{pmatrix} 0.35 \\ -0.73 \\ 1.2 \end{pmatrix} \times \begin{pmatrix} 0.85 \\ 0.32 \\ -0.77 \end{pmatrix} = \begin{pmatrix} 0.178\,1 \\ 1.\,289\,5 \\ 0.732\,5 \end{pmatrix}$$

$$\begin{bmatrix} 1 & -2 & 5 \end{bmatrix} \times \begin{bmatrix} 5 & 3 & -5 \end{bmatrix} = \begin{bmatrix} -5 & 30 & 13 \end{bmatrix}$$

Three-dimensional vectors are often written in terms of the standard basis:

$$\mathbf{i} = (1, 0, 0)$$
$$\mathbf{j} = (0, 1, 0)$$
$$\mathbf{k} = (0, 0, 1)$$

The cross product of the two vectors $a_1 \mathbf{i} + a_2 \mathbf{j} + a_3 \mathbf{k}$ and $b_1 \mathbf{i} + b_2 \mathbf{j} + b_3 \mathbf{k}$ can then be computed by using the determinant

$$\begin{vmatrix} \mathbf{i} & \mathbf{j} & \mathbf{k} \\ a_1 & a_2 & a_3 \\ b_1 & b_2 & b_3 \end{vmatrix} = \mathbf{i} (a_2 b_3 - a_3 b_2) - \mathbf{j} (a_1 b_3 - b_1 a_3) + \mathbf{k} (a_1 b_2 - a_2 b_1)$$

Triple Cross Product

Since the cross product of two vectors produces another vector, it is possible to string cross products together. Use the same vectors a, b, c, and d as before for these *triple vector products*. Note that different choices of position for parentheses generally produce different results. This demonstrates that the cross product is not an associative operation.

The default order of operations for cross products is from left to right.

▶ **Evaluate**

$$a \times b \times c = [8, -4, -16] \qquad\qquad c \times a \times b = \begin{pmatrix} 8 & 8 & 8 \end{pmatrix}$$
$$a \times (b \times c) = \begin{pmatrix} 16 & 4 & -8 \end{pmatrix} \qquad (a \times b) \times c = \begin{pmatrix} 8 & -4 & -16 \end{pmatrix}$$
$$a \times ((b \times c) \times d) = \begin{pmatrix} 0 & 0 & 0 \end{pmatrix} \qquad (a \times (b \times c)) \times d = \begin{pmatrix} -8 & -16 & -24 \end{pmatrix}$$

To obtain intermediate results, select a subexpression that is surrounded by parentheses and hold the CTRL key down while evaluating. This technique does an in-place computation, as illustrated in the following examples.

▶ CTRL + Evaluate, Evaluate

$$([\ 1\quad -2\quad 5\ ] \times [\ 5\quad 3\quad -5\ ]) \times [\ -7\quad 2\quad 8\ ]$$
$$= [\ -5\quad 30\quad 13\ ] \times [\ -7\quad 2\quad 8\ ] = [\ 214\quad -51\quad 200\ ]$$

$$[\ 1\quad -2\quad 5\ ] \times ([\ 5\quad 3\quad -5\ ] \times [\ -7\quad 2\quad 8\ ])$$
$$= [\ 1\quad -2\quad 5\ ] \times [\ 34\quad -5\quad 31\ ] = [\ -37\quad 139\quad 63\ ]$$

Tip Parentheses are important. As always, careful and consistent use of mathematical notation is in order. When in doubt, add extra parentheses to clarify an expression.

Triple Scalar Product

When mixing cross products with scalar products, use parentheses for clarity.

▶ Evaluate

$$(1,0,1) \cdot ((1,2,3) \times (3,2,1)) = -8 \qquad ((1,0,1) \times (1,2,3)) \cdot (3,2,1) = -8$$

Without these parentheses, you obtain different results for the first product.

▶ Evaluate

$$(1,0,1) \cdot (1,2,3) \times (3,2,1) = (\ 12\quad 8\quad 4\ )$$

$$(1,0,1) \times (1,2,3) \cdot (3,2,1) = -8 \qquad (1,2,3) \times (3,2,1) \cdot (1,0,1) = 64$$

Note The triple scalar product has an interesting geometric interpretation. The volume of the parallelepiped spanned by three vectors A, B, and C is equal to $|A \cdot (B \times C)|$.

Example The volume of the parallelepiped spanned by $(1,1,0)$, $(1,0,1)$, and $(0,1,1)$ is given by
$$|(1,1,0) \cdot [(1,0,1) \times (0,1,1)]| = 2$$
In particular, this value does not depend on the order of the vectors in the triple scalar product.
$$|(1,0,1) \cdot [(1,1,0) \times (0,1,1)]| = 2$$
$$|(0,1,1) \cdot [(1,1,0) \times (1,0,1)]| = 2$$

The parallelepiped can be viewed by drawing the visible edges. Plot the vector $(0, 0, 0, 1, 1, 0, 2, 1, 1)$ and drag to the plot each of $(0, 1, 1, 1, 2, 1, 2, 2, 2, 1, 1, 2, 0, 1, 1)$, $(0, 0, 0, 1, 1, 0, 2, 1, 1)$, $(0, 0, 0, 0, 1, 1)$, $(1, 1, 0, 1, 2, 1)$, and $(2, 1, 1, 2, 2, 2)$.

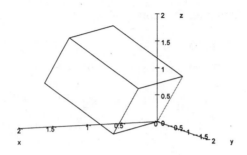

Turn: 70   Tilt: 80

The triple product $(a_1, a_2, a_3) \cdot [(b_1, b_2, b_3) \times (c_1, c_2, c_3)]$ can also be interpreted as the determinant

$$\begin{vmatrix} a_1 & a_2 & a_3 \\ b_1 & b_2 & b_3 \\ c_1 & c_2 & c_3 \end{vmatrix} = a_1 b_2 c_3 - a_1 b_3 c_2 - b_1 a_2 c_3 + b_1 a_3 c_2 + c_1 a_2 b_3 - c_1 a_3 b_2$$

It is clear from this equation that if all the vertices of a parallelepiped have integer coordinates, then the volume is also an integer.

Vector Norms

You can compute vector norms $\|v\|_n$ for every positive integer n and for ∞, where

$$\|v\|_n = \left(\sum |v_i|^n \right)^{\frac{1}{n}} \qquad \qquad \|v\|_\infty = \max \left(|v_i| \right)$$

with entries v_i either real or complex, as illustrated by the following examples.

▶ **To compute a vector norm**

1. Select the vector. Click the Brackets button ![button] on the Math Objects toolbar or choose Insert + Brackets, and choose the norm symbols. Choose OK.

2. Click ![Nx button] or choose Insert + Subscript, and enter a positive integer or the symbol ∞.

3. Place the insertion point in the vector and choose Evaluate.

▶ Evaluate

$$\|(a,b,c)\|_1 = |a| + |b| + |c| \qquad\qquad \|(-1,2,1)\|_5 = \sqrt[5]{34}$$

$$\|(a,b,c)\|_3 = \sqrt[3]{|a|^3 + |b|^3 + |c|^3} \qquad \|(5,-1.9,7)\|_8 = 7.0576$$

$$\|(a,b,c)\|_\infty = \max(|a|,|b|,|c|) \qquad \|[8,-10,2+i]\|_\infty = 10$$

$$\|(\ a \quad b \quad c\ )\|_4 = \sqrt[4]{|a|^4 + |b|^4 + |c|^4} \qquad \|(\ 2+3i \quad 4-i\ )\|_4 = \|\sqrt[4]{458}\|$$

The default $\|v\|$ is the 2-norm, which is also known as the *Euclidean norm*. This norm does not require a subscript.

You can also obtain the 2-norm from a command on the **Matrices** submenu.

▶ Matrices + Norm

$$(\ a \quad b \quad c\ ), \text{2-norm: } \sqrt{aa^* + bb^* + cc^*} \qquad [8,-10,2+i], \text{2-norm: } 13$$

Note the differences between real and complex environments. The standard default is produces complex results.

▶ Evaluate

$$\text{assume}(\text{real}) = \text{real}$$
$$\|(\ a \quad b\ )\| = \sqrt{a^2 + b^2}$$
$$\text{assume}(\text{complex}) = \text{complex}$$
$$\|(\ a \quad b\ )\| = \sqrt{aa^* + bb^*}$$

Before doing the next set of examples, make the following definition.

▶ Definitions + New Definition

$$v = [3,2,1]$$

▶ Evaluate, Evaluate Numerically

$$\|v\|_1 = 6 \qquad\qquad \|v\|_2 = \sqrt{14} = 3.7417 \qquad\qquad \|v\|_6 = \sqrt[6]{794} = 3.043$$

$$\|v\|_{10} = \sqrt[10]{60\,074} = 3.0052 \qquad \|v\|_{20} = \sqrt[20]{34878\,32978} = 3.000045103$$

$$\|v\|_\infty = 3$$

This series of examples suggests that for a vector v,
$$\lim_{n\to\infty} \|v\|_n = \|v\|_\infty$$

Example   The area of the parallelogram in the plane with vertices $(0,0)$, (a_1, a_2), (b_1, b_2), and $(a_1 + b_1, a_2 + b_2)$ is given by

$$\| (a_1, a_2, 0) \times (b_1, b_2, 0) \|$$

In particular, the area of the parallelogram spanned by the two vectors $(1, 2)$ and $(2, 1)$ is given by

$$\|(1, 2, 0) \times (2, 1, 0)\| = 3$$

This parallelogram appears in the following plot.

▶ **Plot 2D + Rectangular, Equal Scaling Along Each Axis**

$(0, 0, 1, 2, 3, 3, 2, 1, 0, 0)$

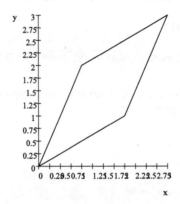

Since $A \cdot B = \|A\| \, \|B\| \cos \theta$, where θ is the angle between the vectors A and B, you can use the dot product to find the angle between two vectors.

▶ **Evaluate**

$$(1, 2, -3) \cdot (-2, 1, 2) = -6 \qquad \|(1, 2, -3)\| \, \|(-2, 1, 2)\| \cos \theta = 3 \, (\cos \theta) \sqrt{14}$$

▶ **Solve + Exact**

$-6 = 3 \, (\cos \theta) \sqrt{14}$, Solution is: $\left\{ \pi + 2X_2 \pi - \left(\arccos \tfrac{1}{7} \sqrt{14} \right) \mid X_2 \in \mathbb{Z} \right\} \cup$
$\left\{ 2X_4 \pi - \pi + \left(\arccos \tfrac{1}{7} \sqrt{14} \right) \mid X_4 \in \mathbb{Z} \right\}$

Planes and Lines in $\mathbb{R}^3$

A vector equation of the plane through the point (x_0, y_0, z_0) and orthogonal to the vector (a, b, c) is given by

$$[(x, y, z) - (x_0, y_0, z_0)] \cdot (a, b, c) = 0$$

▶ **To find the plane through three given points**

(x_0, y_0, z_0), (x_1, y_1, z_1), **and** (x_2, y_2, z_2)

1. Compute the differences $u = (x_0, y_0, z_0) - (x_1, y_1, z_1)$ and $v = (x_0, y_0, z_0) - (x_2, y_2, z_2)$

2. Compute the cross product $n = u \times v$

3. Simplify the equation $[(x, y, z) - (x_0, y_0, z_0)] \cdot n = 0$

Example   To find an equation of the plane through the points $(1, 1, 0)$, $(1, 0, 1)$, and $(0, 1, 1)$, we first compute the vectors

$$\begin{aligned} u &= (1, 1, 0) - (1, 0, 1) = (0, 1, -1) \\ v &= (1, 1, 0) - (0, 1, 1) = (1, 0, -1) \end{aligned}$$

and the cross product

$$n = (0, 1, -1) \times (1, 0, -1) = (-1, -1, -1)$$

and simplify the equation

$$\begin{aligned} [(x, y, z) - (1, 1, 0)] \cdot (-1, -1, -1) &= 0 \\ -x + 2 - y - z &= 0 \\ x + y + z &= 2 \end{aligned}$$

We plot this plane by first solving for z.

▶ Solve + Exact, Variable to Solve For: z

$x + y + z = 2$, Solution is: $2 - y - x$

▶ Plot 3D + Rectangular

$2 - y - x$ (View: $0 \leq x \leq 2, 0 \leq y \leq 2, 0 \leq z \leq 2$)

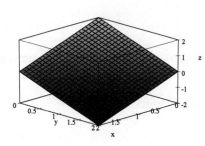

A vector equation of the line through the point (a, b, c) in the direction of $m = (u_1, u_2, u_3)$ is given by

$$(x, y, z) = (a, b, c) + t\,(u_1, u_2, u_3)$$

This is equivalent to the system of three parametric equations

$$
\begin{aligned}
x &= a + tu_1 \\
y &= b + tu_2 \\
z &= c + tu_3
\end{aligned}
$$

Example   To find an equation of the line through the two points $(1, 2, 3)$ and $(2, 1, 2)$, we first compute a vector

$$m = (1, 2, 3) - (2, 1, 2) = (-1, 1, 1)$$

that is parallel to the line, then simplify the equation

$$(x, y, z) = (1, 2, 3) + t\,(-1, 1, 1) = (1 - t, 2 + t, 3 + t)$$

The line can now be plotted.

▶ **Plot 3D + Rectangular**

$(1 - t, 2 + t, 3 + t)$

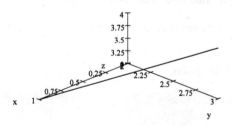

Lines and other curves in space can sometimes be more easily visualized by using a fat curve.

▶ **Plot 3D + Tube (Radius = 0.05)**

$(1 - t, 2 + t, 3 + t)$

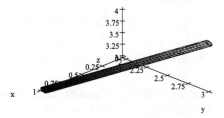

Gradient, Divergence, and Curl

Gradient, divergence, and curl are implemented with their usual notation "∇", "$\nabla\cdot$", and "$\nabla\times$" followed by **Evaluate**. These operations also appear as special commands on the **Vector Calculus** menu. To enter the nabla symbol ∇, click $\boxed{\nabla}$ on the **Symbol Cache** toolbar, or select the nabla from the **Miscellaneous Symbols** panel under $\boxed{\infty \partial}$.

Gradient

If $f(x_1, x_2, \ldots, x_n)$ is a scalar function of n variables, then the vector

$$\left(\frac{\partial f}{\partial x_1}(c_1, c_2, \ldots, c_n), \frac{\partial f}{\partial x_2}(c_1, c_2, \ldots, c_n), \ldots, \frac{\partial f}{\partial x_{n1}}(c_1, c_2, \ldots, c_n) \right)$$

is the *gradient* of f at the point $(c_1, c_2, \ldots, c_n)$ and is denoted ∇f. For $n = 3$, the vector ∇f at (a, b, c) is normal to the level surface $f(x, y, z) = f(a, b, c)$ at the point (a, b, c).

▶ **To compute the gradient of a function** $f(x, y, z)$

- Place the insertion point in the expression $\nabla f(x, y, z)$ and choose **Evaluate**.
  or

- Place the insertion point in the expression $f(x, y, z)$ and choose **Vector Calculus + Gradient**.

▶ **Evaluate**

$$\nabla xyz = \begin{pmatrix} yz \\ xz \\ xy \end{pmatrix}$$

▶ Vector Calculus + Gradient

$$xyz, \text{Gradient is} \begin{pmatrix} yz \\ xz \\ xy \end{pmatrix}$$

You can also operate on the function name after defining a function. For example, if f is defined by the equation $f(x, y, z) = xyz$, then you can evaluate $\nabla f(x, y, z)$.

▶ Evaluate

$$\nabla f(x, y, z) = \begin{bmatrix} yz \\ xz \\ xy \end{bmatrix}$$

The default basis variables are x, y, z. If you use a different set of basis variables, choose **Vector Calculus + Set Basis Variables** and enter the new set of basis variables. The basis variables should appear in red, separated by red commas.

To specify a different variable list, reset the variables.

▶ Vector Calculus + Set Basis Variables

- u, v, w

After setting u, v, w as basis variables, the computing engine regards c as a constant.

▶ Evaluate

$$\nabla \left(cuv + v^2 w \right) = \begin{bmatrix} cv \\ cu + 2vw \\ v^2 \end{bmatrix}$$

In the following example, we regard xy as the value of a function of three variables.

▶ Vector Calculus + Set Basis Variables

- x, y, z

▶ Evaluate

$$\nabla xy = \begin{bmatrix} y \\ x \\ 0 \end{bmatrix}$$

Note   In physics, f represents potential energy, and ∇f represents force.

Divergence

A *vector field* is a vector-valued function. If
$$F(x, y, z) = [p(x, y, z), q(x, y, z), r(x, y, z)]$$
is a vector field, then the scalar
$$\nabla \cdot F = \frac{\partial p}{\partial x}(a, b, c) + \frac{\partial q}{\partial y}(a, b, c) + \frac{\partial r}{\partial z}(a, b, c)$$
is the *divergence* of F at the point (a, b, c). The dot product notation is used because the symbol ∇ can be thought of as the vector operator
$$\nabla = \left(\frac{\partial}{\partial x}, \frac{\partial}{\partial y}, \frac{\partial}{\partial z} \right)$$

▶ **To compute the divergence of a vector field $F(x, y, z)$**

- Place the insertion point in the expression $\nabla \cdot F(x, y, z)$ and choose **Evaluate**.

The default is that the field variables are x, y, and z, in that order. If you wish to label the field variables differently, reset the default with **Set Basis Variables** on the **Vector Calculus** submenu.

For the following example, use **Definitions + New Definition** to define the following vector fields

$$F = [yz, 2xz, xy] \qquad\qquad G = (xz, 2yz, z^2)$$

$$H = \begin{bmatrix} yz & 2xz & xy \end{bmatrix} \qquad J = \begin{pmatrix} x^2 \\ xy \\ 2xz \end{pmatrix}$$

where F and G are represented as 3-tuples, H is represented as a 1×3 matrix, and J as a 3×1 matrix. Compute divergence with **Evaluate** or **Vector Calculus + Divergence**.

▶ **Evaluate**

$\nabla \cdot F = 0$ $\qquad\qquad$ $\nabla \cdot G = 5z$ $\qquad$ $\nabla \cdot (xz, 2iyz + x, z^2) = (3 + 2i) z$

$\nabla \cdot (xy, x, 0) = y$ $\qquad$ $\nabla \cdot H = 0$ $\qquad$ $\nabla \cdot J = 5x$

$\nabla \cdot (a, b, c) = 0$ $\qquad$ $\nabla \cdot [ax, bxy, cz^2] = a + bx + 2cz$

▶ **Vector Calculus + Divergence**

$[yz, 2xz, xy]$, Divergence is 0 $\qquad\qquad$ $\begin{bmatrix} yz & 2xz & xy \end{bmatrix}$, Divergence is 0

$(xz, 2yz, z^2)$, Divergence is $5z$ $\qquad\qquad$ $\begin{pmatrix} x^2 \\ xy \\ 2xz \end{pmatrix}$, Divergence is $5x$

Curl

If $F(x, y, z) = (p(x, y, z), q(x, y, z), r(x, y, z))$ is a vector field, then the vector

$$\nabla \times F = \left(\frac{\partial r}{\partial y} - \frac{\partial q}{\partial z}, \frac{\partial p}{\partial z} - \frac{\partial r}{\partial x}, \frac{\partial q}{\partial x} - \frac{\partial p}{\partial y} \right)$$

is called the *curl* of F. The default is that the field variables are x, y, and z, in that order. If you wish to label the field variables differently, reset the default with **Set Basis Variables** on the **Vector Calculus** submenu. The vector field F in the following example is defined as in the previous section. Compute the curl with **Evaluate** or **Vector Calculus + Curl**.

▶ Evaluate

$$\nabla \times \begin{pmatrix} yz & 2xz & xy \end{pmatrix} = \begin{pmatrix} -x \\ 0 \\ z \end{pmatrix} \qquad \nabla \times F = \begin{pmatrix} -x \\ 0 \\ z \end{pmatrix}$$

$$\nabla \times \begin{pmatrix} x^2 \\ xy \\ 2xz \end{pmatrix} = \begin{pmatrix} 0 \\ -2z \\ y \end{pmatrix} \qquad \nabla \times \begin{bmatrix} ax^2 \\ bxy \\ 2icxz \end{bmatrix} = \begin{bmatrix} 0 \\ -2icz \\ by \end{bmatrix}$$

▶ Vector Calculus + Curl

$$(yz, 2xz, xy), \text{Curl is } \begin{bmatrix} -x \\ 0 \\ z \end{bmatrix} \qquad \begin{pmatrix} yz & 2xz & xy \end{pmatrix}, \text{Curl is } \begin{pmatrix} -x \\ 0 \\ z \end{pmatrix}$$

$$\begin{pmatrix} x^2 \\ xy \\ 2xz \end{pmatrix}, \text{Curl is } \begin{pmatrix} 0 \\ -2z \\ y \end{pmatrix} \qquad \begin{bmatrix} ax^2 \\ bxy \\ 2cxz \end{bmatrix}, \text{Curl is } \begin{bmatrix} 0 \\ -2cz \\ by \end{bmatrix}$$

In terms of the basis

$$\begin{aligned} \mathbf{i} &= (1, 0, 0) \\ \mathbf{j} &= (0, 1, 0) \\ \mathbf{k} &= (0, 0, 1) \end{aligned}$$

the curl of F can be interpreted as the determinant

$$\nabla \times F = \begin{vmatrix} \mathbf{i} & \mathbf{j} & \mathbf{k} \\ \frac{\partial}{\partial x} & \frac{\partial}{\partial y} & \frac{\partial}{\partial z} \\ p(x, y, z) & q(x, y, z) & r(x, y, z) \end{vmatrix}$$

$$= \mathbf{i} \left(\frac{\partial r}{\partial y} - \frac{\partial q}{\partial z} \right) + \mathbf{j} \left(\frac{\partial p}{\partial z} - \frac{\partial r}{\partial x} \right) + \mathbf{k} \left(\frac{\partial q}{\partial x} - \frac{\partial p}{\partial y} \right)$$

Laplacian

The *Laplacian* of a scalar field $f(x, y, z)$ is the divergence of ∇f and is written

$$\begin{aligned}
\nabla^2 f &= \nabla \cdot \nabla f \\
&= \nabla \cdot \left(\frac{\partial f}{\partial x}, \frac{\partial f}{\partial y}, \frac{\partial f}{\partial z} \right) \\
&= \frac{\partial^2 f}{\partial x^2} + \frac{\partial^2 f}{\partial y^2} + \frac{\partial^2 f}{\partial z^2}
\end{aligned}$$

The name of this operator comes from its relation to Laplace's equation

$$\frac{\partial^2 f}{\partial x^2} + \frac{\partial^2 f}{\partial y^2} + \frac{\partial^2 f}{\partial z^2} = 0$$

The default field variables for the Laplacian are x, y, and z, in that order. If you wish to label the field variables differently, reset the default with **Set Basis Variables** on the **Vector Calculus** submenu. Compute the Laplacian with **Evaluate** or **Vector Calculus + Laplacian**.

▶ **Evaluate**

$$\nabla^2 \left(x + y^2 + 2z^3 \right) = 12z + 2 \qquad \nabla \left(x + y^2 + 2z^3 \right) = \begin{bmatrix} 1 \\ 2y \\ 6z^2 \end{bmatrix}$$

$$\nabla \cdot \nabla \left(x + y^2 + 2z^3 \right) = 12z + 2 \qquad \nabla \cdot \begin{bmatrix} 1 \\ 2y \\ 6z^2 \end{bmatrix} = 12z + 2$$

▶ **Vector Calculus + Laplacian**

$x + y^2 + 2z^3$, Laplacian is $12z + 2$

$1 - 2y + 6z^2$, Laplacian is 12

Directional Derivatives

The *directional derivative* of a function f at the point (a, b, c) in the direction $\mathbf{u} = (u_1, u_2, u_3)$ is given by the inner product of ∇f and $\mathbf{u}$ at the point (a, b, c). That is, for a vector $\mathbf{u}$ of unit length and a scalar function f,

$$\begin{aligned}
D_{\mathbf{u}} f(a, b, c) &= \nabla f(a, b, c) \cdot \mathbf{u} \\
&= \frac{\partial f}{\partial x}(a, b, c)\, u_1 + \frac{\partial f}{\partial y}(a, b, c)\, u_2 + \frac{\partial f}{\partial z}(a, b, c)\, u_3
\end{aligned}$$

▶ **To compute the directional derivative of** $f(x, y, z) = xyz$ **in the direction**

$$\mathbf{u} = \left(\cos\frac{\pi}{8} \sin\frac{\pi}{9}, \sin\frac{\pi}{8} \sin\frac{\pi}{9}, \cos\frac{\pi}{9} \right)$$

1. Enter the dot product $(\nabla xyz) \cdot \left(\cos\frac{\pi}{8} \sin\frac{\pi}{9}, \sin\frac{\pi}{8} \sin\frac{\pi}{9}, \cos\frac{\pi}{9} \right)$. Note that the expression ∇xyz is enclosed in parentheses.

2. With the insertion point in the expression, choose **Evaluate** or **Evaluate Numerically**.

▶ **Evaluate, Evaluate Numerically**

$(\nabla xyz) \cdot \left(\cos\frac{\pi}{8} \sin\frac{\pi}{9}, \sin\frac{\pi}{8} \sin\frac{\pi}{9}, \cos\frac{\pi}{9} \right)$ $= xy \cos\frac{1}{9}\pi + \frac{1}{2}yz \left(\sin\frac{1}{9}\pi \right) \sqrt{\sqrt{2} + 2} +$
$\frac{1}{2}xz \left(\sin\frac{1}{9}\pi \right) \sqrt{2 - \sqrt{2}}$

$\qquad = 0.939\,69xy + 0.130\,89xz + 0.315\,99yz$

Plots of Vector Fields and Gradients

A function that assigns a vector to each point of a region in two- or three-dimensional space is called a *vector field*. The gradient of a scalar-valued function of two variables is a vector field.

Plots of Vector Fields

The operation **Plot 2D + Vector Field** requires a pair of expressions in two variables representing the horizontal and vertical components of the vector field.

▶ **To plot a two-dimensional vector field**

1. Type a pair of two-variable expressions, representing the horizontal and vertical components of a vector field, into a vector.

2. Leave the insertion point in the vector and, from the **Plot 2D** submenu, choose **Vector Field**.

Example   To visualize the vector field $F(x, y) = [x + y, x - y]$
Place the insertion point in the vector $[x + y, x - y]$, and from the **Plot 2D** submenu, choose **Vector Field**. From the **Items Plotted** page of the Plot Properties dialog, choose Intervals and increase the **Sample Size** to 20×20.

▶ Plot 2D + Vector Field

$[x + y, x - y]$

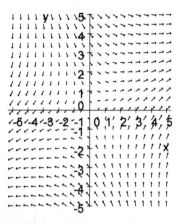

At a point (x, y) on a solution curve of a differential equation of the form $\frac{dy}{dx} = f(x, y)$, the curve has slope $f(x, y)$. You can get an idea of the appearance of the graphs of the solution of a differential equation from the *direction field*—that is, a plot depicting short line segments with slope $f(x, y)$ at points (x, y). This can be done using Plot 2D + Vector Field and the vector-valued function

$$F(t, y) = \frac{\left(1, \frac{dy}{dt}\right)}{\left\|\left(1, \frac{dy}{dt}\right)\right\|}$$

that assigns to each point (t, y) a vector of length one in the direction of the derivative at the point (t, y).

Example The direction field for the differential equation $\frac{dy}{dt} = y^2 - t^2$ is the two-dimensional vector field plot of the vector valued function

$$F(t, y) = \frac{\left(1, y^2 - t^2\right)}{\left\|\left(1, y^2 - t^2\right)\right\|}$$

▶ Plot 2D + Vector Field

$$\left(\frac{1}{\sqrt{\left(1 + |y^2 - t^2|^2\right)}}, \frac{y^2 - t^2}{\sqrt{\left(1 + |y^2 - t^2|^2\right)}}\right)$$

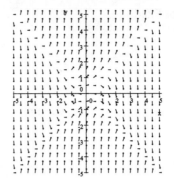

Several of the solution curves are depicted in the following plot.

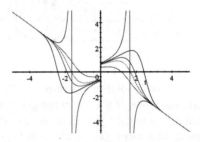

The operation **Plot 3D + Vector Field** requires a triple of expressions in three variables representing the rectangular components of the vector field.

▶ **To plot a three-dimensional vector field**

1. Type three three-variable expressions, representing the x-, y-, and z-components of a vector field, into a vector.

2. Leave the insertion point in the vector.

3. From the **Plot 3D** submenu, choose **Vector Field**.

▶ Plot 3D + Vector Field

$$[-y/z, x/z, z]$$

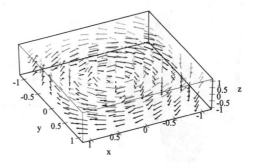

Three-dimensional vector fields can be challenging to visualize. Multiple views are helpful. Three or four views of the vector field $f(x, y, z) = (-y/z, x/z, z)$ provide a reasonable graphical representation. Boxed axes can also help the visualization. This type of vector field has been used to study circular wind patterns.

▶ **To change the view**

1. Click the frame until a small box ⬛ appears in the upper-right corner of the frame.

2. With the left mouse button held down, rotate the plot.

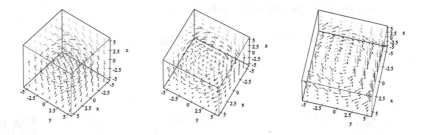

Plots of Gradient Fields

Scalar-valued functions of two variables can be visualized in several ways. Given the function $f(x, y) = xy \sin xy$, choosing **Rectangular** from the **Plot 3D** submenu produces a surface represented by the function values. Another way to visualize such a function is to choose **Gradient** from the **Plot 2D** submenu. This choice produces a plot of the vector field that is the gradient of this expression, plotting vectors at grid points whose magnitude and direction indicate the steepness of the surface and the direction of steepest ascent. The vector field that assigns to each point (x, y) the gradient of f at (x, y) is called the *gradient field* associated with the function f.

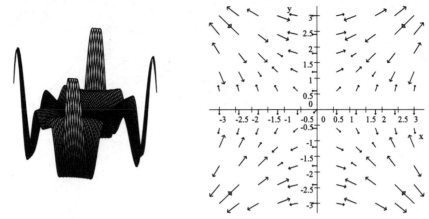

$xy \sin xy$ Gradient field of $xy \sin xy$

▶ **To plot a gradient field**

1. Type an expression $f(x, y)$.

2. Leave the insertion point in the expression, and from the **Plot 2D** submenu, choose **Gradient**.

For example, type the expression $x^2 + 2y^2$, and choose **Gradient** from the **Plot 2D** submenu. This procedure produces a plot of the vector field that is the gradient of this expression. The following plots show the relative steepness on the left, the surface in the middle, and contours on the right.

▶ Plot 2D + Gradient, Plot 3D + Rectangular

$x^2 + 2y^2$

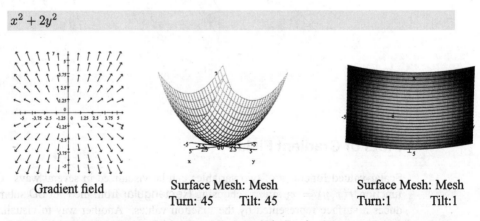

Gradient field   Surface Mesh: Mesh   Surface Mesh: Mesh
                  Turn: 45   Tilt: 45   Turn:1   Tilt:1

The gradient field for a scalar-valued function $f(x, y, z)$ of three variables is a three-dimensional vector field in which each vector represents the direction of maximal increase. The surface represented by the function values is embedded in four-dimensional space, so you must use indirect methods such as plotting the gradient field to help you visualize this surface.

▶ Plot 3D + Gradient

$$xz + xy + yz$$

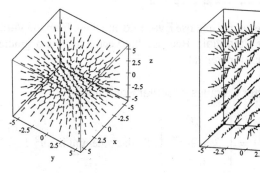

Another way to visualize the function $f(x, y, z)$ is to plot a series of implicit plots of surfaces of constant values. This is available with *Scientific WorkPlace* only. The gradient field points from surfaces of lower constant values to surfaces with higher constant values. As with other three-dimensional vector fields, multiple views convey graphical information more effectively than does a single view.

Scalar and Vector Potentials

The Scalar Potential command on the Vector Calculus menu produces the *inverse* of the gradient in the sense that it finds a scalar function whose gradient is the given vector field, or it tells you that such a function does not exist. The vector potential has an analogous interpretation in terms of the curl.

Scalar Potentials

A scalar potential exists for a vector field F if and only if
$$\operatorname{curl} F = \nabla \times F = 0$$
That is, the vector field is *irrotational*.

The following are examples of scalar potential with the standard basis variables.

▶ Vector Calculus + Scalar Potential

(x, y, z), Scalar potential is $\frac{1}{2}x^2 + \frac{1}{2}y^2 + \frac{1}{2}z$

(x, z, y), Scalar potential is $yz + \frac{1}{2}x^2$

(y, z, x), Scalar potential does not exist.

The vector field (y, z, x) does not have a scalar potential because its curl is not 0.

▶ Evaluate

$$\nabla \times (y, z, x) = \begin{bmatrix} -1 \\ -1 \\ -1 \end{bmatrix} \neq \begin{bmatrix} 0 \\ 0 \\ 0 \end{bmatrix}$$

In the next example, choose Evaluate and then, from the Vector Calculus submenu, choose Scalar Potential. Because the vector field is a gradient, it has the original function as a scalar potential.

▶ Evaluate, Vector Calculus + Scalar Potential

$$\nabla \left(xy^2 + yz^3 \right) = \begin{bmatrix} y^2 \\ 2xy + z^3 \\ 3yz^2 \end{bmatrix}, \text{ Scalar potential is } xy^2 + yz^3$$

You would normally expect the scalar potential of the vector field $\left(cv, cu + 2vw, v^2 \right)$ to be $ucv + v^2w$; that is, you expect c to be treated as a constant. When the number of variables differs from the number of components in the field vector, a dialog box asks for the field variables. In this case, you can enter **u,v,w** to get the expected result.

The dialog box also appears when you ask for the scalar potential of a vector field that specifies fewer than three variables, such as $(y, x, 0)$. Enter **x,y,z** in the dialog box to get the expected result xy for the scalar potential of this vector field.

Vector Potential

A vector potential exists for a vector field $\vec{F}$ if and only if
$$\text{div } F = \nabla \cdot F = 0$$
That is, the vector field is *solenoidal*.

Unless otherwise specified, the operators *curl* and *vector potential* apply to scalar or vector functions of a set of exactly three *standard basis variables*. The default is x, y, z, but you can use other sets of basis variables by choosing Set Basis Variables from the Vector Calculus submenu and changing the default variables.

Start with $\nabla \times (xy, yz, zx) = \begin{bmatrix} -y \\ -z \\ -x \end{bmatrix}$ to get the following vector potential.

▶ Vector Calculus + Vector Potential

$$\begin{bmatrix} -y \\ -z \\ -x \end{bmatrix}, \text{ Vector potential is } \begin{bmatrix} xy - \frac{1}{2}z^2 \\ yz \\ 0 \end{bmatrix}$$

Notice that we did not get the original vector field when we asked for a vector potential of its curl. That is because the vector potential is determined only up to a field whose curl is zero. You can verify that this is the case. First, calculate the difference of the two vectors. Then compute the curl of the difference.

▶ Evaluate

$$
\begin{bmatrix} xy \\ yz \\ zx \end{bmatrix} - \begin{bmatrix} xy - \frac{1}{2}z^2 \\ yz \\ 0 \end{bmatrix} = \begin{bmatrix} \frac{1}{2}z^2 \\ 0 \\ xz \end{bmatrix} \qquad \nabla \times \begin{bmatrix} \frac{1}{2}z^2 \\ 0 \\ xz \end{bmatrix} = \begin{bmatrix} 0 \\ 0 \\ 0 \end{bmatrix}
$$

Try the same experiment after changing the basis variables to u, v, w with Vector Calculus + Set Basis Variables. Note that a vector field can be written either as the triple (v, w, u) or as a column matrix.

▶ Vector Calculus + Vector Potential

$$
\begin{bmatrix} -v \\ -w \\ -u \end{bmatrix}, \text{Vector potential is} \quad \begin{bmatrix} uv - \frac{1}{2}w^2 \\ vw \\ 0 \end{bmatrix}
$$

$$
(v, w, u), \text{Vector potential is} \quad \begin{bmatrix} -uv + \frac{1}{2}w^2 \\ -vw \\ 0 \end{bmatrix}
$$

Matrix-Valued Operators

Matrix-valued operators include the Hessian, the Jacobian, and the Wronskian matrix.

Hessian

The *Hessian* is the $n \times n$ matrix

$$
\begin{bmatrix} \dfrac{\partial^2 f}{\partial x_1^2} & \dfrac{\partial^2 f}{\partial x_1 \partial x_2} & \cdots & \dfrac{\partial^2 f}{\partial x_1 \partial x_n} \\[2ex] \dfrac{\partial^2 f}{\partial x_2 \partial x_1} & \dfrac{\partial^2 f}{\partial x_2^2} & \cdots & \dfrac{\partial^2 f}{\partial x_2 \partial x_n} \\[2ex] \vdots & \vdots & \ddots & \vdots \\[2ex] \dfrac{\partial^2 f}{\partial x_n \partial x_1} & \dfrac{\partial^2 f}{\partial x_n \partial x_2} & \cdots & \dfrac{\partial^2 f}{\partial x_n^2} \end{bmatrix}
$$

of second partial derivatives of a scalar function $f(x_1, x_2, \ldots, x_n)$ of n variables.

The order of the set of basis variables affects the ordering of the rows and columns of the Hessian. For the following examples, the set of basis variables is x, y, z.

▶ **Vector Calculus + Hessian**

xyz, Hessian is $\begin{bmatrix} 0 & z & y \\ z & 0 & x \\ y & x & 0 \end{bmatrix}$ $x^2 + y^3$, Hessian is $\begin{bmatrix} 2 & 0 & 0 \\ 0 & 6y & 0 \\ 0 & 0 & 0 \end{bmatrix}$

$wxyz$, Hessian is $\begin{bmatrix} 0 & wz & wy \\ wz & 0 & wx \\ wy & wx & 0 \end{bmatrix}$ $a^3 + b^3$, Hessian is $\begin{bmatrix} 0 & 0 & 0 \\ 0 & 0 & 0 \\ 0 & 0 & 0 \end{bmatrix}$

To specify a different variable list, reset the variables.

▶ **Vector Calculus + Set Basis Variables:** a, b, c

▶ **Vector Calculus + Hessian**

$a^3 + b^3$, Hessian is $\begin{bmatrix} 6a & 0 & 0 \\ 0 & 6b & 0 \\ 0 & 0 & 0 \end{bmatrix}$

▶ **Vector Calculus + Set Basis Variables:** a, b

▶ **Vector Calculus + Hessian**

$b^3 + a^3$, Hessian is $\begin{bmatrix} 6a & 0 \\ 0 & 6b \end{bmatrix}$

▶ **Vector Calculus + Set Basis Variables:** x, y, z, w

▶ **Vector Calculus + Hessian**

$x^2 z + y^3 w$, Hessian is $\begin{bmatrix} 2z & 0 & 2x & 0 \\ 0 & 6wy & 0 & 3y^2 \\ 2x & 0 & 0 & 0 \\ 0 & 3y^2 & 0 & 0 \end{bmatrix}$

▶ **Vector Calculus + Set Basis Variables:** y, x

▶ **Vector Calculus + Hessian**

$x^2 + y^3$, Hessian is $\begin{bmatrix} 6y & 0 \\ 0 & 2 \end{bmatrix}$

After defining $f(x, y, z) = 3xy^2z$, you can find the Hessian for the expression $f(x, y, z)$.

▶ Vector Calculus + Set Basis Variables: x, y, z

▶ Vector Calculus + Hessian

$$f(x, y, z), \text{Hessian is} \begin{bmatrix} 0 & 6yz & 3y^2 \\ 6yz & 6xz & 6xy \\ 3y^2 & 6xy & 0 \end{bmatrix}$$

Jacobian

The *Jacobian* is the $n \times n$ matrix

$$\begin{bmatrix} \dfrac{\partial f_1}{\partial x_1} & \dfrac{\partial f_1}{\partial x_2} & \cdots & \dfrac{\partial f_1}{\partial x_n} \\ \dfrac{\partial f_2}{\partial x_1} & \dfrac{\partial f_2}{\partial x_2} & \cdots & \dfrac{\partial f_2}{\partial x_n} \\ \vdots & \vdots & \ddots & \vdots \\ \dfrac{\partial f_n}{\partial x_1} & \dfrac{\partial f_n}{\partial x_2} & \cdots & \dfrac{\partial f_n}{\partial x_n} \end{bmatrix}$$

of partial derivatives of the entries in a vector field

$$(f_1(x_1, x_2, \ldots, x_n), f_2(x_1, x_2, \ldots, x_n), \ldots, f_n(x_1, x_2, \ldots, x_n))$$

Jacobians resemble Hessians in that the order of the variables in the variable list determines the order of the columns of the matrix. In the following examples, the variable list is x, y, z. To verify these examples, choose **Jacobian** while the insertion point is in the given vector field.

▶ Vector Calculus + Jacobian

$$(yz, xz, xy), \text{Jacobian is} \begin{bmatrix} 0 & z & y \\ z & 0 & x \\ y & x & 0 \end{bmatrix}$$

$$(x^2z, x + z, xz^2), \text{Jacobian is} \begin{bmatrix} 2xz & 0 & x^2 \\ 1 & 0 & 1 \\ z^2 & 0 & 2xz \end{bmatrix}$$

$$(x^2z, y + c, yz^2), \text{Jacobian is} \begin{bmatrix} 2xz & 0 & x^2 \\ 0 & 1 & 0 \\ 0 & z^2 & 2yz \end{bmatrix}$$

To specify a different variable list, reset the variables.

▶ Vector Calculus + Set Basis Variables: a, b, c

▶ Vector Calculus + Jacobian

$$(x^2 z, y + c, yz^2), \text{ Jacobian is } \begin{bmatrix} 0 & 0 & 0 \\ 0 & 0 & 1 \\ 0 & 0 & 0 \end{bmatrix}$$

Wronskian

The *Wronskian* with respect to functions $f_1, f_2, \ldots, f_n$ defined on an interval I, often denoted by $W(f_1(x), f_2(x), \ldots, f_n(x))$, is defined as

$$\det \begin{pmatrix} f_1(x) & f_2(x) & \cdots & f_n(x) \\ f_1^{(1)}(x) & f_2^{(1)}(x) & \cdots & f_n^{(1)}(x) \\ \vdots & \vdots & \cdots & \vdots \\ f_1^{(n-1)}(x) & f_2^{(n-1)}(x) & \cdots & f_n^{(n-1)}(x) \end{pmatrix}$$

Observe that $W(f_1(x), f_2(x), \ldots, f_n(x))$ is a function defined on the interval I. To compute the Wronskian, take the determinant of the Wronskian matrix.

▶ Vector Calculus + Wronskian

$$(x^3 - 3x^2, 3x^2 - 7, x^4 + 5x^2),$$
$$\text{Wronskian matrix is } \begin{bmatrix} -3x^2 + x^3 & 3x^2 - 7 & 5x^2 + x^4 \\ -6x + 3x^2 & 6x & 10x + 4x^3 \\ 6x - 6 & 6 & 12x^2 + 10 \end{bmatrix}$$

▶ Matrices + Determinant

$$\begin{bmatrix} -3x^2 + x^3 & 3x^2 - 7 & 5x^2 + x^4 \\ -6x + 3x^2 & 6x & 10x + 4x^3 \\ 6x - 6 & 6 & 12x^2 + 10 \end{bmatrix}, \text{ determinant: } 84x^4 - 336x^3 - 210x^2 - 6x^6$$

It follows that the Wronskian of the functions

$$\begin{aligned} f_1(x) &= x^3 - 3x^2 \\ f_2(x) &= 3x^2 - 7 \\ f_3(x) &= x^4 + 5x^2 \end{aligned}$$

is given by

$$W(x^3 - 3x^2, 3x^2 - 7, x^4 + 5x^2) = -6x^6 + 84x^4 - 210x^2 - 336x^3$$

Consider the special case where there are two functions. We define two generic functions f_1 and f_2.

▶ **Definitions + New Definition**

$f_1(x)$

$f_2(x)$

▶ **Vector Calculus + Wronskian**

$(f_1(x), f_2(x))$, Wronskian matrix is $\begin{bmatrix} f_1(x) & f_2(x) \\ \frac{\partial f_1(x)}{\partial x} & \frac{\partial f_2(x)}{\partial x} \end{bmatrix}$

▶ **Matrices + Determinant**

$\begin{bmatrix} f_1(x) & f_2(x) \\ \frac{\partial f_1(x)}{\partial x} & \frac{\partial f_2(x)}{\partial x} \end{bmatrix}$, determinant: $f_1(x) \frac{\partial f_2(x)}{\partial x} - f_2(x) \frac{\partial f_1(x)}{\partial x}$

Note that since

$$\frac{d}{dx}\left(\frac{f_2(x)}{f_1(x)}\right) = \frac{f_1(x) \frac{\partial f_2(x)}{\partial x} - f_2(x) \frac{\partial f_1(x)}{\partial x}}{(f_1(x))^2}$$

it follows that two functions are proportional if and only if $\frac{f_2(x)}{f_1(x)}$ is a constant, which is equivalent to $\frac{d}{dx}\left(\frac{f_2(x)}{f_1(x)}\right) = 0$; that is, their Wronskian is zero.

Plots of Complex Functions

A complex function $F(z)$ (z and $F(z)$ are both complex) is a challenge to graph, because the natural graph would require four dimensions.

Conformal Plots

A *conformal plot* of a complex function $F(z)$ is the image of a two-dimensional rectangular grid of horizontal and vertical line segments. The default is an 11×11 grid, with each of the intervals $0 \le \operatorname{Re}(z) \le 1$ and $0 \le \operatorname{Im}(z) \le 1$ subdivided into 10 equal subintervals. If $F(z)$ is analytic, then it preserves angles at every point at which $F'(z) \ne 0$; hence, the image is a grid composed of two families of curves that intersect at right angles.

To create a conformal plot of $F(z) = \dfrac{z-1}{z+1}$, put the insertion point in the expression, and choose **Conformal** from the **Plot 2D** submenu. The number of grid lines and the view can be changed in the **Plot Properties** tabbed dialogs.

▶ Plot 2D + Conformal

$$\frac{z-1}{z+1}$$

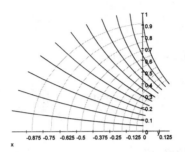

In the following example,

- $\text{Re}(z)$ and $\text{Im}(z)$ both range from -3 to 3.
- The **View Intervals** are set at $-2 \le \text{Re}(z) \le 4$ and $-3 \le \text{Im}(z) \le 3$.
- The **Grid Size** has been increased to 40 by 40.
- **Samples per Horizontal Grid Line** and **Samples per Vertical Grid Line** have both been increased to 60.

▶ Plot 2D + Conformal

$$\frac{z-1}{z+1}$$

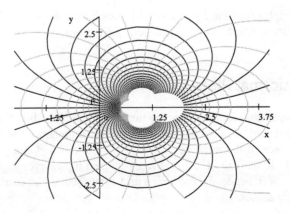

Exercises

1. Evaluate the directional derivative of $f(x,y,z) = 3x - 5y + 2z$ at $(2,2,1)$ in the direction of the outward normal to the sphere $x^2 + y^2 + z^2 = 9$.

2. Find a vector v normal to the surface $z = \sqrt{x^2 + y^2} + \left(x^2 + y^2\right)^{3/2}$ at the point $(x,y,z) \neq (0,0,0)$ on the surface.

3. Let $f(x,y,z) = \dfrac{mM}{\sqrt{x^2 + y^2 + z^2}}$ denote Newton's gravitational potential. Show that the gradient is given by

$$\nabla f(x,y,z) = -\frac{mM}{(x^2 + y^2 + z^2)^{3/2}} \begin{bmatrix} x \\ y \\ z \end{bmatrix}$$

4. Let $u_1(t), u_2(t), u_3(t)$ be three functions having third-order derivatives. Show that the derivative of the Wronskian $W(u_1(t), u_2(t), u_3(t))$ is the determinant

$$\begin{vmatrix} u_1(t) & u_2(t) & u_3(t) \\ \frac{d}{dt}u_1(t) & \frac{d}{dt}u_2(t) & \frac{d}{dt}u_3(t) \\ \frac{d^3}{dt^3}u_1(t) & \frac{d^3}{dt^3}u_2(t) & \frac{d^3}{dt^3}u_3(t) \end{vmatrix}$$

In other words, the derivative of $W(u_1(t), u_2(t), u_3(t))$ may be obtained by first differentiating the elements in the last row of the Wronskian matrix of $(u_1(t), u_2(t), u_3(t))$ and then taking the derivative of the resulting matrix.

5. Starting with the function $f(x,y) = \sin xy$, observe connections between the surface $z = f(x,y)$, the gradient of $f(x,y)$, and the vector field of $\nabla f(x,y)$.

6. Observe the vector field of $(\sin xy, \cos xy)$ and describe the flow. Is there a function $g(x,y)$ whose gradient is $(\sin xy, \cos xy)$?

Solutions

1. The directional derivative is given by $D_u f(x,y,z)) = \nabla f(x,y,z) \cdot u$, where u is a unit vector in the direction of the outward normal to the sphere $x^2 + y^2 + z^2 = 9$. The vector

$$\nabla\left(x^2 + y^2 + z^2\right) = \begin{bmatrix} 2x \\ 2y \\ 2z \end{bmatrix}$$

is normal to the sphere $x^2 + y^2 + z^2 = 9$, and at $(2,2,1)$ this normal is $(4,4,2)$. A unit vector in the same direction is given by

$$u = \begin{bmatrix} 4 \\ 4 \\ 2 \end{bmatrix} \div \left\| \begin{bmatrix} 4 \\ 4 \\ 2 \end{bmatrix} \right\| = \frac{1}{6} \begin{bmatrix} 4 \\ 4 \\ 2 \end{bmatrix} = \begin{bmatrix} 2/3 \\ 2/3 \\ 1/3 \end{bmatrix}$$

and is shown in the following figure.

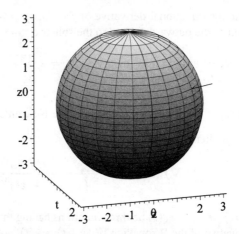

Since $\nabla f\left(x, y, z\right) = (3, -5, 2)$ it follows that $\nabla f\left(x, y, z\right) \cdot \left(\frac{2}{3}, \frac{1}{3}, \frac{2}{3}\right) = \frac{5}{3}$.

2. A normal vector is given by

$$\nabla \left(\sqrt{x^2 + y^2} + \left(x^2 + y^2\right)^{3/2} - z\right)$$

$$= \left(\frac{1}{\sqrt{(x^2 + y^2)}} x + 3\sqrt{(x^2 + y^2)}x, \; \frac{1}{\sqrt{(x^2 + y^2)}} y + 3\sqrt{(x^2 + y^2)}y, -1\right)$$

$$= \left(x \frac{1 + 3x^2 + 3y^2}{\sqrt{(x^2 + y^2)}}, y \frac{1 + 3x^2 + 3y^2}{\sqrt{(x^2 + y^2)}}, -1\right)$$

$$= \frac{1}{\sqrt{(x^2 + y^2)}} \left(x \left(1 + 3x^2 + 3y^2\right), y \left(1 + 3x^2 + 3y^2\right), -\sqrt{(x^2 + y^2)}\right)$$

Hence, any scalar multiple of

$$\left[x \left(1 + 3x^2 + 3y^2\right), y \left(1 + 3x^2 + 3y^2\right), -\sqrt{(x^2 + y^2)}\right]$$

is also normal to the given surface.

3. Evaluate the expression $\nabla \left(\dfrac{mM}{\sqrt{x^2 + y^2 + z^2}}\right)$. Then delete the first two rows of the vector, because m and M are constant parameters.

$$\nabla \left(\frac{mM}{\sqrt{x^2 + y^2 + z^2}}\right)$$

$$= \left(-m \frac{M}{\left(\sqrt{(x^2+y^2+z^2)}\right)^3} x, \; -m \frac{M}{\left(\sqrt{(x^2+y^2+z^2)}\right)^3} y, \; -m \frac{M}{\left(\sqrt{(x^2+y^2+z^2)}\right)^3} z\right)$$

$$= -\frac{mM}{\left(x^2 + y^2 + z^2\right)^{3/2}} \left(x, y, z\right)$$

This gives the Newtonian gravitational force between two objects of masses m and M, with one object at the origin and the other at the point (x, y, z).

4. Evaluate each of

$$\frac{d}{dt}\begin{vmatrix} u_1\left(t\right) & u_2\left(t\right) & u_3\left(t\right) \\ \frac{d}{dt}u_1\left(t\right) & \frac{d}{dt}u_2\left(t\right) & \frac{d}{dt}u_3\left(t\right) \\ \frac{d^2}{dt^2}u_1\left(t\right) & \frac{d^2}{dt^2}u_2\left(t\right) & \frac{d^2}{dt^2}u_3\left(t\right) \end{vmatrix} \quad \text{and} \quad \begin{vmatrix} u_1\left(t\right) & u_2\left(t\right) & u_3\left(t\right) \\ \frac{d}{dt}u_1\left(t\right) & \frac{d}{dt}u_2\left(t\right) & \frac{d}{dt}u_3\left(t\right) \\ \frac{d^3}{dt^3}u_1\left(t\right) & \frac{d^3}{dt^3}u_2\left(t\right) & \frac{d^3}{dt^3}u_3\left(t\right) \end{vmatrix}$$

Each gives

$$u_1\left(t\right)\frac{\partial u_2\left(t\right)}{\partial t}\frac{\partial^3 u_3\left(t\right)}{\partial t^3} - u_1\left(t\right)\frac{\partial u_3\left(t\right)}{\partial t}\frac{\partial^3 u_2\left(t\right)}{\partial t^3} - \frac{\partial u_1\left(t\right)}{\partial t}u_2\left(t\right)\frac{\partial^3 u_3\left(t\right)}{\partial t^3} +$$

$$\frac{\partial u_1\left(t\right)}{\partial t}u_3\left(t\right)\frac{\partial^3 u_2\left(t\right)}{\partial t^3} + \frac{\partial^3 u_1\left(t\right)}{\partial t^3}u_2\left(t\right)\frac{\partial u_3\left(t\right)}{\partial t} - \frac{\partial^3 u_1\left(t\right)}{\partial t^3}u_3\left(t\right)\frac{\partial u_2\left(t\right)}{\partial t}$$

5. The surface $z = \sin xy$ has ridges along the hyperbolas $xy = \frac{\pi}{2} + 2n\pi$ and valleys along the hyperbolas $xy = \frac{3\pi}{2} + 2n\pi$ in the first quadrant. The gradient $\nabla \sin xy = (y\cos xy, x\cos xy)$ produces a vector field whose vectors show the steepness of the surface $z = \sin xy$. Note that plotting the gradient of $f(x, y)$ is the same as plotting the vector field of $\nabla f(x, y)$. The ridges and valleys are indicated by vectors of zero length.

Surface Mesh: Mesh
Turn: -66   Tilt: 43

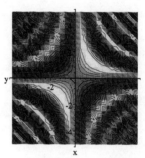

Surface Mesh: VLines
Turn: -90   Tilt: 1

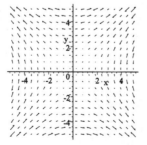

gradient of $\sin xy$

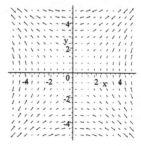

vector field of $(y\cos xy, x\cos xy)$

6. A plot of the vector field $(\sin xy, \cos xy)$ suggests an interesting pattern of flow. However, a search for a scalar potential fails.

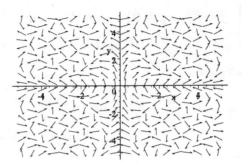

10 Differential Equations

A *differential equation* is an equation that includes differentials or derivatives of an unknown function. A *solution* to a differential equation is any function that satisfies the given equation. For example, $y = \sin x$ is a solution to the differential equation $y'' + y = 0$, because if $y = \sin x$, then $y' = \cos x$ and $y'' = -\sin x$, and hence $y'' + y = \sin x - \sin x = 0$. Differential equations are encountered in the study of problems in both pure and applied mathematics, in the sciences, in engineering, and in business and the social sciences.

Ordinary Differential Equations

With the choices on the Solve ODE submenu you will be able to find closed-form solutions to many differential equations. The solution is generally returned as an equation in $y(x)$ and x (or whatever variables were specified) with any arbitrary constants represented as $C_1, C_2, \ldots, C_n$.

▶ **To solve a differential equation**

1. Type the differential equation using standard mathematical notation.

2. With the insertion point in the equation, from the Solve ODE submenu, choose Exact or Laplace.

   or

   Put the differential equation in a vector or display, with initial condition(s) in separate rows; place the insertion point in the equation and, from the Solve ODE submenu, choose Numeric.

These different choices are explained in more detail in the ensuing sections.

Exact Solutions

Two methods, Exact and Laplace, return exact solutions to a linear differential equation. Laplace, which, as its name suggests, uses the Laplace transform to derive solutions, works for either homogeneous or nonhomogeneous linear differential equations with constant coefficients. Initial conditions are displayed in the solution. Exact is more general in the sense that it works for some nonlinear differential equations as well.

Exact Method

When a notation is used for differentiation that names the independent variable, the variable is taken from context.

▶ Solve ODE + Exact

$$\frac{dy}{dx} = xy, \text{ Exact solution is}: C_1 e^{\frac{1}{2}x^2}$$

To check this result, define $y(x) = e^{\frac{1}{2}x^2} C_1$. Replace y by $y(x)$ in the differential equation and evaluate both sides.

▶ Evaluate

$$\frac{dy(x)}{dx} = xe^{\frac{1}{2}x^2} C_1 \qquad xy(x) = xe^{\frac{1}{2}x^2} C_1$$

For any given number C_1, the solution describes a curve. Since C_1 may, in general, take on infinitely many values, there are an infinity of solution curves—or a *one-parameter family* of solution curves—for this equation.

When a prime indicates differentiation, the independent variable will be named if it is unambiguous; otherwise, a variable name must be specified. In the equations $y' = y$, $y' = \sin x$ and $y' = \sin x + t$, the independent variable is ambiguous and a dialog box appears asking for the independent variable.

▶ Solve ODE + Exact

$y' = \sin x$ (Specify x), Exact solution is: $C_1 - \cos x$

$y' = \sin x$ (Specify t), Exact solution is: $C_1 + (\sin x) t$

$y' = y$ (Specify t), Exact solution is: $C_1 e^t$

There is a family of solutions, one for each choice of the constant C_1. The following figure shows solutions for $y' = y$ corresponding to the choices $C_1 = \frac{1}{2}$, 1, 2, 3, and 4. To replicate this plot, drag solutions to the frame one at a time.

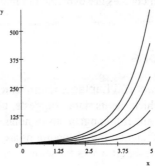

Solutions to $y' = y$: $\frac{1}{2}e^t$, e^t, $2e^t$, $3e^t$, $4e^t$

The program recognizes a variety of notations for a differential equation. The following examples illustrate some of this variety. The Leibniz notation $\frac{dy}{dx}$ and the D_x notation provide enough information so the independent variable can be determined by the computational engine.

▶ Solve ODE + Exact

$\frac{dy}{dx} = y + x$, Exact solution is: $e^{x - C_1} - x - 1$

$D_x y - y = \sin x$, Exact solution is: $C_1 e^x - \frac{1}{2}\sin x - \frac{1}{2}\cos x$

$y' + xy = ax$ (Independent Variable: x), Exact solution is: $\left(\dfrac{a}{a + e^{-C_2 - \frac{1}{2}x^2}} \right)$

Following is a plot of three particular solutions for $D_x y - y = \sin x$ corresponding to $C_1 = 1, 2, 3$. To replicate this plot, drag solutions to the frame one at a time. To distinguish the solutions, edit the colors or line styles on the Items Plotted page of the Plot Properties dialog.

▶ Plot 2D + Rectangular

$e^x - \frac{1}{2}\cos x - \frac{1}{2}\sin x$

• Select and drag to the frame each of the expressions $2e^x - \frac{1}{2}\cos x - \frac{1}{2}\sin x$ and $3e^x - \frac{1}{2}\cos x - \frac{1}{2}\sin x$

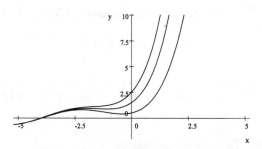

Three solutions for $D_x y - y = \sin x$

The three solutions can be distinguished by evaluation at 0. For example, the solution with $C_1 = 1$ crosses the y-axis at $y = \frac{1}{2}$.

▶ Solve ODE + Exact

$y'' + y = x^2$ (Specify x), Exact solution is: $C_1 \cos x - C_2 \sin x + x^2 - 2$

The following plot shows three solutions generated with constants $(C_1, C_2) = (1, 1)$, $(C_1, C_2) = (5, 1)$, and $(C_1, C_2) = (1, 5)$. To replicate this plot, drag solutions to the frame one at a time and then edit the colors or line styles on the **Items Plotted** page of the **Plot Properties** dialog.

▶ **Plot 2D + Rectangular**

$$\sin x + \cos x + x^2 - 2$$

- Select and drag to the frame the expressions $5 \sin x + \cos x + x^2 - 2$ and $\sin x + 5 \cos x + x^2 - 2$.

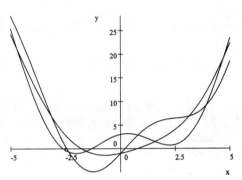

Three solutions to $y'' + y = x^2$

The particular solution $y(x) = \sin x + 5 \cos x + x^2 - 2$ is the one whose graph crosses the y-axis at $y = 3$.

▶ **Solve ODE + Exact**

$xy' - y = x^2$ (Specify x), Exact solution is: $C_1 x + x^2$

$\dfrac{d^2 y}{dx^2} + \dfrac{dy}{dx} = x + y$, Exact solution is: $C_1 e^{x\left(\frac{1}{2}\sqrt{5} - \frac{1}{2}\right)} - x + C_2 e^{x\left(-\frac{1}{2}\sqrt{5} - \frac{1}{2}\right)} - 1$

Some differential equations that are not readily solvable by this method can be solved after rewriting the equation, as shown in the following two examples.

Example The differential equation

$$\frac{dy}{dx} = \frac{1}{x^2 \sin y - xy}$$

can be solved by reversing the role of the two variables.

- **Solve ODE + Exact** $\dfrac{dx}{dy} = x^2 \sin y - xy$, Exact solution is:

$$\left(\begin{array}{c} 0 \\ \dfrac{1}{C_1 e^{\frac{1}{2}y^2} - e^{\frac{1}{2}y^2} \int \frac{\sin y}{e^{\frac{1}{2}y^2}} \, dy} \end{array} \right)$$

Example   The differential equation

$$\frac{dy}{dx} + \frac{xy}{1 - x^2} = x\sqrt{y}$$

can be solved after a change of variable. Take $\sqrt{y} = z$ to get $2z\frac{dz}{dx} + \frac{xz^2}{1 - x^2} = xz.$

- Solve ODE + Exact, Simplify $\frac{dz}{dx} + \frac{xz}{2(1 - x^2)} = \frac{x}{2},$

  Exact solution is: $\frac{1}{3}x^2 + C_1\sqrt[4]{x^2 - 1} - \frac{1}{3}$

Thus $y(x) = \left(\frac{1}{3}x^2 + C_1\sqrt[4]{x^2 - 1} - \frac{1}{3}\right)^2.$

Laplace Method

Laplace transforms solve either homogeneous or nonhomogeneous linear systems in which the coefficients are all constants. Initial conditions appear explicitly in the solution.

▶ Solve ODE + Laplace

$\frac{dy}{dx} = y$, Laplace solution is: $y(x) = e^x y(0)$

$y' + y = x + \sin x$ (Specify x),

$\qquad$ Laplace solution is: $x - \frac{1}{2}\cos x + \frac{1}{2}\sin x + e^{-x}\left(y(0) + \frac{3}{2}\right) - 1$

The following examples compare exact and Laplace solutions.

Equation	Exact	Laplace
$y' = \sin x$	$y(x) = C_1 - \cos x$	$y(x) = y(0) - \cos x + 1$
$D_x y = x + t$	$y(x) = C_1 + tx + \frac{1}{2}x^2$	$y(x) = tx + y(0) + \frac{1}{2}x^2$
$\frac{dy}{dx} = y$	$y(x) = C_1 e^x$	$y(x) = e^x y(0)$

Equation (Independent Variable: x)	Exact	Laplace
$y' = y^2 + 1$	$\begin{cases} \tan\left(\frac{1}{2}\pi + C_{35} + x\right) \\ -i \\ i \end{cases}$	Fails
$(y')^3 - 3(y')^2 + 2y' = 0$	$\begin{cases} C_1 \\ C_1 + x \\ C_1 + 2x \end{cases}$	Fails

Series Solutions

For many applications requiring a solution to a differential equation, a few terms of a Taylor series solution are sufficient. You can control the number of terms that appear in the solution by changing Series Order for ODE Solutions in the Engine Setup dialog. For most of the following examples, the series order has been set at 6.

▶ **To specify the order of terms in a series solution**

1. Choose Tools + Engine Setup, General page.

2. Set Series Order for ODE Solutions to the desired order.

3. Choose OK.

In the following examples, notice that the initial condition $y(0)$ appears explicitly in each solution. Choose Series from the Solve ODE submenu to produce the following solutions.

▶ Solve ODE + Series

$D_x y = y$, Series solution is : $y(x) = y(0) + y(0)x + \left(\frac{1}{2}y(0)\right)x^2 + \left(\frac{1}{6}y(0)\right)x^3 + \left(\frac{1}{24}y(0)\right)x^4 + \left(\frac{1}{120}y(0)\right)x^5 + O\left(x^6\right)$

$y' = \dfrac{\sin x}{x}$ (Specify x), Series solution is : $y(x) = y(0) + x - \frac{1}{18}x^3 + \frac{1}{600}x^5 + O\left(x^6\right)$

Heaviside and Dirac Functions

Laplace and Fourier transforms interact closely with the Heaviside unit-step function and the Dirac unit-impulse function. The Dirac and Heaviside functions are related by

$$\int_{-\infty}^{x} \text{Dirac}(t)\, dt = \text{Heaviside}(x) \quad \text{and} \quad \frac{d}{dx}\text{Heaviside}(x) = \text{Dirac}(x)$$

The Dirac function is not a function in the usual sense. It represents an infinitely short, infinitely strong unit-area impulse. It satisfies $\text{Dirac}(x) = 0$ if $x \neq 0$ and can be obtained as the limit of functions $f_n(x)$ satisfying $\int_{-\infty}^{\infty} f_n(x)\, dx = 1$. The Heaviside function equals 0 for $x < 0$ and 1 for $x \geq 0$.

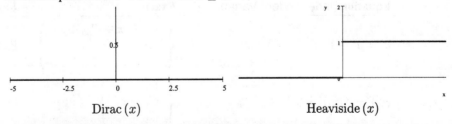

$\text{Dirac}(x)$ $\text{Heaviside}(x)$

▶ **To enter the name of the Dirac or Heaviside function**

1. Click [sin cos], or choose Insert + Math Name.

2. Enter the function name in the Name box using upper- and lowercase letters as they appear above, and choose OK.

If you prefer to work with shorter names, you can define them as follows. Results of computations will, however, return the long name.

▶ Definitions + New Definition

$$\delta(x) = \text{Dirac}(x) \qquad\qquad H(x) = \text{Heaviside}(x)$$

You can test your definition of $\delta(x)$ by calculating an appropriate integral:

▶ Evaluate

$$\int_{-1}^{1} \delta(x)\ dx = 1$$

You can test your definition of $H(x)$ by calculating an appropriate derivative:

▶ Evaluate

$$\tfrac{d}{dx} H(x) = \text{Dirac}(x)$$

You can create characteristic functions with the Heaviside function. For example, Heaviside $(1-x)$ Heaviside$(2+x)$ gives the function that is 1 on the interval $[-2, 1]$ and 0 elsewhere.

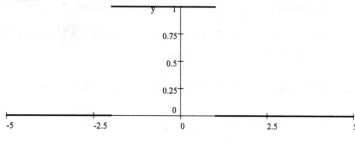

$$\text{Heaviside}(1-x)\,\text{Heaviside}(2+x)$$

Laplace Transforms

If f is a function on $[0, \infty]$, the function $\mathcal{L}(f) = \hat{f}$ defined by the integral

$$\hat{f}(s) = \mathcal{L}(f(t), t, s) = \int_{0}^{\infty} e^{-st} f(t)\, dt$$

for those values of s for which the integral converges is the *Laplace transform* of f. The Laplace transform depends on the function f and the number s. The equation above also defines the Laplace operator $\mathcal{L}$.

A constant coefficient linear differential equation in $f(t)$ is transformed into an algebraic equation in $\hat{f}(s)$ by the operator $\mathcal{L}$. A solution can be found to the differential equation by first solving the algebraic equation to find $\hat{f}(s)$ and then applying the *inverse Laplace transform* to determine $f(t)$ from $\hat{f}(s)$.

Expressions involving exponentials, polynomials, trigonometrics (sin, cos, sinh, cosh) with linear arguments, and Bessel functions (BesselJ, BesselI) with linear arguments can be transformed. The Laplace transform also recognizes derivatives and integrals, the Heaviside unit-step function Heaviside (x) and the Dirac-delta unit impulse function Dirac (x).

Computing Laplace Transforms

You can compute a Laplace transform of an expression in the variable t using a command on the Transforms submenu.

▶ Transforms + Laplace

t, Laplace transform is: $\frac{1}{s^2}$

$t^{\frac{3}{2}} - e^t + \sinh at$, Laplace transform is: $\frac{a}{s^2-a^2} - \frac{1}{s-1} + \frac{3}{4}\frac{\sqrt{\pi}}{s^{\frac{5}{2}}}$

te^{-at} Heaviside(t), Laplace transform is: $\frac{1}{(s+a)^2}$

You can also compute a Laplace transform using the symbol $\mathcal{L}$.

▶ **To compute a Laplace transform using the symbol $\mathcal{L}$**

1. From the Miscellaneous Symbols panel under $\boxed{\infty\partial}$, choose $\mathcal{L}$, select the symbol, and change it to mathematics mode.

2. Click $\boxed{(\square)}$, or from the Insert + Brackets panel choose ( ).

3. Inside the parentheses, enter an expression in the variable t

   or

   Enter, separated by commas,

   a. An expression for the object function.
   b. The variable of integration.
   c. The variable for the transform.

4. Choose Evaluate.

The default variable of integration is t and the default transform variable is s. The computing engine evaluates a Laplace transform with t as input variable and produces a solution using the variable s.

▶ Evaluate

$\mathcal{L}(t) = \frac{1}{s^2}$ $\mathcal{L}(t^3) = \frac{6}{s^4}$ $\mathcal{L}(\delta(t)) = 1$

$\mathcal{L}(3\sin t) = \frac{3}{s^2+1}$ $\mathcal{L}(t^5) = \frac{120}{s^6}$ $\mathcal{L}(H(t)) = \frac{1}{s}$

You can use other variable names by specifying the variable of integration and the transform variable as in the following examples.

$$\mathcal{L}(x, x, y) = \frac{1}{y^2} \qquad \mathcal{L}(e^{-\alpha t} H(t), t, s) = \frac{1}{s+\alpha}$$

$$\mathcal{L}(3\sin x, x, s) = \frac{3}{s^2+1} \qquad \mathcal{L}(te^{-\alpha t} H(t), t, s) = \frac{1}{(s+\alpha)^2}$$

Computing Inverse Laplace Transforms

You can compute an inverse Laplace transform of an expression in the variable s from the Transforms submenu.

▶ Transforms + Inverse Laplace

$\frac{2}{s^3}$, Is Laplace transform of t^2 $\frac{1}{s+\alpha}$, Is Laplace transform of $e^{-t\alpha}$

1, Is Laplace transform of Dirac (t) $\frac{120}{s^6}$, Is Laplace transform of t^5

You can also compute an inverse Laplace transform by evaluating an expression of the form $\mathcal{L}^{-1}(f(s))$.

▶ **To compute a Laplace transform using the symbol $\mathcal{L}$**

1. From the Miscellaneous Symbols panel under [∞∂], choose $\mathcal{L}$, select the symbol, and change it to mathematics mode.

2. Insert the superscript -1.

3. Click [(□)] or, from the Insert + Brackets panel, choose ( ).

4. Inside the parentheses, enter an expression in the variable s

   or

   Enter, separated by commas,

   a. An expression for the transform.
   b. The variable of the transform.
   c. The variable for the object function.

5. Choose Evaluate.

The default variable for the transform is s, and the default object function variable is t. The inverse Laplace transform will correctly interpret an expression with these variables as input.

▶ Evaluate

$$\mathcal{L}^{-1}\left(\tfrac{1}{s^2}\right)=t \qquad \mathcal{L}^{-1}(1)=\mathrm{Dirac}\,(t) \qquad \mathcal{L}^{-1}\left(\tfrac{3}{s^2+1}\right)=3\sin t$$

For other variable names, the variable of integration and the transform variable must be specified, as in the following examples.

▶ Evaluate

$$\mathcal{L}^{-1}\left(\tfrac{1}{x+a},x,y\right)=e^{-\alpha y} \qquad \mathcal{L}^{-1}\left(\tfrac{2}{s^3},s,x\right)=x^2 \qquad \mathcal{L}^{-1}\left(\tfrac{120}{y^6},y,x\right)=x^5$$

If the range of parameters must be restricted, you can use the functions assume and additionally. (See page 260 for details.)

▶ Evaluate

$\mathrm{assume}\,(n>0)$

$\mathrm{additionally}\,(n,\mathrm{integer})$

$$\mathcal{L}\left(\tfrac{t^{n-1}}{(n-1)!}H\,(t)\,,t,s\right)=\tfrac{1}{(n-1)!}s^{-n}\Gamma\,(n)$$

The following two examples demonstrate the use of the Laplace Transform to solve a differential equation.

Example In order to solve the problem
$$f'+af=0,\,f\,(0)=b$$
use Definitions + New Definition to define $f\,(t)$ as a generic function and both a and b as a generic constants. Then evaluate both sides of the equation:
$$\mathcal{L}\,(f'+af)=\mathcal{L}\,(0)$$
to get
$$s\mathcal{L}\,(f)-f\,(0)+a\mathcal{L}\,(f)=0$$
Solve this equation for $\mathcal{L}\,(f)$ to get
$$\mathcal{L}\,(f)=\frac{b}{s+a}$$
Now use the inverse Laplace transform to get
$$f\,(t)=\mathcal{L}^{-1}\left(\frac{b}{s+a}\right)=be^{-ta}$$
Check: Define $f\,(t)=be^{-ta}$ and evaluate $f'\,(t)+af\,(t)$ and $f\,(0)$ to get
$$f'\,(t)+af\,(t)=0$$
$$f\,(0)=b$$

Example   Consider the second-order differential equation
$$y'' + y = 0$$
with the initial conditions $y(0) = 1$ and $y'(0) = -2$. Define $y(t)$ as a generic function and apply Evaluate to $\mathcal{L}(y''(t) + y(t), t, s)$ to get
$$\mathcal{L}(y''(t) + y(t), t, s) = s(s\mathcal{L}(y) - y(0)) - y'(0) + \mathcal{L}(y) = 0$$
Solve the equation
$$s(s\mathcal{L}(y) - y(0)) - y'(0) + \mathcal{L}(y) = 0$$
for $\mathcal{L}(y)$ with Solve + Exact to get
$$\mathcal{L}(y) = \frac{sy(0) + y'(0)}{s^2 + 1}$$
Replace $y(0)$ with 1 and $y'(0)$ with -2 to get
$$\mathcal{L}(y) = \frac{s - 2}{s^2 + 1}$$
Now take the inverse Laplace transform by applying Evaluate to $\mathcal{L}^{-1}\left(\frac{s-2}{s^2+1}\right)$ (or Transforms + Inverse Laplace to $\frac{s-2}{s^2+1}$) to get
$$\mathcal{L}^{-1}\left(\frac{s-2}{s^2+1}\right) = \cos t - 2\sin t$$
Check: If $y(t) = \cos t - 2\sin t$, then $y''(t) = -\cos t + 2\sin t$, and $y''(t) + y(t) = 0$, $y(0) = 1$, and $y'(0) = -2$.

Fourier Transforms

Fourier transforms provide techniques for solving problems in linear systems and provide a unifying mathematical approach to the study of diverse fields including electrical networks and information theory.

If f is a real-valued function on $[-\infty, \infty]$, the function $\hat{f} = \mathcal{F}(f)$ defined by the integral
$$\hat{f}(w) = \mathcal{F}(f(x), x, w) = \int_{-\infty}^{\infty} e^{iwx} f(x)\, dx$$
for those values of w for which the integral converges is the *Fourier transform* of f; that is, it is the integral transform with kernel $K(w, t) = e^{-iwt}$ or $K(w, t) = e^{iwt}$. The Fourier transform depends on the function f and the number w.

Computing Fourier Transforms

You can compute a Fourier transform of an expression in the variable x from the Transforms submenu.

▶ Transforms + Fourier

1, Fourier transform is: $2\pi \operatorname{Dirac}(w)$

$\operatorname{Dirac}(x)$, Fourier transform is: 1

$\operatorname{Heaviside}(x)$, Fourier transform is: $\pi \operatorname{Dirac}(w) + \frac{i}{w}$

e^{-ix}, Fourier transform is: $2\pi \operatorname{Dirac}(w-1)$

You can also compute a Fourier transform using the symbol $\mathcal{F}$.

▶ **To compute a Fourier Transform using the symbol $\mathcal{F}$**

1. From the Symbol Panels toolbar under ⬛, choose $\mathcal{F}$, select the symbol, and change it to mathematics mode.

2. Click (□) or, from the Insert + Brackets panel, choose ().

3. Inside the parentheses,
   - Enter an expression for the object function in terms of the variable x.
     or
   - Enter, separated by commas,
       i   An expression for the object function.
       ii The variable of integration (that is, the variable of the object function).
       iii The variable for the transform.

4. Choose Evaluate.

▶ Evaluate

$$\mathcal{F}(1) = 2\pi \operatorname{Dirac}(w) \qquad\qquad \mathcal{F}(\operatorname{Dirac}(x)) = 1$$

$$\mathcal{F}(1, t, w) = 2\pi \operatorname{Dirac}(w) \qquad\quad \mathcal{F}(\operatorname{Dirac}(t), t, w) = 1$$

$$\mathcal{F}(\operatorname{Heaviside}(t), t, w) = \pi \operatorname{Dirac}(w) + \frac{i}{w} \qquad \mathcal{F}(1/x, t, w) = 2\frac{\pi}{x} \operatorname{Dirac}(w)$$

$$\mathcal{F}(1/x) = i\pi (2 \operatorname{Heaviside}(w) - 1)$$

Computing Inverse Fourier Transforms

You can compute an inverse Fourier transform of an expression in the variable w from the Transforms submenu.

▶ Transforms + Inverse Fourier

$2\pi\,\mathrm{Dirac}\,(w)$, Is Fourier transform of 1 $\qquad$ 1, Is Fourier transform of $\mathrm{Dirac}\,(x)$

$\pi\,\mathrm{Dirac}\,(w) + \frac{i}{w}$, Is Fourier transform of $\frac{1}{2\pi}\,(\pi - \pi\,(2\,\mathrm{Heaviside}\,(-x) - 1))$

$2\pi\,\mathrm{Dirac}\,(w + 1)$, Is Fourier transform of e^{ix}

$2\pi\,\mathrm{Dirac}\,(w - 2\pi)$, Is Fourier transform of $e^{-2i\pi x}$

You can also compute an inverse Fourier transform using the symbol $\mathcal{F}^{-1}$.

▶ **To compute an inverse Fourier transform using the symbol $\mathcal{F}^{-1}$**

1. From the **Symbol Panels** toolbar under $\boxed{\infty\!\partial}$, choose $\mathcal{F}$, select the symbol and put it in mathematics mode.

2. Click $\boxed{\mathsf{N}^{\times}}$ or from the **Insert** menu choose **Superscript**.

   a. In the input box, enter -1.

   b. Press the SPACE BAR to move from the input box.

3. Click $\boxed{(\square)}$ or, from the **Insert + Brackets** panel, choose ( ).

4. Inside the parentheses,

   • Enter an expression for the transform in terms of the variable w.

   or

   • Enter, separated by commas,

       i An expression for the transform.

       ii The variable of the transform.

       iii The variable for the object function.

5. Choose **Evaluate**.

The transform variable must be w or the variable of integration and the transform variable must be specified, as in the following examples.

▶ Evaluate

$\mathcal{F}^{-1}\,(2\pi\,\mathrm{Dirac}\,(w)) = 1$ $\qquad\qquad$ $\mathcal{F}^{-1}\,(2\pi\,\mathrm{Dirac}\,(h)\,,h,s) = 1$

$\mathcal{F}^{-1}\,\left(\pi\,\mathrm{Dirac}\,(s) - \frac{i}{s}, s, h\right) = \frac{1}{2\pi}\,(\pi + \pi\,(2\,\mathrm{Heaviside}\,(-h) - 1))$

For some of these expressions, **Simplify** gives a better form for the solution.

▶ Simplify

$$\mathcal{F}^{-1}\left(-i\pi\delta\left(-\omega+\omega_0\right)+i\pi\delta\left(\omega+\omega_0\right),\omega,t\right)=\tfrac{1}{2}e^{it\omega_0}\left(i-ie^{-2it\omega_0}\right)$$

To compute the transforms and inverse transforms of multiple expressions, enter the expressions in a single column matrix and evaluate.

▶ Evaluate

$$\mathcal{F}\left(\begin{array}{c}e^{2\pi ix}\\2\pi\,\mathrm{Dirac}\left(x-2\pi\right)\end{array}\right)=\left(\begin{array}{c}2\pi\,\mathrm{Dirac}\left(w-2\pi\right)\\2\pi e^{-2i\pi w}\end{array}\right)$$

Systems of Ordinary Differential Equations

Systems consisting of more than one equation are handled in a consistent manner. Such problems include initial-value problems and systems of differential equations.

Exact Solutions

The statement of some problems requires more than one equation. You enter systems with initial conditions, systems of differential equations, boundary-value problems, or a mixture of these problems using $n \times 1$ matrices, where n is the number of equations and conditions involved. You can also click the Display button ▤ and enter such systems in a multiline display.

▶ **To enter and solve a system of differential equations**

1. Click ▦ or, from the Insert menu, choose Matrix.

2. Select 1 column, and set the number of rows equal to the number of equations.

3. Choose OK.

4. From the View menu, choose Matrix Lines and Input Boxes (unless Matrix Lines and Input Boxes are already visible) to show where to enter the required equations.

5. Enter the equations.

6. Leave the insertion point in the matrix.

7. From the Solve ODE submenu, choose Exact or Laplace.

▶ Solve ODE + Exact, Expand

$$y' + y = x$$
$$y(0) = 1 \quad \text{(Specify } x\text{)},$$

Exact solution is : $x + e^{\ln 2 - x} - 1 = x + \frac{2}{e^x} - 1$

To solve the second-order initial-value problem $y'' + y = x^2$, $y(0) = 1$, $y'(0) = 1$, enter these three equations into a 3×1 matrix and choose **Laplace** from the **Solve ODE** submenu.

▶ Solve ODE + Laplace

$$y'' + y = x^2$$
$$y(0) = 1 \quad \text{(Specify } x\text{)},$$
$$y'(0) = 1$$

$$x^2 - 2 + \sin x + 3 \cos x$$

The following examples illustrate some of the different notations you can use for entering and solving systems of differential equations.

▶ Solve ODE + Laplace

$$\frac{dy}{dx} = \sin x \quad \text{, Laplace solution is: } 2 - \cos x$$
$$y(0) = 1$$

$$D_{xx}y - y = 0$$
$$y(0) = 1 \text{ , Laplace solution is: } \cosh x$$
$$y'(0) = 0$$

A new independent variable is introduced in certain instances where none is provided.

▶ Solve ODE + Exact

$$y' = x$$
$$x' = -y \quad \text{(Specify } t\text{)},$$

Exact solution is: $\left[x(t) = C_6 e^{it} + C_7 e^{-it}, y(t) = iC_7 e^{-it} - iC_6 e^{it} \right]$

Notice that an exact solution to this problem involves a two-parameter family of solutions.

▶ Solve ODE + Laplace

$y' = x$
$x' = -y$
$x(0) = 0$ (Specify t), Laplace solution is: $[y(t) = \cos t, x(t) = -\sin t]$
$y(0) = 1$

Subscripted dependent variables are allowed.

▶ Solve ODE + Laplace

$D_{xx}y_1 - y_1 = 0$
$\quad y_1(0) = 1$, Laplace solution is: $\cosh x$
$\quad y_1'(0) = 0$

The next two examples show solutions using **Exact** for nonlinear equations. The command **Laplace** produces no result for these equations, as Laplace transforms are appropriate for linear equations only. The command **Series** also fails with the second example, because $\ln x$ does not have a series expansion about $x = 0$ in powers of x.

▶ Solve ODE + Exact

$y' = y^2 + 4$
$y(0) = -2$ (Specify t), Exact solution is: $2\tan\left(2t - \frac{1}{4}\pi\right)$

$(x+1)y' + y = \ln x$
$\quad y(1) = 10$ (Specify x), Exact solution is: $-\frac{1}{x+1}(x - x\ln x - 21)$

Series Solutions

The following examples illustrate series solutions to two types of systems of differential equations. You can control the number of terms that appear in the solution by changing Series Order for ODE Solutions in the Engine Setup dialog under Tools. For the following examples, the series order has been set at 6.

▶ Solve ODE + Series

$D_{xx}y_1 - y_1 = 0$
$\quad y_1(0) = 1$, Series solution is: $1 + \frac{1}{2}x^2 + \frac{1}{24}x^4 + O(x^5)$
$\quad y_1'(0) = 0$

$y' = y^2 + 4$
$y(0) = -2$ (Specify t), Series solution is: $-2 + 8t - 16t^2 + \frac{128}{3}t^3 - \frac{320}{3}t^4 + O(t^5)$

Numerical Methods For ODE's

Appropriate systems can be solved numerically. These numeric solutions are functions that can be evaluated at points or plotted.

Initial-Value Problems

▶ **To solve an initial-value problem numerically**

1. Start with a column matrix and enter an initial-value problem, such as

$$y' = -y$$
$$y(0) = 1$$

with one equation per row.

2. From the Solve ODE submenu, choose Numeric.

▶ Solve ODE + Numeric

$\begin{aligned} y' &= -y \\ y(0) &= 1 \end{aligned}$, Functions defined: y

This calculation defines a function y that can be evaluated at given arguments. You can use the function to generate a table of values, and as you will see in the next section, the function can be plotted.

▶ Evaluate

$y(1) = 0.36788$ $y(10.7) = 2.2543 \times 10^{-5}$

▶ **To generate a table of function values for a function y**

1. Define the function $g(i) = 0.1i$ and, from the Matrix submenu, choose Fill Matrix.

2. In the dialog box, select 3 rows and 1 column, select Defined by function and enter the function name g.

3. Select the column and enclose it with brackets and place y at the left of the column and outside of the brackets.

4. Choose Evaluate.

$$y \begin{bmatrix} 0.1 \\ 0.2 \\ 0.3 \end{bmatrix} = \begin{bmatrix} 0.90484 \\ 0.81873 \\ 0.74082 \end{bmatrix}$$

This calculation generates a list of function values for y as x varies from 0.1 to 1. As a check, solve the initial-value problem exactly.

▶ Solve ODE + Exact

$$y' + y = 0$$
$$y(0) = 1$$ (Specify t), Exact solution is: e^{-t}

Compute the function values for $y(t) = e^{-t}$ for the same arguments as before. You will find that they agree exactly, at least to the indicated precision.

$$y \begin{bmatrix} 0.1 \\ 0.2 \\ 0.3 \end{bmatrix} = \begin{bmatrix} 0.904\,84 \\ 0.818\,73 \\ 0.740\,82 \end{bmatrix}$$

Graphical Solutions to Initial-Value Problems

Numerical solutions can be plotted. To view the solution to the initial-value problem $y' = \sin xy$, $y(0) = 3$, enter the two equations into a 2×1 matrix, and choose Numeric from the Solve ODE submenu.

▶ Solve ODE + Numeric

$$y' = \sin xy$$
$$y(0) = 3$$, Functions defined: y

Now plot y by choosing Rectangular or ODE from the Plot 2D submenu.

▶ Plot 2D + Rectangular

y

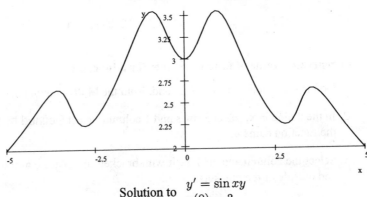

Solution to $\begin{aligned} y' &= \sin xy \\ y(0) &= 3 \end{aligned}$

▶ Plot 2D + ODE

y

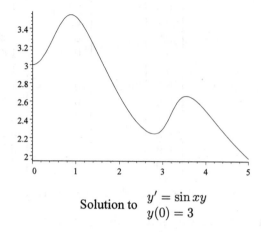

Solution to $\quad \begin{array}{l} y' = \sin xy \\ y(0) = 3 \end{array}$

Numerical Solutions to Systems of Differential Equations

You can solve the following system numerically by entering the equations into a 6×1 matrix and choosing **Numeric** from the **Solve ODE** submenu. Three functions x, y, and z are returned as output.

▶ Solve ODE + Numeric

$$\begin{array}{l} x' = x + y - z \\ y' = -x + y + z \\ z' = -x - y + z \\ x(0) = 1 \\ y(0) = 1 \\ z(0) = 1 \end{array}$$, Functions defined: x, y, z

The following table lists values of x, y, and z as the independent variable t varies from 0 to 1.

t	x	y	z
0	1.0000	1.0000	1.0000
0.1	1.1158	1.0938	0.8842
0.2	1.2668	1.1695	0.7332
0.3	1.4582	1.2173	0.5418
0.4	1.6953	1.2253	0.3047
0.5	1.9830	1.1791	0.0170
0.6	2.3256	1.0619	−0.3256
0.7	2.7265	0.8542	−0.7265
0.8	3.1873	0.5344	−1.1873
0.9	3.7077	0.0777	−1.7077
1.0	4.2842	−0.5424	−2.2842

You can create a matrix with these values as follows.

1. Apply **Evaluate** to x
$$\begin{bmatrix} 0.0 \\ 0.1 \\ 0.2 \\ 0.3 \\ 0.4 \\ 0.5 \\ 0.6 \\ 0.7 \\ 0.8 \\ 0.9 \\ 1.0 \end{bmatrix} \text{to get } x \begin{bmatrix} 0.0 \\ 0.1 \\ 0.2 \\ 0.3 \\ 0.4 \\ 0.5 \\ 0.6 \\ 0.7 \\ 0.8 \\ 0.9 \\ 1.0 \end{bmatrix} = \begin{bmatrix} 1.0 \\ 1.1158 \\ 1.2668 \\ 1.4582 \\ 1.6953 \\ 1.983 \\ 2.3256 \\ 2.7265 \\ 3.1873 \\ 3.7077 \\ 4.2843 \end{bmatrix}$$

2. Similarly, apply **Evaluate** to get the y and z columns.

3. To create a matrix with all four columns, place the **t, x, y,** and **z** columns next to one another and, from the **Matrices** submenu, choose **Concatenate**.

4. To add a row at the top for labels, select the matrix and choose **Edit + Insert Rows**.

5. To line up entries, select a column, choose **Edit + Properties**, and change **Column Alignment** to **Aligned Left** or **Aligned Right**.

Tip   You must select only the matrix, not including brackets, to have **Insert Rows** appear on the **Edit** menu.

You can also take advantage of the fact that you are using a word processor to put the values into a 12×4 table. To make a table that will print with lines, click ⊞ on the **Standard** toolbar or choose **Insert + Table**. Copy the information into the table by selecting, clicking and dragging each piece of data. Choose **Edit + Properties** and add lines according to instructions in the **Table Properties** dialog. This is only for purpose of creating a special appearance—a table does not behave mathematically as a matrix.

Graphical Solutions to Systems of ODEs

You can create and plot matrices for x, y, and z as follows.

1. Concatenate the columns for t and x to generate the matrix
$$\begin{bmatrix} 0.0 & 1.0 \\ 0.1 & 1.1158 \\ 0.2 & 1.2668 \\ 0.3 & 1.4582 \\ 0.4 & 1.6953 \\ 0.5 & 1.983 \\ 0.6 & 2.3256 \\ 0.7 & 2.7265 \\ 0.8 & 3.1873 \\ 0.9 & 3.7077 \\ 1.0 & 4.2843 \end{bmatrix}.$$

2. Plot the matrix using Plot 2D + Rectangular.

3. Generate a similar matrix using t and y and drag it to the plot frame.

4. Generate a similar matrix using t and z and drag it to the plot frame.

5. Open the Plot Properties dialog, and choose the Items Plotted page. Note that a matrix for x appears as Item number 1.

6. Choose Item Number 1, which contains a matrix for x, and change Line Thickness to Medium.

7. Choose Item Number 2, which contains a matrix for y, and change Line Style to Dash. Click OK.

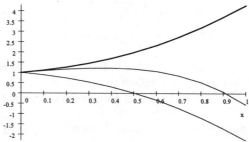

Note that the numeric output for t between 0 and 1 does not predict long-range behavior. This system of differential equations describes a highly unstable system.

Bessel Functions

The Bessel functions are rather complicated oscillatory functions with many interesting properties. The functions $I_v(z)$ and $K_v(z)$ (or BesselI$_v(z)$ and BesselK$_v(z)$) are solutions known as first and second kind, respectively, to the modified Bessel equation

$$z^2 \frac{d^2 w}{dz^2} + z \frac{dw}{dz} - \left(z^2 + v^2\right) w = 0$$

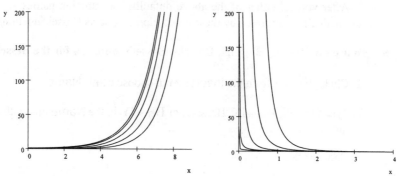

BesselI$_v(z)$, $\nu = 0, 1, 2, 3, 4$ BesselK$_v(z)$, $v = 0, 1, 2, 3, 4$

The functions $J_v(z)$ and $Y_v(z)$ (or BesselJ$_v(z)$ and BesselY$_v(z)$) are solutions of the first and second kind, respectively, to the Bessel equation

$$z^2 \frac{d^2 w}{dz^2} + z \frac{dw}{dz} + \left(z^2 - v^2\right) w = 0$$

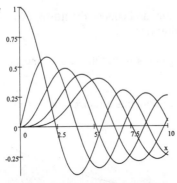

BesselJ$_v(z)$, $v = 0, 1, 2, 3, 4$ BesselY$_v(z)$, $v = 0, 1, 2, 3, 4$

You can reset the default so that I_v, J_v, K_v, Y_v will be interpreted as Bessel functions, either globally (for all documents) or locally (for one document).

▶ **To compute with the I, J, K, Y notation globally**

1. From the Tools menu, choose Computation Setup and select the General page.

2. Under Bessel Function Notation, check Use I, J, K, and Y with Subscripts.

This action sets a default that affects all documents.

▶ **To compute with the I, J, K, Y notation locally**

1. From the Compute menu, choose Settings and select the General page.

2. Check Set Document Values, and under Bessel Function Notation, check Use I, J, K, and Y with Subscripts.

This default will be saved with your document, and will not affect the behavior of the system for other documents.

After you set either of the above defaults, the function names I_v, J_v, K_v, Y_v are automatically interpreted by the computation engine as Bessel functions.

▶ **To use the BesselI, BesselK, BesselJ, BesselY notation for the Bessel functions**

1. Click [sin cos] or, from the Insert menu, choose Math Name.

2. Type BesselI, BesselK, BesselJ, or BesselY in the Name box with capital letters as indicated.

3. Choose OK.

4. Enter a subscript. Enter an argument enclosed in parentheses.

These custom names are automatically interpreted as Bessel functions.
Following are 3D plots of the Bessel functions.

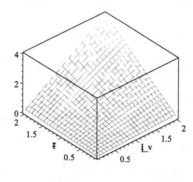

$$\mathrm{BesselI}_v\,(z) = I_v(z)$$

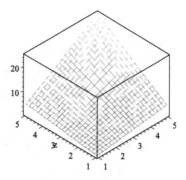

$$\mathrm{BesselK}_v\,(z) = K_v(z)$$

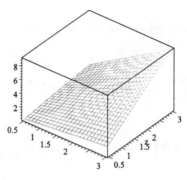

$$\mathrm{BesselJ}_v\,(z) = J_v(z)$$

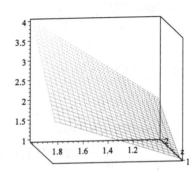

$$\mathrm{BesselY}_v\,(z) = Y_v\,(z)$$

Exercises

1. Find the general solution of the equation $y'' - 6y' + 5y = 0$.

2. Find the general solution of the equation $x^2 y'' - 3xy' - 6y = 0$.

3. Find the general solution of the equation $2x^2 y' = xy + 3y^2$.

4. Solve the initial-value problem $y' + y = 2$, $y(0) = 0$.

5. Solve the initial-value problem $\frac{dy}{dx} - y + 3 = 0$, $y(0) = 1$.

6. Newton's law of cooling states that the rate of change in the temperature of an object is given by $\frac{dT}{dt} = k(T - R)$, where k is a constant that depends on how well insulated the object is, T is the temperature of the object, and R is room temperature. A cup of coffee is initially $160°$; 10 minutes later, it is $120°$. Assuming the room temperature

is a constant $70°$, give a formula for the temperature at any time t. What will the temperature of the coffee be after 20 minutes?

Solutions

1. Solve ODE + Exact

$$y'' - 6y' + 5y = 0, \text{ Exact solution is: } C_1 e^{5t} + C_2 e^t$$

2. Solve ODE + Exact (Specify x)

$$x^2 y'' - 3xy' - 6y = 0, \text{ Exact solution is: } y(x) = C_1 x^{2+\sqrt{10}} + C_2 x^{2-\sqrt{10}}$$

3. Solve ODE + Exact (Specify x)2

$$2x^2 y' = xy + 3y^2, \text{ Exact solution is: } \left[\begin{array}{c} \frac{x}{9C_{16}-9x} \left(3C_{16} - 3\sqrt{C_{16}x} \right) \\ \frac{x}{9C_{16}-9x} \left(3C_{16} + 3\sqrt{C_{16}x} \right) \end{array} \right]$$

4. Solve ODE + Laplace

$$\begin{array}{c} y' + y = 2 \\ y(0) = 0 \end{array}, \text{ Laplace solution is: } y(t) = 2 - 2e^{-t}$$

5. Solve ODE + Exact

$$\begin{array}{c} \frac{dy}{dx} - y + 3 = 0 \\ y(0) = 1 \end{array} \quad \text{Exact solution is: } 3 - 2e^x$$

6. Solve ODE + Exact

$$\frac{dT}{dt} = k(T - 70), \text{ Exact solution is: } T(t) = \left[\begin{array}{c} 70 \\ C_{18} e^{kt} + 70 \end{array} \right]$$

Solve + Exact

$$\begin{array}{c} 160 = 70 + e^{k(0)} C_1 \\ 120 = 70 + e^{k(10)} C_1 \end{array}, \text{ Solution is: } \left[C_1 = 90, k \in \left\{ \frac{1}{5} i\pi X_{219} + \frac{1}{10} \left(\ln \frac{5}{9} \right) \mid X_{219} \in \mathbb{Z} \right\} \right]$$

Definitions + New Definition

$$\begin{array}{rcl} C_1 & = & 90 \\ k & = & \frac{1}{10} \ln \frac{5}{9} \\ T(t) & = & 70 + e^{kt} C_1 \end{array}$$

Evaluate

$$T(t) = 90 \exp \left(t \left(\frac{1}{10} \ln 5 - \frac{1}{10} \ln 9 \right) \right) + 70$$

Evaluate Numerically

$$T(20) = 97.8°$$

11 Statistics

Statistics is the science and art of obtaining and analyzing quantitative data in order to make sound inferences in the face of uncertainty. The word *statistics* is used both to refer to a set of quantitative data and to a field of study. The field includes the development and application of effective methods for obtaining and using quantitative data.

Introduction to Statistics

You can perform statistical operations on data with the various items on the Statistics submenu. In addition to the menu items, you will find a number of the standard statistical distribution functions and densities available as functions, either built in or definable.

The items Mean, Median, Mode, Moment, Quantile, Mean Deviation, Standard Deviation, and Variance on the Statistics submenu take a single argument that can be presented as a list of data or as a matrix. The result of an operation is a number, or, in the case of a matrix or vector, a number for each column.

The items Correlation, Covariance, and Fit Curve to Data on the Statistics submenu take a single argument that must be a matrix. For the multiple regression curve-fitting commands, the columns must be labeled with variable names. The menu item Random Numbers on the Statistics submenu allows you to get random samples from families of distributions listed in the dialog box that appears when you choose Random Numbers.

Lists and Matrices

You can store data in lists or in matrices. Numbers in a list should be separated by commas, with the numbers and commas both in mathematics mode. Lists can be plain or enclosed in brackets. A list of data is also referred to as a *set* of data. A list (not enclosed by brackets) can be reshaped into a matrix.

▶ **To reshape a list or set of data into a matrix**

1. If the list is in text, select the list and click T to change it to mathematics.

1. Place the insertion point in the list or set.

2. From the Matrices submenu, choose Reshape.

3. In the dialog box that appears, enter the Number of Columns.

411

4. Choose OK.

5. Select the matrix with the mouse and click 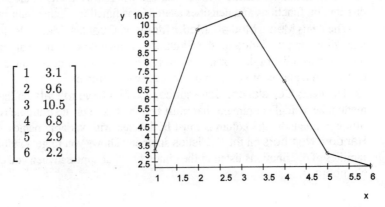.

▶ **Matrices + Reshape**

$$2, 4, 7, 15, -8, 0, -10, \text{(Number of Columns: 4)} \begin{bmatrix} 2 & 4 & 7 & 15 \\ -8 & 0 & -10 \end{bmatrix}$$

$$1, 3.1, 2, 9.6, 3, 10.5, 4, 6.8, 5, 2.9, 6, 2.2, \text{(Number of Columns: 2)} \begin{bmatrix} 1 & 3.1 \\ 2 & 9.6 \\ 3 & 10.5 \\ 4 & 6.8 \\ 5 & 2.9 \\ 6 & 2.2 \end{bmatrix}$$

▶ **Plot 2D + Rectangular**

$$\begin{bmatrix} 1 & 3.1 \\ 2 & 9.6 \\ 3 & 10.5 \\ 4 & 6.8 \\ 5 & 2.9 \\ 6 & 2.2 \end{bmatrix}$$

For this plot, the matrix of points was used for two items. Item 1 is a **Point Plot** with **Point Marker** set to **Circle**. Item 2 has **Line Style** set to **Dash**. (See page 311 for further examples.)

Importing Data from an ASCII File

Both **File + Open** and **File + Import Contents** allow you to read text files. One rather general method for converting data from a text file to a mathematical list or matrix is outlined below. Look under **Help + Search + importing data** and **Help + Search + calculators** for additional options for importing data.

If the data is a list of numbers separated by commas, select the data with click and drag, then click $\boxed{T}$ to change it to mathematics. Then use the data as a list, or use the techniques described in the previous section to convert the data to a vector or matrix.

If the data is stored as a column of numbers, it will come in as in-line numbers separated by spaces.

▶ **To change numbers separated by spaces to a list**

1. Select the data with click and drag.

2. Choose Edit + Replace.

3. In the Search for box, enter a space.

4. In the Replace with box, enter a comma followed by an Allowbreak (choose Insert + Spacing + Break, check Allowbreak, check OK).

5. Choose Replace All.

6. Select the data with click and drag.

7. Click ⊞T or choose Insert + Mathematics.

Note   The order in which these operations are carried out is very important. If you change to mathematics before replacing spaces by commas, the spaces will not be retained and the digits will move together to form one number.

Example   File + Import contents: 345 26 14 8 19 36 32 14 9 4 20

Edit + Replace + Replace All: 345,26,14,8,19,36,32,14,9,4,20

Insert + Mathematics: $345, 26, 14, 8, 19, 36, 32, 14, 9, 4, 20$

Compute + Matrices + Reshape (Choose 2 columns):
$$\begin{bmatrix} 345 & 26 \\ 14 & 8 \\ 19 & 36 \\ 32 & 14 \\ 9 & 4 \\ 20 & \end{bmatrix}$$

The following example illustrates a method for working with one-dimensional data.

Example   File + Import contents: 345,26,14,8,19,36
Insert + Mathematics: $345, 26, 14, 8, 19, 36$

Compute + Matrices + Reshape (Choose 1 column):
$$\begin{bmatrix} 345 \\ 26 \\ 14 \\ 8 \\ 19 \\ 36 \end{bmatrix}$$

Definitions + New Definition $g(i) = i$

Matrices + Fill Matrix (Defined by Function g, 6 Rows, 1 Column) $\begin{bmatrix} 1 \\ 2 \\ 3 \\ 4 \\ 5 \\ 6 \end{bmatrix}$

Matrices + Concatenate $\begin{bmatrix} 1 \\ 2 \\ 3 \\ 4 \\ 5 \\ 6 \end{bmatrix} \begin{bmatrix} 345 \\ 26 \\ 14 \\ 8 \\ 19 \\ 36 \end{bmatrix}$, concatenate: $\begin{bmatrix} 1 & 345 \\ 2 & 26 \\ 3 & 14 \\ 4 & 8 \\ 5 & 19 \\ 6 & 36 \end{bmatrix}$

Measures of Central Tendency

You can compute ordinary measures of central tendency. Several of these, such as Mean, Median, Mode, Geometric Mean, and Harmonic Mean are items on the Statistics submenu.

Arithmetic Mean

The *mean* (arithmetic mean, average) of the numbers $x_1, x_2, \ldots, x_n$ is the most commonly used measure of central tendency. It is the sum of the numbers divided by the number of numbers.

$$\frac{\sum_{i=1}^{n} x_i}{n}$$

▶ **To find the mean of the numbers in a list**

1. Place the insertion point in the list.

2. From the Statistics submenu, choose Mean.

▶ Statistics + Mean

a, b, c, Mean(s): $\frac{1}{3}a + \frac{1}{3}b + \frac{1}{3}c$

$23, 5, -6, 18, 23, -22, 5$, Mean(s): $\frac{46}{7}$

$16.5, 22.1, 6.9, 14.2, 9.0$, Mean(s): 13.74

Applying Mean from the Statistics submenu to a matrix gives the means of the columns. Applying Mean again, this time to the list of column means, gives the mean of the matrix entries.

▶ Statistics + Mean, Statistics + Mean

$$\begin{bmatrix} 23 & 5 & -6 \\ 18 & 23 & -22 \\ 5 & 0 & 0 \end{bmatrix}, \text{Mean(s): } \left[\frac{46}{3}, \frac{28}{3}, -\frac{28}{3}\right], \text{Mean(s): } \frac{46}{9}$$

$$\begin{bmatrix} x & y & z \\ 1 & 1 & 4 \\ 3 & 2 & 5 \\ 5 & 3 & 6 \\ 7 & 4 & 7 \end{bmatrix}, \text{Mean(s): } \left[4, \frac{5}{2}, \frac{11}{2}\right], \text{Mean(s): } 4$$

$$\begin{bmatrix} a & b \\ c & d \\ f & g \end{bmatrix}, \text{Mean(s): } \left[\frac{1}{2}c + \frac{1}{2}f, \frac{1}{2}d + \frac{1}{2}g\right], \text{Mean(s): } \frac{1}{4}c + \frac{1}{4}d + \frac{1}{4}f + \frac{1}{4}g$$

Notice that these final two matrices were interpreted as labeled matrices, so the first row was ignored.

Median

A *median* of a finite list of numbers is a number such that at least half the numbers in the set are equal to or less than it, and at least half the numbers in the set are equal to or greater than it. If two different numbers satisfy this criterion, MuPAD takes the smaller number as the median. The value computed for a median may vary according to different conventions. The median is interpreted here according to the algorithms implemented by the computational engine you are using.

You do not have to arrange the numbers in increasing order before computing the median. Leave the insertion point in a list or set of data, a vector, or a matrix and, from the Statistics submenu, choose Median.

▶ Statistics + Median

$1, 5, 2$, Median(s): 2 $\qquad\qquad$ $1, 2, 3, 4$, Median(s): 2

$2, 3, 3, 3$, Median(s): 3 $\qquad\qquad$ $23, 5, -6, 18, 23, -22, 5, 7$, Median(s): 5

For a matrix, you obtain the medians of the columns. The second of the following matrices is interpreted as a labeled matrix, and the first row is ignored.

▶ Statistics + Median

$$\begin{bmatrix} 23 & 5 & -6 \\ 18 & 23 & -22 \\ 5 & 0 & 0 \end{bmatrix}, \text{Median(s): } [18, 5, -6] \qquad \begin{bmatrix} a & b \\ 1 & 2 \\ 5 & 6 \\ 3 & 4 \end{bmatrix}, \text{Median(s): } [3, 4]$$

Quantile

The qth quantile of a set, where q is a number between zero and one, is a number Q satisfying the condition that the fraction q of the numbers falls below Q and the fraction $1 - q$ lies above Q. The 0.5th quantile is a median or 50th percentile, whereas the 0.25th quantile is a first quartile or 25th percentile, and so forth. Take the qth quantile of a matrix to find the qth quantiles of the columns. The value of a quantile of a finite set of numbers may vary according to different conventions. The quantile is interpreted here according to the algorithms implemented by the computational engine you are using.

You can find quantiles of a list of numbers, a set of numbers, a vector, or columns of a matrix, as shown below.

▶ Statistics + Quantile

$1, 2, 3, 4, 5, 6, 7, 8, 9, 10, 0.87$th Quantile(s): 9

$\{5.6, 7, 8.3, 57, 1.4, 37, 2\}, 0.25$th Quantile(s): 2

$\begin{bmatrix} 23 & 5 & -6 \\ 9 & -3 & 7 \\ 18 & 23 & -22 \end{bmatrix}, 0.5$th Quantile(s): $[18, 5, -6]$

$\begin{bmatrix} 23 & 5 & -6 \\ 18 & 23 & -22 \\ 5 & 0 & 0 \end{bmatrix}, 0.33$th Quantile(s): $[5, 0, -22]$

$\begin{bmatrix} \dfrac{1137}{100}, \dfrac{49}{20}, -\dfrac{354}{25} \end{bmatrix}, 0.75$th Quantile(s): $\frac{1137}{100}$

Mode

A *mode* is a value that occurs with maximum frequency. To find the modes of a list of numbers or of the columns of a matrix, leave the insertion point in the list or matrix and, from the Statistics submenu, choose Mode. The computational engine also returns the multiplicity of the mode or modes.

▶ Statistics + Mode

$23, 5, -6, 18, 23, -22, 5$, Mode(s): $[23, 5], 2$

$1, 1, 5, 5, 5, 7, 7, 8, 9, 9, 9$, Mode(s): $[5, 9], 3$

$\begin{bmatrix} 23 & 5 & -6 \\ 18 & 23 & -22 \\ 5 & 23 & 0 \end{bmatrix}$, Mode(s): $[[23, 18, 5], 1, [23], 2, [-6, -22, 0], 1]$

▶ **Statistics + Mode**

$$\begin{bmatrix} a & b \\ 3 & 4 \\ 1 & 2 \end{bmatrix}, \text{Mode(s): } [[3,1],1,[4,2],1]$$

The previous matrix is interpreted as a labeled matrix, and the modes returned are the modes of the matrix $\begin{bmatrix} 3 & 4 \\ 1 & 2 \end{bmatrix}$.

Geometric Mean

The *geometric mean* of n nonnegative numbers $x_1, x_2, \ldots, x_n$ is the nth root of the product of the numbers

$$\sqrt[n]{x_1 x_2 \cdots x_n}$$

The geometric mean is useful with data for which the ratio of any two consecutive numbers is nearly constant, such as money invested with compound interest.

To find the geometric mean of a set of nonnegative numbers, leave the insertion point in a list, set, vector, or matrix of numbers, and from the Statistics submenu choose Geometric Mean. For a matrix, the result is a list of geometric means of the columns.

▶ **Statistics + Geometric Mean**

$$3, 56, 14, 2 = \sqrt[4]{4704} \qquad\qquad 5.19, 7.3, 2.77, 3.67, 8 = 4.985\,9$$

$$\begin{bmatrix} 2.9 & 5.2 & 9.7 \\ 6.2 & 8.8 & 1.1 \end{bmatrix} = [4.240\,3, 6.764\,6, 3.266\,5]$$

You can also find the geometric mean directly from the defining formula, as follows.

▶ **Evaluate Numerically**

$$\sqrt[4]{3 \times 56 \times 14 \times 2} = 8.2816 \qquad\qquad \sqrt[5]{(5.19)(7.3)(2.77)(3.67)(8)} = 4.985\,9$$

Another way you can compute the geometric mean is by defining the function

$$G(z, n) = \sqrt[n]{\prod_{i=1}^{n} z_i}$$

and a vector $z = [z_1, z_2, \ldots, z_n]$ and then evaluating $G(z, n)$.

▶ **Definitions + New Definition**

$$G(z, n) = \sqrt[n]{\prod_{i=1}^{n} z_i}$$

$$s = [3, 56, 14, 2] \qquad\qquad t = [5.19, 7.3, 2.77, -3.67, -8]$$
$$u = [4, 7, 18] \qquad\qquad v = [4, 7, 13, 18]$$

▶ Evaluate, Evaluate Numerically

$$G(s,4) = \sqrt[4]{4704} = 8.2816 \qquad\qquad G(t,5) = 4.9859$$
$$G(u,3) = \sqrt[3]{504} = 7.9581 \qquad\qquad G(v,4) = \sqrt[4]{6552} = 8.9969$$

Example If you invest \$1 and earn 10% per year for six years, the value of your investment in this and the succeeding years is

$$1.00, 1.10, 1.21, 1.33, 1.46, 1.61, 1.77$$

The geometric average of these seven numbers is 1.33.

Harmonic Mean

The *harmonic mean* of n positive numbers $x_1, x_2, \ldots, x_n$ is the reciprocal of the mean of the reciprocals.

$$\frac{n}{\sum_{i=1}^{n} \frac{1}{x_i}}$$

The harmonic mean can be used in averaging speeds, where the distances applying to each speed are the same.

To find the harmonic mean of a set of positive numbers, leave the insertion point in a list, set, vector, or matrix of numbers, and from the Statistics submenu choose Harmonic Mean. For a matrix or vector, the result is a list of harmonic means of the columns.

▶ Statistics + Harmonic Mean

$$a, b, c = \frac{3}{\frac{1}{a} + \frac{1}{b} + \frac{1}{c}} \qquad\qquad 2,4,6,8 = \frac{96}{25} \qquad\qquad 0.67, 1.9, 6.2, 5.8, 4.7 = 1.9491$$

You can also compute a harmonic mean directly from the defining formula. Following are the harmonic mean of 2, 4, 6, and 8, and the harmonic mean of 0.67, 1.9, 6.2, 5.8, and 4.7, respectively.

▶ **Evaluate**

$$\frac{4}{\frac{1}{2} + \frac{1}{4} + \frac{1}{6} + \frac{1}{8}} = \frac{96}{25}$$

$$5\left((0.67)^{-1} + (1.9)^{-1} + (6.2)^{-1} + (5.8)^{-1} + (4.7)^{-1}\right)^{-1} = 1.9491$$

Another way to compute the harmonic mean is by defining the function

$$H(z,n) = \frac{n}{\sum_{k=1}^{n} \frac{1}{z_k}}$$

and the vector $z = [z_1, z_2, \ldots, z_n]$ and then evaluating $H(z,n)$.

▶ Definitions + New Definition

$$H(z,n) = \frac{n}{\sum_{k=1}^{n} \frac{1}{z_k}}$$

$s = [2,4,6,8]$ $t = [0.67, 1.9, 6.2, 5.8, 4.7]$
$u = [4,7,18]$ $v = [4,7,13,18]$

▶ Evaluate, Evaluate Numerically

$H(s,4) = \frac{96}{25} = 3.84$ $H(t,5) = 1.9491$
$H(u,3) = \frac{756}{113} = 6.6903$ $H(v,4) = \frac{13\,104}{1721} = 7.6142$

Example   If you average 20 MPH driving from your home to a friend's home and 30 MPH driving back home over the same route, then your "average" speed for the round trip is the harmonic mean

$$\frac{2}{\frac{1}{20} + \frac{1}{30}} = 24 \text{ MPH}$$

This computation gives the speed that you would have to travel if you did the round trip at a constant speed, taking the same total amount of time.

Measures of Dispersion

The various measures of dispersion describe different aspects of the spread, or dispersion, of a set of variates about their mean.

Mean Deviation

The *mean deviation* is the mean of the distances of the data from the data mean. The mean deviation of $x_1, x_2, \ldots, x_n$ is

$$\frac{\sum_{i=1}^{n} \left| x_i - \frac{\sum_{j=1}^{n} x_j}{n} \right|}{n}$$

where the vertical bars denote absolute value. (Without the absolute values, this sum would be zero.) For example, the mean deviation of $\{1,2,3,4,5\}$ is

$$\frac{|1-3| + |2-3| + |3-3| + |4-3| + |5-3|}{5} = \frac{6}{5}$$

You can present the data as a list, vector, or matrix. In the latter case, you get the mean deviations of the columns.

▶ **Statistics + Mean Deviation**

$1, 2, 3, 4, 5,$ Mean deviation(s): $\frac{6}{5}$

$\left[25, 76, \frac{87}{2}\right]$

$\begin{bmatrix} -85 & -55 & -37 \\ -35 & 97 & 50 \end{bmatrix}$, Mean deviation(s):

Variance and Standard Deviation

The *sample variance* for $x_1, x_2, \ldots, x_n$ is the sum of the squares of differences with the mean, divided by $n - 1$.

$$\frac{\sum_{i=1}^{n} \left(x_i - \frac{\sum_{j=1}^{n} x_j}{n} \right)^2}{n - 1}$$

▶ **To compute variance**

1. Place the insertion point in a list of data, in a vector, or in a matrix.

2. From the Statistics submenu, choose Variance.

▶ **Statistics + Variance**

$5, 1, 89, 4, 29, 47, 18,$ Variance(s): $\frac{21\,055}{21}$

$\begin{bmatrix} 18.1 \\ 5.3 \\ 7.6 \end{bmatrix}$, Variance(s): 46.563

$\begin{bmatrix} 23 & 5 & -6 \\ 18 & 23 & -22 \\ 5 & 0 & 0 \end{bmatrix}$, Variance(s): $\left[\frac{259}{3}, \frac{439}{3}, \frac{388}{3}\right]$

$\begin{pmatrix} x & y \\ a & b \\ c & d \end{pmatrix}$, Variance(s): $\left[\left(\frac{1}{2}c - \frac{1}{2}a\right)^2 + \left(\frac{1}{2}a - \frac{1}{2}c\right)^2, \left(\frac{1}{2}d - \frac{1}{2}b\right)^2 + \left(\frac{1}{2}b - \frac{1}{2}d\right)^2\right]$

The square root of the variance is called the *standard deviation*. It is the most commonly used measure of dispersion.

$$\sqrt{\frac{\sum_{i=1}^{n} \left(x_i - \frac{\sum_{j=1}^{n} x_j}{n} \right)^2}{n - 1}}$$

▶ Statistics + Standard Deviation

$[5, 1, 89, 4, 29, 47, 18]$, Standard deviation(s): $\frac{1}{21}\sqrt{21}\sqrt{21\,055}$

$\begin{pmatrix} 18.1 \\ 5.3 \\ 7.6 \end{pmatrix}$, Standard deviation(s): 6.8237

$\begin{pmatrix} x & y \\ a & b \\ c & d \end{pmatrix}$, Standard deviation(s):

$$\left[\sqrt{\left(\tfrac{1}{2}c - \tfrac{1}{2}a\right)^2 + \left(\tfrac{1}{2}a - \tfrac{1}{2}c\right)^2}, \; \sqrt{\left(\tfrac{1}{2}d - \tfrac{1}{2}b\right)^2 + \left(\tfrac{1}{2}b - \tfrac{1}{2}d\right)^2} \right]$$

Note that the preceding matrix was treated as a labeled matrix, and the first row was ignored.

▶ Statistics + Standard Deviation

$\begin{bmatrix} 23 & 5 & -6 \\ 18 & 23 & -22 \\ 5 & 0 & 0 \end{bmatrix}$, Standard deviation(s): $\left[\tfrac{1}{3}\sqrt{3}\sqrt{259}, \; \tfrac{1}{3}\sqrt{3}\sqrt{439}, \; \tfrac{2}{3}\sqrt{3}\sqrt{97} \right]$

$\begin{bmatrix} -8.5 & 5.0 & 5.7 \\ -5.5 & 7.9 & -5.9 \\ -3.7 & 5.6 & 4.5 \\ -3.5 & 4.9 & -8.0 \\ 9.7 & 6.3 & -9.3 \end{bmatrix}$, Standard deviation(s): $[7.0014, 1.23, 7.1456]$

Covariance

The *covariance matrix* of an $m \times n$ matrix $X = [x_{ij}]$ is an $n \times n$ matrix with (i, j)th entry

$$\frac{\sum_{k=1}^{m} \left(x_{ki} - \frac{\sum_{s=1}^{m} x_{si}}{m} \right) \left(x_{kj} - \frac{\sum_{t=1}^{m} x_{tj}}{m} \right)}{m - 1}$$

Note that for each i, the (i, i)th entry is the *variance* of the data in the ith column, making the variances of the column vectors occur down the main diagonal of the covariance matrix. The definition of covariance matrix is symmetric in i and j, so the covariance matrix is always a symmetric matrix.

▶ Statistics + Mean, Statistics + Variance, Statistics + Covariance

$$\begin{bmatrix} 1 & 2 \\ 3 & 5 \\ 4 & 3 \end{bmatrix}, \text{Mean(s): } 2.7, 3.3, \text{Variance(s): } [2.3333, 2.3333],$$

$$\text{Covariance matrix: } \begin{bmatrix} 2.3333 & 1.1667 \\ 1.1667 & 2.3333 \end{bmatrix}$$

$$\begin{bmatrix} 8.5 & -5.5 & -3.7 \\ -3.5 & 9.7 & 5.0 \\ 7.9 & 5.6 & 4.9 \end{bmatrix}, \begin{array}{l} \text{Mean(s): } [4.3, 3.2667, 2.0667] \\ \text{Variance(s): } [45.72, 61.843, 24.943] \end{array},$$

$$\text{Covariance matrix: } \begin{bmatrix} 45.72 & -39.3 & -18.45 \\ -39.3 & 61.843 & 38.018 \\ -18.45 & 38.018 & 24.943 \end{bmatrix}$$

Moment

The rth *moment* of a set $\{x_1, x_2, \ldots, x_n\}$ about the point a is the following sum:

$$\frac{1}{n} \sum_{i=1}^{n} (x_i - a)^r$$

Thus, the mean is also known as the *first moment about zero*. The second moment about zero is the quantity $\mu^2 + \sigma^2$, where μ is the mean and σ^2 is the variance of the data. The rth *moment about the mean* is the sum

$$\frac{1}{n} \sum_{i=1}^{n} \left(x_i - \frac{1}{n} \sum_{j=1}^{n} x_j \right)^r$$

Example The 3rd and 4th moments of the set $\{2, 4, 6, 8, 10, 12, 14, 16, 18\}$ about the mean are

$$\frac{1}{9} \sum_{i=1}^{9} \left(2i - \frac{1}{9} \sum_{j=1}^{9} 2j \right)^3 = 0$$

$$\frac{1}{9} \sum_{i=1}^{9} \left(2i - \frac{1}{9} \sum_{j=1}^{9} 2j \right)^4 = \frac{3776}{3} \approx 1258.7$$

▶ Statistics + Moment

$$\begin{bmatrix} 8.5 \\ -5.5 \\ -3.7 \\ 3.5 \end{bmatrix}$$ (Moment Number: 1, Moment Origin: About 0),
Moment(s): 0.7
(Moment Number: 2, Moment Origin: About 0),
Moment(s): 32.11

$(\begin{array}{cccc} 0.123 & 0.703 & 0.445 & 0.284 \end{array})$,

(Moment Number: 1, Moment Origin: About 0.5),
Moment(s): -0.11125
(Moment Number: 3, Moment Origin: About 0.5),
Moment(s): -1.3865×10^{-2}

$(\begin{array}{cccc} 0.123 & 0.703 & 0.445 & 0.284 \end{array})$,

(Moment Number: 1, Moment Origin: About Mean),
Moment(s): 0
(Moment Number: 2, Moment Origin: About Mean),
Moment(s): 4.5878×10^{-2}

Correlation

In dealing with two random variables, we refer to the measure of their linear correlation as the *correlation coefficient*. When two random variables are independent, this measure is 0. If two random variables X and Y are linearly related in the sense $Y = a + bX$ for some constants a and b, then the coefficient of correlation reaches one of the extreme values $+1$ or -1. In either of these cases, X and Y are referred to as *perfectly* correlated. The formula for the coefficient of correlation for two random variables is

$$\rho = \rho(X, Y) = \frac{\text{Cov}(X, Y)}{\sigma_x \sigma_y} = \frac{\sigma_{xy}}{\sigma_x \sigma_y}$$

where σ_x and σ_y are the standard deviations of the two random variables.

To compute the coefficient of correlation between two samples, enter the data as two columns of a matrix and, from the **Statistics** submenu, choose **Correlation**. You can apply this operation to any size matrix to get the coefficient of correlation for each pair of columns: the number in the i, j position is the coefficient of correlation between column i and column j. A correlation matrix is always symmetric, with ones on the main diagonal.

▶ Statistics + Correlation

$$\begin{bmatrix} 43 & -62 \\ 77 & 66 \\ 54 & -5 \\ 99 & -61 \end{bmatrix}, \text{Correlation matrix:} \begin{bmatrix} 1.0 & 7.4831 \times 10^{-2} \\ 7.4831 \times 10^{-2} & 1.0 \end{bmatrix}$$

▶ Statistics + Correlation

$$\begin{bmatrix} -50 & -12 & -18 \\ 31 & -26 & -62 \\ 1 & -47 & -91 \end{bmatrix}, \text{Correlation matrix:} \begin{bmatrix} 1.0 & -0.52883 & -0.71054 \\ -0.52883 & 1.0 & 0.97297 \\ -0.71054 & 0.97297 & 1.0 \end{bmatrix}$$

The relationship $\dfrac{\mathrm{Cov}\,(X,Y)}{\sigma_x \sigma_y} = \rho\,(X,Y)$ among correlation, covariance, and the standard deviations is illustrated in the following example.

$$\begin{bmatrix} -50 & -12 \\ 31 & -26 \\ 1 & -47 \end{bmatrix}, \begin{cases} \text{Correlation matrix:} \begin{bmatrix} 1.0 & -0.52883 \\ -0.52883 & 1.0 \end{bmatrix} \\[2mm] \text{Covariance matrix:} \begin{bmatrix} 1677.0 & -381.5 \\ -381.5 & 310.33 \end{bmatrix} \\[2mm] \text{Standard deviation(s): } [40.951, 17.616] \\[2mm] \dfrac{-381.5}{40.951 \times 17.616} = -0.52884 \end{cases}$$

Distributions and Densities

A *cumulative distribution function* $F\,(x)$ of a random variable X is the function $F\,(x) = P\,(X \leq x)$, the probability that $X \leq x$. If $F(x)$ has a derivative $f(x)$, then $f(x)$ is non-negative and is called the *probability density function* of x. The *inverse distribution function* $G(\alpha)$ satisfies $G\,(F\,(x)) = x$ and $F\,(G\,(\alpha)) = \alpha$. The names for these functions are obtained by adding Dist, Den, or Inv to the name of the distribution. For example, NormalDist, NormalDen, and NormalInv are the three functions for the normal distribution. Such function names will automatically turn gray when typed in mathematics mode.

Cumulative Distribution Functions

A *cumulative distribution function* is a nondecreasing function defined on the interval $(-\infty, \infty)$, with values in the interval $[0, 1]$. The definition of a distribution function generally describes only the values where the function is positive, the implicit assumption being that the distribution function is zero up to that point. For discrete cumulative distribution functions, the definition also gives only the values where the function changes, the implicit assumption being that the cumulative distribution function is a step function. Commonly, definitions of these functions are stated only for integers. The definition of a density function also generally describes only the values where the function is positive, the implicit assumption being that the function is zero elsewhere.

These distribution and density functions satisfy the relationships

$$f(x) = \frac{d}{dx}F(x)$$

$$F(x) = \int_{-\infty}^{x} f(u)du$$

Also note that the cumulative distribution function satisfies $\lim_{x \to \infty} F(x) = 1$ and $\lim_{x \to -\infty} F(x) = 0$. Cumulative distribution functions are named FunctionDist, and the density functions are named FunctionDen. For example, the probability density functions for the normal distributions are called NormalDen.

You can compute with several families of distributions: Normal, Cauchy, Student's t, Chi-Square, F, Exponential, Weibull, Gamma, Beta, Uniform, Binomial, Poisson, and Hypergeometric.

Inverse Distribution Functions

For a distribution function F mapping $(-\infty, \infty)$ into $[0, 1]$, the *inverse distribution function* G performs the corresponding inverse mapping from (a subset of) $[0, 1]$ into $(-\infty, \infty)$; that is, $G(F(x)) = x$ and $F(G(\alpha)) = \alpha$. Equivalently, $\text{Prob}[X \leq G(\alpha)] = F(x) = \alpha$. Note that the value that is *exceeded* with probability α is given by the function $G(1 - \alpha)$. This function is also of interest.

$$\text{Prob}[X \leq G(1 - \alpha)] = F(x) = 1 - \alpha = 1 - \text{Prob}[X \leq G(\alpha)]$$

When cumulative distribution functions are named FunctionDist, then the inverse cumulative distribution functions are named FunctionInv. For example, NormalInv is the name of the inverse cumulative distribution function for the normal distribution.

Distribution Tables

Depending on the particular family of distributions, the distribution tables in statistics books list function values for selected parameters of one of the functions described earlier—either the cumulative distribution, the inverse cumulative distribution, or the probability density function. With access to these functions, not only can you compute the tabular entries easily and accurately, but you can also find the corresponding values directly for any variables and parameters to any degree of accuracy you wish.

If the *Reference Library* is installed on your system, you can find interactive distribution tables. Choose Help + Search and look under tables, reference: Statistical distributions. (A "Typical" installation does not include the *Reference Library* for space considerations, but if you have space on your system, you can copy the files from the installation CD or add them with a "Custom" install.)

Families of Continuous Distributions

The relationship of the various distribution, inverse distribution, and density functions to the entries in standard statistical tables is explained in the following sections for each of the families of distributions available.

Gamma Function

The *Gamma function* $\Gamma(t)$ that appears in the definition of the Student's t distribution and the gamma distribution is the continuous function $\Gamma(t) = \int_0^\infty e^{-x} x^{t-1} dx$ defined for positive real numbers t. The Gamma function satisfies

$$\Gamma(1) = 1 \text{ and } \Gamma(t+1) = t\Gamma(t)$$

and for positive integers k, it is the familiar factorial function

$$\Gamma(k) = (k-1)!$$

The Gamma function is active. For example, place the insertion point in the expression $\Gamma(5)$ and choose Evaluate to get $\Gamma(5) = 24$. Note that $24 = 4 \times 3 \times 2 \times 1$. See page 169) for a plot of the Gamma function.

Use Rewrite + Factorial to convert the Gamma function to a factorial expression. (Here it is assumed that x is an integer.)

▶ Rewrite + Factorial

$\Gamma(x) = (x-1)!$

Use Rewrite + Gamma to convert factorials, binomials, and multinomial coefficients to an expression in the Gamma function.

▶ Rewrite + Gamma

$(x-1)! = \Gamma(x)$ $\binom{m}{n} = \frac{\Gamma(m+1)}{\Gamma(n+1)\Gamma(m-n+1)}$ $x!y!z! = \Gamma(x+1)\,\Gamma(y+1)\,\Gamma(z+1)$

Normal Distribution

The *normal cumulative distribution function* is defined for all real numbers μ and for positive σ by the integral

$$\text{NormalDist}(x; \mu, \sigma) = \frac{1}{\sigma\sqrt{2\pi}} \int_{-\infty}^{x} e^{-\frac{(u-\mu)^2}{2\sigma^2}} du$$

of the *normal probability density function*

$$\text{NormalDen}(u; \mu, \sigma) = \frac{1}{\sigma\sqrt{2\pi}} e^{-\frac{(u-\mu)^2}{2\sigma^2}}$$

The inverse of the normal cumulative distribution function, NormalInv, is also available. All three of these functions can be typed in mathematics, and they will automatically turn gray as you type the final letter.

The parameters μ and σ are optional parameters for mean and standard deviation, with the default values 0 and 1 defining the *standard normal distribution*

$$\text{NormalDist}(x) = \frac{1}{\sqrt{2\pi}} \int_{-\infty}^{x} e^{-\frac{u^2}{2}} du$$

A normal distribution table, as found in the back of a typical statistics book, lists some values of the standard normal cumulative distribution function. Certain versions of the table list the values $1 - \text{NormalDist}(x)$.

Note that the function NormalDist can be evaluated as a function of one variable (with default parameters $(0, 1)$) or as a function of one variable and two parameters.

▶ Evaluate Numerically

- NormalDist$(2.44) = 0.992\,66$
- NormalDist$(2.44; 0, 1) = 0.99266$
- NormalDist$(2.44; 1, 2) = 0.76424$
- NormalDen$(2.44; 1, 2) = 0.153\,93$

Graphs of the normal density functions are the familiar bell-shaped curves. The following plots show the density functions NormalDen $(x; \mu, \sigma)$ and distribution functions NormalDist $(x; \mu, \sigma)$ for the parameters $(\mu, \sigma) = (0, 1), (0, 5), (0, 0.5), (1, 1)$.

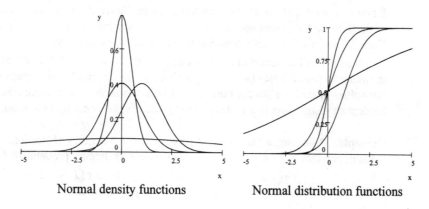

Normal density functions   Normal distribution functions

Student's t Distribution

The *Student's t cumulative distribution function* TDist$(x; v)$ is defined by the integral

$$TDist(x; v) = \frac{\Gamma(\frac{v+1}{2})}{\Gamma(\frac{v}{2})\sqrt{\pi v}} \int_{-\infty}^{x} \left(1 + \frac{1}{v}u^2\right)^{-\frac{v+1}{2}} du$$

of the density function

$$TDen(u; v) = \frac{\Gamma(\frac{v+1}{2})}{\Gamma(\frac{v}{2})\sqrt{\pi v}} \left(1 + \frac{1}{v}u^2\right)^{-\frac{v+1}{2}}$$

with shape parameter v, called degrees of freedom, that ranges over the positive integers. The variance for a Student's t distribution is $\frac{v}{v-2}$, provided $v > 2$.

The function TInv$(p; v)$ is the value of x for which the integral has the value p, as demonstrated here:

TDist$(63.66; 1) = 0.995$ TDist$(-0.97847; 3) = 0.2$
TInv$(0.995; 1) = 63.657$ TInv$(0.2; 3) = -0.97847$

The following plots display the density and distribution functions TDen $(x; v)$ and TDist $(x; v)$ for the parameters $v = 1$ and $v = 15$ with $-5 \leq x \leq 5$.

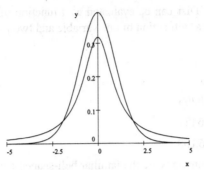

Student's t density functions

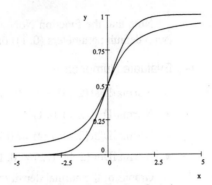

Student's t distribution functions

Note that the Student's t density functions resemble the standard normal density function in shape, although these curves are a bit flatter at the center. It is not difficult to show, using the definitions of the two density functions, that $\lim_{v \to \infty} \text{TDen}(u; v) = \text{NormalDen}(u)$, the density function for the standard normal distribution.

Student's t distribution tables list values of the inverse distribution function corresponding to probabilities (values of the distribution function) and degrees of freedom. For values of v above 30, the normal distribution is such a close approximation for the Student's t distribution that tables usually provide values only up to $v = 30$.

Example Assuming a Student's t distribution with 5 degrees of freedom, determine a value c such that $\Pr(-c < T < c) = 0.90$, where Pr denotes probability. Now

$$\Pr(-c \;<\; T < c) = \Pr(T \le c) - \Pr(T \le -c)$$
$$= \text{TDist}(c; 5) - \text{TDist}(-c; 5)$$

So, you need to solve

$$\text{TDist}(c; 5) - \text{TDist}(-c; 5) = 0.90$$

The Student's t distribution satisfies

$$\text{TDist}(c; 5) + \text{TDist}(-c; 5) = 1$$

So, the problem reduces to

$$2\,\text{TDist}(c; 5) - 1 \;=\; 0.90$$
$$\text{TDist}(c; 5) \;=\; \frac{0.90 + 1}{2} = 0.95$$

The problem is solved by

$$\text{TInv}(0.95; 5) = 2.015$$

Chi-Square Distribution

The *chi-square cumulative distribution function* is defined for nonnegative x and μ by the integral

$$\text{ChiSquareDist}(x; \mu) = \frac{1}{\Gamma(\frac{\mu}{2})2^{\frac{\mu}{2}}} \int_0^x u^{\frac{\mu}{2}-1} e^{-\frac{u}{2}}\, du$$

The integrand is the *chi-square probability density function*

$$\text{ChiSquareDen}(u; \mu) = \frac{1}{\Gamma(\frac{\mu}{2})2^{\frac{\mu}{2}}} u^{\frac{\mu}{2}-1} e^{-\frac{u}{2}}$$

The indexing parameter $\mu > 0$ is the mean of the distribution; it is referred to as the *degrees of freedom*.

The following plots show density functions $\text{ChiSquareDen}(x; \mu)$ and distribution functions $\text{ChiSquareDist}(x; \mu)$ for $\mu = 1, 5, 10, 15$ and $0 \leq x \leq 25$.

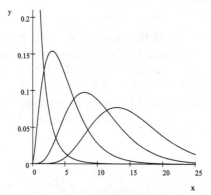

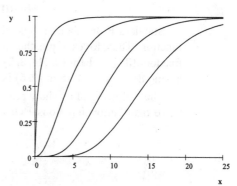

Chi-square density functions   Chi-square distribution functions

The function, $\text{ChiSquareInv}(t; \nu)$ gives the value of x for which $\text{ChiSquareDist}(x; \nu) = t$. This relationship is demonstrated in the following examples:

$$
\begin{aligned}
\text{ChiSquareDist}(1.6103; 5) &= 9.9999 \times 10^{-2} \approx 0.1 \\
\text{ChiSquareInv}(0.1; 5) &= 1.6103 \\
\text{ChiSquareDist}(2.366; 3) &= 0.5 \\
\text{ChiSquareInv}(0.5; 3) &= 2.366
\end{aligned}
$$

A chi-square distribution table shows values of ν down the left column and values u of ChiSquareDist across the top row. The entry in row ν and column u is $\text{ChiSquareInv}(u; \nu)$.

F Distribution

The *F cumulative distribution function* is given by the integral

$$\text{FDist}(x; n, m) = \frac{\Gamma(\frac{n+m}{2})}{\Gamma(\frac{n}{2})\Gamma(\frac{m}{2})} \left(\frac{n}{m}\right)^{\frac{n}{2}} \int_0^x u^{\frac{n-2}{2}} \left(1 + \frac{n}{m}u\right)^{-\frac{n+m}{2}} du$$

of the probability density function

$$\text{FDen}(u; n, m) = \frac{\Gamma(\frac{n+m}{2})}{\Gamma(\frac{n}{2})\Gamma(\frac{m}{2})} \left(\frac{n}{m}\right)^{\frac{n}{2}} u^{\frac{n-2}{2}} \left(1 + \frac{n}{m}u\right)^{-\frac{n+m}{2}}$$

The variable x can be any positive number, and n and m can be any positive integers. The F distribution is used to determine the validity of the assumption of identical standard deviations of two normal populations. It is the distribution on which the analysis of variance procedure is based.

The inverse distribution function $FInv(p; n, m)$ gives the value of x for which the integral $FDist(x; n, m)$ has the value p. These function names automatically turn gray when they are entered in mathematics mode. The relationship between these two functions is illustrated in the following examples.

$$
\begin{aligned}
FDist(0.1; 3, 5) &= 4.3419 \times 10^{-2} \\
FInv(0.043419; 3, 5) &= 0.1 \\
FDist(3.7797; 2, 5) &= 0.900\,00 \\
FInv(0.9; 2, 5) &= 3.7797
\end{aligned}
$$

Standard F distribution tables list some of the values of the inverse F distribution function. Thus, for example, the 4.4th percentile for the F distribution having degrees of freedom $(3, 5)$ is $FInv(0.044; 3, 5) = 0.1$, and the 90th percentile for the F distribution having degrees of freedom $(2, 5)$ is $FInv(0.90; 2, 5) = 3.7797$.

The following plots show probability density functions $FDen(x; n, m)$ and cumulative distribution functions $FDist(x; n, m)$ for $(n, m) = (1, 1), (2, 5), (3, 15)$, and $0 \leq x \leq 5$.

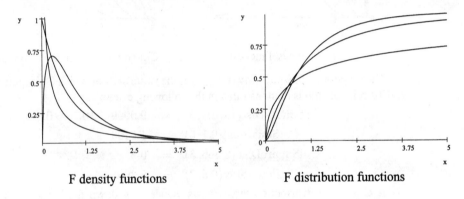

F density functions   F distribution functions

Exponential Distribution

The *exponential cumulative distribution function with parameter* μ, *or mean* μ, is defined by the integral

$$
\text{ExponentialDist}(x; \mu) = \frac{1}{\mu} \int_0^x e^{\frac{-u}{\mu}} \, du = 1 - e^{\frac{-x}{\mu}}
$$

of the exponential density function

$$
\text{ExponentialDen}(u; \mu) = \frac{1}{\mu} e^{\frac{-u}{\mu}}
$$

for $x \geq 0$, and is 0 otherwise.

The *inverse exponential distribution function*

$$
\text{ExponentialInv}(\alpha; \mu) = \mu \ln \frac{1}{1 - \alpha}
$$

is the value of x for which the integral has the value α, as illustrated by the following:

► **Evaluate**

$\mathrm{ExponentialInv}\,(0.73;0.58) = 0.75941$

$\mathrm{ExponentialDist}\,(0.75941;0.58) = 0.73000$

$\mathrm{ExponentialDen}\,(0.75941;0.58) = 0.465\,52$

The following plots show density functions $\mathrm{ExponentialDen}(x;\mu)$ and distribution functions $\mathrm{ExponentialDist}\,(x;\mu)$, for the parameters $\mu = 1,3,5$ and $0 \le x \le 25$.

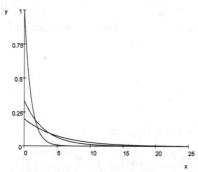

Exponential density functions   Exponential distribution functions

Weibull Distribution

The *Weibull distribution* with scale parameter $b > 0$ and shape parameter $a > 0$ is defined by the integral

$$\mathrm{WeibullDist}\,(x;a,b) = ab^{-a}\int_0^x u^{a-1}e^{-u^a b^{-a}}\,du = 1 - e^{-x^a b^{-a}}$$

of the density function

$$\mathrm{WeibullDen}(u;a,b) = ab^{-a}u^{a-1}e^{-u^a b^{-a}}$$

for $x \ge 0$, and is 0 otherwise.

The *inverse Weibull distribution function*

$$\mathrm{WeibullInv}\,(\alpha;a,b) = b\left(\ln\frac{1}{1-\alpha}\right)^{\frac{1}{a}}$$

is the value of x for which the integral has the value α, as illustrated by the following:

► **Evaluate**

$\mathrm{WeibullInv}\,(0.73;0.5,0.3) = 0.51431$

$\mathrm{WeibullDist}\,(0.51431;0.5,0.3) = 0.73$

The following plots show the probability density functions $\mathrm{WeibullDen}(x;a,b)$ and cumulative distribution functions $\mathrm{WeibullDist}\,(x;a,b)$ for parameters $(a,b) = (0.5,1)$, $(1,1)$, $(3,0.5)$, and $(3,1)$, and $0 \le x \le 3$.

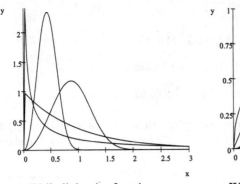

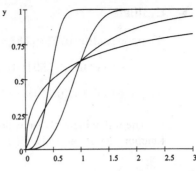

Weibull density functions   Weibull distribution functions

Gamma Distribution

The *gamma distribution* is defined for $x > 0$ by the integral

$$\text{GammaDist}(x; a, b) = \frac{1}{b^a \Gamma(a)} \int_0^x u^{a-1} e^{-\frac{u}{b}} \, du$$

where $\Gamma(t) = \int_0^\infty e^{-u} u^{t-1} \, du$ is the Gamma function. The parameters a and b are called the shape parameter and scale parameter, respectively. The mean of this distribution is ab and the variance is ab^2. The probability density function for the gamma distribution is

$$\text{GammaDen}(u; a, b) = \frac{1}{b^a \Gamma(a)} u^{a-1} e^{-\frac{u}{b}}$$

The following plots show probability density functions $\text{GammaDen}(x; a, b)$ and cumulative distribution functions $\text{GammaDist}(x; a, b)$ for $(a, b) = (1, 0.5), (1, 1)$, and $(2, 1)$ and $0 \le x \le 4$.

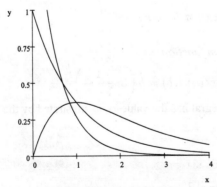

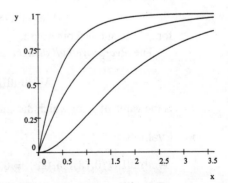

Gamma density functions   Gamma distribution functions

Beta Distribution

The *beta distribution* is defined for $0 \leq x \leq 1$ by the integral

$$\mathrm{BetaDist}\,(x; v, w) = \frac{1}{B\,(v, w)} \int_0^x u^{v-1}\,(1-u)^{w-1}\,du$$

where $B(v, w) = \int_0^1 u^{v-1}\,(1-u)^{w-1}\,du$ is the Beta function with parameters v and w.

The probability density function for the beta distribution is

$$\mathrm{BetaDen}\,(u; v, w) = \frac{u^{v-1}\,(1-u)^{w-1}}{B\,(v, w)}$$

The parameters v and w are positive real numbers called *shape parameters*, and $0 \leq u \leq 1$. The mean of the beta distribution is $\dfrac{v}{v+w}$.

▶ Evaluate Numerically

$\mathrm{BetaDist}\,(0.5; 2, 3) = 0.6875$

$\mathrm{BetaDen}\,(0.5; 2, 3) = 1.5$

$\mathrm{BetaInv}\,(0.6875; 2, 3) = 0.5$

The following plots show probability density functions $\mathrm{BetaDen}(x; b, c)$ and cumulative distribution functions $\mathrm{BetaDist}\,(x; b, c)$ for $(b, c) = (2, 3)$, $(5, 1)$, $(3, 8)$, and $0 \leq x \leq 1$.

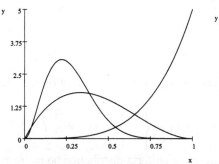

Beta density functions

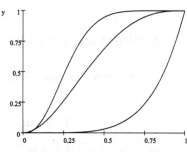

Beta distribution functions

Cauchy Distribution

The *Cauchy cumulative distribution function* is defined for all real numbers α, and for positive β, by the integral

$$\mathrm{CauchyDist}(x; \alpha, \beta) = \frac{1}{\pi\beta} \int_{-\infty}^x \left(1 + \left(\frac{u-\alpha}{\beta}\right)^2\right)^{-1} du$$

The integrand is the *Cauchy probability density function*

$$\text{CauchyDen}(u; \alpha, \beta) = \frac{1}{\pi \beta \left(1 + \left(\frac{u - \alpha}{\beta}\right)^2\right)}$$

The median of this distribution is α. The Cauchy probability density function is symmetric about α and has a unique maximum at α.

The following plots show probability density functions $\text{CauchyDen}(x; \alpha, \beta)$ and cumulative distribution functions $\text{CauchyDist}(x; \alpha, \beta)$ for the parameters $(\alpha, \beta) = (-3, 1)$, $(0, 1.5)$, and $(3, 1)$, and $-5 \leq x \leq 5$.

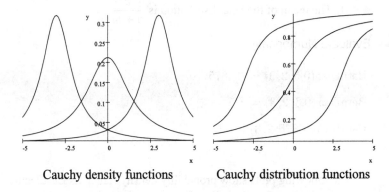

Cauchy density functions   Cauchy distribution functions

Uniform Distribution

The *uniform cumulative distribution function* $\text{UniformDist}(x; a, b)$ for $a < b$ is the function

$$\text{UniformDist}(x; a, b) = \begin{cases} 0 & if & x \leq a \\ \frac{x - a}{b - a} & if & a \leq x \leq b \\ 1 & if & b \leq x \end{cases}$$

The *probability density function* of the uniform distribution on an interval $[a, b]$, where $a < b$, is the function

$$\text{UniformDen}(x; a, b) = \begin{cases} 0 & if & x \leq a \\ \frac{1}{b - a} & if & a \leq x \leq b \\ 0 & if & b \leq x \end{cases}$$

The uniform random variable is the continuous version of "choosing a number at random." The probability that a uniform random variable on $[a, b]$ will have a value in either of two subintervals of $[a, b]$ of equal length is the same.

The following plots show probability density functions $\text{UniformDen}(x; a, b)$ and cumulative distribution functions $\text{UniformDist}(x; a, b)$ for $(a, b) = (0, 1)$, $(1.5, 5)$, $(3, 15)$ and $-5 \leq x \leq 20$.

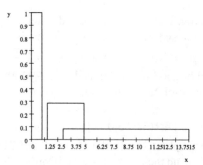

Uniform density functions

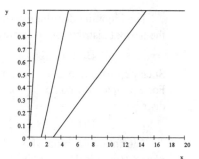

Uniform distribution functions

Families of Discrete Distributions

Several of the standard distributions are functions of a discrete variable, usually the integers. They are commonly plotted with bar graphs or broken line (polygonal) graphs.

Binomial Distribution

The *binomial distribution functions* are functions of a nonnegative integer x,

$$\text{BinomialDist}(x; n, p) = \sum_{k=0}^{x} \binom{n}{k} p^k q^{n-k}$$

with Bernoulli trial parameter (or sample size) a positive integer n, Bernoulli probability parameter a real number p with $0 < p < 1$ and $q = 1-p$. (To enter *binomial coefficients*,

$\binom{n}{k} = \frac{n!}{k!(n-k)!}$, click the binomial fraction and choose None for line.) The corresponding *binomial probability density function* is

$$\text{BinomialDen}(x; n, p) = \binom{n}{x} p^x q^{n-x}$$

for the same conditions on x, n, and p. The mean for this distribution is np, and the variance is npq.

Binomial distribution tables found in statistics books give selected values of either the binomial probability density function $\text{BinomialDen}(x; n, p)$ or the cumulative distribution function $\text{BinomialDist}(x; n, p)$.

The binomial density $\text{BinomialDen}(x; n, p)$ gives the probability of x successes in n independent Bernoulli trials, when the probability of success at each trial is p. It is by far the most common discrete distribution, since people deal with many experiments in which a dichotomous classification of the result is of primary interest. The name *binomial distribution* comes from the fact that the coefficients $\binom{n}{k} = \frac{n!}{k!(n-k)!}$ are commonly called binomial coefficients.

Example   Compute the probability that, in 100 tosses of a coin with $\Pr(\text{heads}) = 0.55$, no more than 54 heads turn up, assuming a binomial distribution.
Solution: $\Pr(X \le 54) = \text{BinomialDist}(54; 100, 0.55) = 0.45868$.

The binomial distribution function with parameters n and p can be approximated by the normal distribution with mean np and variance $np(1-p)$; that is,

$$\text{BinomialDist}(x; n, p) \approx \text{NormalDist}(x; np, \sqrt{np(1-p)})$$

Such approximations are reasonably good if both np and $n(1-p)$ are greater than 5. For example, to find an approximate solution to the preceding problem using a normal distribution, use

$$\Pr(X \le 54) \approx \text{NormalDist}(54; 55.0, 4.9749) = 0.42035$$

The following plots show the graph of $\text{NormalDist}(x; 55.0, 4.9749)$ with a point plot of $\text{BinomialDist}(x; 100, 0.55)$, and the graph of $\text{NormalDen}(x; 55.0, 4.9749)$ with a point plot of $\text{BinomialDen}(x; 100, 0.55)$ for $0 \le x \le 100$.

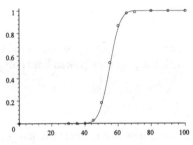   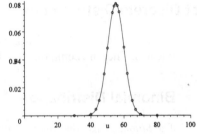

Normal and Binomial distributions   Normal and Binomial densities

Poisson Distribution

The *Poisson cumulative distribution function* is a discrete function defined for non-negative integers. The Poisson distribution with mean $\mu > 0$ is given by the summation

$$\text{PoissonDist}(x; \mu) = \sum_{k=0}^{x} \frac{\mu^k e^{-\mu}}{k!}$$

The *Poisson probability density function* is

$$\text{PoissonDen}(k; \mu) = \frac{\mu^k e^{-\mu}}{k!}$$

for nonnegative integers k and real numbers $\mu > 0$. A Poisson distribution table lists selected values of the Poisson probability density function $\text{PoissonDen}(k; \mu)$.

▶ Evaluate Numerically

$\text{PoissonDen}(2; 3) = 0.22404$ $\text{PoissonDen}(5; 0.3) = 1.5002 \times 10^{-5}$

The Poisson distribution can be used to approximate the binomial distribution when the probability is small and n is large; that is,

$$\text{PoissonDist}(k; \mu) \approx \text{BinomialDist}(k; \mu, \mu(1-p))$$

where $\mu = np$. This distribution has been used as a model for a variety of random phenomena of practical importance.

Hypergeometric Distribution

Suppose that, from a population of M elements, of which x possess a certain attribute, you draw a sample of n items without replacement. The number of items that possess the certain attribute in such a sample is a hypergeometric variate. The *hypergeometric cumulative distribution function* is a discrete function defined for nonnegative integers x. The hypergeometric distribution with M elements in the population, K successes in the population, and sample size n is defined by the following summation of quotients of binomial coefficients for $0 \leq x \leq n$:

$$\text{HypergeomDist}\,(x; M, K, n) = \sum_{k=0}^{x} \frac{\binom{K}{k}\binom{M-K}{n-k}}{\binom{M}{n}}$$

For $x < 0$, the distribution function is 0, and for $x \geq n$, the function is 1. The *hypergeometric probability density function* is

$$\text{HypergeomDen}\,(k; M, K, n) = \frac{\binom{K}{k}\binom{M-K}{n-k}}{\binom{M}{n}}$$

for integers k, K, n, and M satisfying $0 \leq k \leq n$, $0 \leq K \leq M$, and $0 < n \leq M$.

Example   What is the probability of at most five successes when you draw a sample of 10 from a population of 100, of which 30 members are identified as successes? The probability of exactly x successes is given by HypergeomDen $(x; 100, 30, 10)$. Thus, the probability of at most five successes is the sum of exactly 0, 1, 2, 3, 4, and 5 successes, or HypergeomDist $(5; 100, 30, 10) = 0.96123$.

The hypergeometric distribution is the model for sampling without replacement. The hypergeometric distribution can be approximated by the binomial distribution when the sample size is relatively small.

These plots (created as polygonal plots) depict HypergeomDen $(x; 100, 30, 10)$ and HypergeomDist $(x; 100, 30, 10)$ for $0 \leq x \leq 10$.

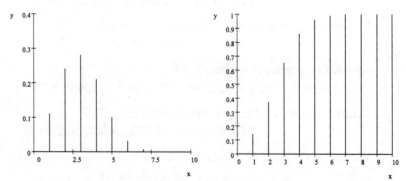

Hypergeometric density function   Hypergeometric distribution function

Random Numbers

The random-number generators on the Statistics submenu give you a set of random numbers from one of several families of distribution functions. The choices in the dialog are Beta, Binomial, Cauchy, Chi-Square, Exponential, F, Gamma, Normal, Poisson, Student's t, Uniform, and Weibull. From the Statistics submenu, choose Random Numbers. Choose a distribution from the dialog, specify how many random numbers you want, and enter appropriate parameters. Following are sample results.

▶ Statistics + Random Numbers

Beta, order 3, order 7:
$$0.31172, 0.28533, 7.8338 \times 10^{-2}, 0.14925, 0.41693$$

Binomial, number of trials 10, probability of success .5:
$$6, 2, 6, 5, 6$$

Cauchy, median 10, shape parameter 5:
$$8.8, 6.9, 7.3, 9.6, 11.$$

Chi-Square, degrees of freedom 3:
$$0.91006, 2.2787, 4.4748, 2.7026, 1.5385$$

Exponential, mean time between arrivals 10:
$$16.851, 16.865, 8.8222, 32.037, 12.434$$

F, degrees of freedom 1 and 3:
$$1.1585 \times 10^{-2}, 1.3279 \times 10^{-2}, 0.18187, 1.5567, 1.8483$$

Gamma, order 5:
$$6.5299, 4.2894, 10.473, 6.9089, 7.9011$$

Normal, mean 3, standard deviation 7:
$$5.223, -4.8075, -5.5782, -1.1218, 1.357$$

Poisson, mean number of occurrences 4:
$$2, 4, 1, 2, 5$$

Student's t, degrees of freedom 7:
$$1.6568 \times 10^{-2}, 0.11424, -1.3886 \times 10^{-2}, -1.8095 \times 10^{-2}, -0.14798$$

Uniform, lower end of range 0, upper end of range 20:
$$5.6016, 16.744, 10.275, 14.057, 10.136$$

Weibull, shape parameter 5, scale parameter 3:
$$3.8, 3.3, 2.7, 3.0, 3.6$$

Curve Fitting

You have the tools to do general curve fitting in an intuitive manner. From the **Statistics** submenu, choose **Fit Curve to Data** and make a choice in the dialog box.

- For straight-line fits, choose **Multiple Regression** or **Multiple Regression (no constant)**.
- For best fits by polynomials, choose **Polynomial of Degree [ ]**.

Linear Regression

Multiple Regression calculates linear-regression equations with *keyed* or labeled data matrices. The result is an equation expressing the variable at the head of the first column as a linear combination of the variables heading the remaining columns, plus a constant (that is missing if **Multiple Regression (no constant)** was chosen).

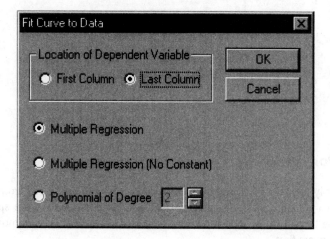

The equation produced is the best fit to the data in the least-squares sense.

▶ **Statistics + Fit Curve to Data + Multiple Regression**

$$\begin{bmatrix} y & x \\ 0 & 1.1 \\ 0.5 & 1.5 \\ 1 & 1.9 \\ 1.5 & 2.4 \end{bmatrix}, \text{ Regression is: } x = 1.08 + 0.86y$$

$$\begin{bmatrix} z & x & y \\ 1 & 0 & 1.1 \\ 2 & 0.5 & 1.1 \\ 4 & 1 & 1.9 \\ 5 & 1.5 & 1.9 \end{bmatrix}, \text{ Regression is: } y = 0.300\,00 + 0.8z - 1.6x$$

The choice **Multiple Regression (no constant)** gives the following linear equations.

▶ **Statistics + Fit Curve to Data + Multiple Regression (no constant)**

$$
\begin{bmatrix}
u & v \\
0 & 1.1 \\
0.5 & 1.5 \\
1 & 1.9 \\
1.5 & 2.4 \\
2 & 2.9
\end{bmatrix}
\text{, Regression is: } u = 0.56733v
$$

$$
\begin{bmatrix}
z & x & y \\
1 & 0 & 1.1 \\
2 & 0.5 & 1.1 \\
4 & 1 & 1.9 \\
5 & 1.5 & 1.9 \\
7 & 2 & 2.9
\end{bmatrix}
\text{, Regression is: } z = 2.1829x + 0.91245y
$$

$$
\begin{pmatrix}
x & y \\
a & b \\
c & d
\end{pmatrix}
\text{, Regression is: } x = y\,\frac{ab + cd}{b^2 + d^2}
$$

Polynomial Fit

Polynomial of Degree [ ] calculates polynomial equations from labeled or unlabeled two-column data matrices. The result is a polynomial of the specified degree that is the best fit to the data in the least-squares sense. For the polynomial fit, the x column appears first.

To find the best fit by a polynomial of second degree to the set of points
$$(0, 0.64), (0.5, 0.09), (1, 0.04), (1.5, 0.49), (2, 1.44)$$
first remove the parentheses and convert the entries into a two-column matrix. To make this conversion, place the insertion point in the list; from the **Matrices** submenu, choose **Reshape**; then specify two columns.

▶ **Matrices + Reshape**

$$
0, 0.64, 0.5, 0.09, 1, 0.04, 1.5, 0.49, 2, 1.44, \quad
\begin{pmatrix}
0 & 0.64 \\
0.5 & 0.09 \\
1 & 0.04 \\
1.5 & 0.49 \\
2 & 1.44
\end{pmatrix}
$$

▶   Statistics + Fit Curve to Data

• Select **Polynomial of Degree [ ]**, enter **2**, and click OK.

$$\begin{pmatrix} 0 & 0.64 \\ 0.5 & 0.09 \\ 1 & 0.04 \\ 1.5 & 0.49 \\ 2 & 1.44 \end{pmatrix}, \text{Polynomial fit: } y = 1.0x^2 - 1.6x + 0.64$$

You can plot the points and polynomial on the same graph. You will notice that these points were chosen such that they lie on the parabola.

▶   Plot 2D + Rectangular

$$\begin{pmatrix} 0 & 0.64 \\ 0.5 & 0.09 \\ 1.0 & 0.04 \\ 1.5 & 0.49 \\ 2.0 & 1.44 \end{pmatrix}$$

1. Change **Plot Style** to **Point** and **Point Marker** to **Circle**.

2. Select and drag $64 - 1.6x + 1.0x^2$ onto the plot.

3. For **Item 2**, set **Plot Intervals** to $-0.05 < x < 2.05$.

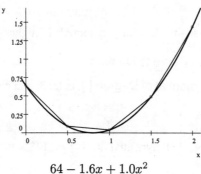

$$64 - 1.6x + 1.0x^2$$

▶ Plot 2D + Rectangular

$$\begin{bmatrix} 0 & 6 \\ 1 & 0.1 \\ 2 & -3 \\ 3 & 2 \\ 4 & 8 \end{bmatrix}$$

1. Select and drag $5.9971 - 8.5243x + 2.2786x^2$ onto the plot.

2. For Item 2, set Plot Intervals to $0 < x < 4$.

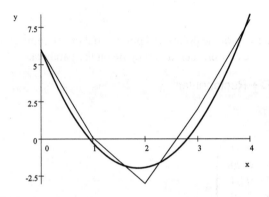

The Fit Curve to Data command operates on labeled matrices.

▶ Statistics + Fit Curve to Data

- Polynomial of Degree 2

$$\begin{pmatrix} x & y \\ 0 & 6 \\ 1 & 0.1 \\ 2 & -3 \\ 3 & 2 \end{pmatrix}, \text{Polynomial fit: } y = 6.265 - 9.685x + 2.725x^2$$

You can also fit data with polynomials of higher degree.

▶ Statistics + Fit Curve to Data

- Select Polynomial of Degree [ ], enter 3, and click OK.

$$\begin{bmatrix} 0 & 0.64 \\ 0.5 & 0.09 \\ 1.0 & 8.04 \\ 1.5 & 0.49 \\ 2.0 & -7.44 \end{bmatrix}, \text{Polynomial fit: } y = 8.1143 \times 10^{-2} + 1.4114x + 9.1143x^2 - 5.92x^3$$

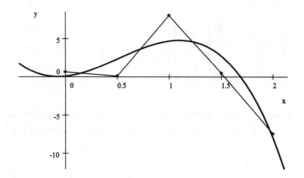

► Statistics + Fit Curve to Data,

    • Select **Polynomial of Degree**, enter **4**, and click **OK**.

$$\begin{bmatrix} 1 & 12 \\ 3 & 4 \\ 5 & 6 \\ 7 & 8 \\ 9 & 18 \end{bmatrix}, \text{Polynomial fit: } y = 7.1563x^2 - 22.042x - 0.95833x^3 + 4.6875 \times 10^{-2}$$
$$x^4 + 27.797$$

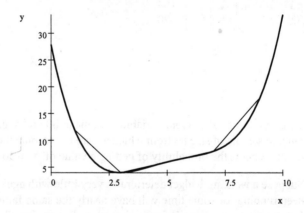

Overdetermined Systems of Equations

The Solve command has been extended to handle overdetermined systems, returning the least-squares solution . Here we give an example of an overdetermined system. Note that (as before) the least-squares solution is the *actual* solution, when an actual solution exists.

▶ Solve + Exact

$$\begin{bmatrix} 1 & 2 \\ 3 & 4 \\ 5 & 6 \\ 7 & 8 \end{bmatrix} \begin{bmatrix} x \\ y \end{bmatrix} = \begin{bmatrix} 3 \\ 7 \\ 11 \\ 15 \end{bmatrix}, \text{Solution is}: \begin{bmatrix} 1 \\ 1 \end{bmatrix}$$

It is easy to multiply both sides of a matrix equation by A^T to check that, when you "solve" $AX = B$, you are actually getting the solution of $\left(A^T A\right) X = A^T B$.

$$\left(A^T A\right) X = \left(\begin{bmatrix} 1 & 3 & 5 & 7 \\ 2 & 4 & 6 & 8 \end{bmatrix} \begin{bmatrix} 1 & 2 \\ 3 & 4 \\ 5 & 6 \\ 7 & 8 \end{bmatrix}\right) \begin{bmatrix} x \\ y \end{bmatrix} = \begin{bmatrix} 84 & 100 \\ 100 & 120 \end{bmatrix} \begin{bmatrix} x \\ y \end{bmatrix}$$

$$A^T B = \begin{bmatrix} 1 & 3 & 5 & 7 \\ 2 & 4 & 6 & 8 \end{bmatrix} \begin{bmatrix} 3 \\ 7 \\ 11 \\ 15 \end{bmatrix} = \begin{bmatrix} 184 \\ 220 \end{bmatrix}$$

This calculation gives the following equation, which has an exact solution.

▶ Evaluate

$$\begin{bmatrix} 84 & 100 \\ 100 & 120 \end{bmatrix} \begin{bmatrix} 1 \\ 1 \end{bmatrix} = \begin{bmatrix} 184 \\ 220 \end{bmatrix}$$

Exercises

1. Consider a normal random variable with mean 50 and standard deviation 10, and a random sample of size 80 from which we are to compute the values of $\bar{X}$, the sample mean. What is the probability of getting a value of $\bar{X}$ as low as 46?

2. Suppose a working widget deteriorates very little with age. That is, a widget that has been running for some time will have nearly the same failure probability during the following hour as it had during its first hour of operation. Then, the failure times have an **exponential distribution** $P(T \leq t)$ of the form $1 - e^{-\frac{x}{\mu}}$. Given that the widget has a mean life of 5 years, what is the probability that the widget will have a lifetime exceeding 7.5 years? If the widget is guaranteed for 2 years, what percentage of such widgets can be expected to need replacement while under warranty?

3. A widget has a mean life of 5 years with a standard deviation of 2 years. Assuming a **normal distribution**, what is the probability that the widget will have a lifetime exceeding 7.5 years? If the widget is guaranteed for 2 years, what percentage of such widgets can be expected to need replacement while under warranty?

4. The mean of a continuous distribution with probability density function $f(u)$ is the integral $\int_{-\infty}^{\infty} uf(u)du = \mu$ of the product of the variable and the probability density function. The variance is the integral $\int_{-\infty}^{\infty} (u - \mu)^2 f(u)du$. Find the mean and variance for each of the continuous distributions discussed in this chapter.

5. The mean of a discrete distribution with probability density function $f(u)$ is the sum $\sum_{-\infty}^{\infty} uf(u) = \mu$, and the variance is $\sum_{-\infty}^{\infty} (u - \mu)^2 f(u) = \sigma^2$.
   - Find the mean and variance for the discrete distributions discussed in this chapter.
   - If the probability density function for a distribution is $f(n) = \left(\frac{1}{2}\right)^n$, what is the mean of the distribution? What is the variance?

6. A die is cast until a 4 appears. What is the probability that it must be cast more than 5 times?

7. A telephone switchboard handles 600 calls on average during a single rush hour. The board can make a maximum of 20 connections per minute. Use the Poisson distribution to evaluate the probability that the board will be overtaxed during any given minute of a rush hour.

8. Find the probability that $x^2 \leq 4$ for a normal distribution with mean 1 and standard deviation 1.

Solutions

1. To solve this problem, you need to know that the distribution of the mean of a sample of size n from a normal distribution of mean μ and standard deviation σ is normal with mean μ and standard deviation $\frac{\sigma}{\sqrt{n}}$. Thus, the probability is

$$\Pr\left(\bar{X} \leq 46\right) = \text{NormalDist}\left(46; 50, \frac{1}{2}\sqrt{5}\right)$$
$$= \text{NormalDist}\left(46; 50, 1.118\right) = 1.7324 \times 10^{-4}$$

2. ExponentialDist $(7.5; 5) = 0.77687 = P(X \leq 7.5)$, so the probability that X is greater than 7.5 is $1 - 0.78 = 0.22$.

   ExponentialDist $(2; 5) = 0.32968 = P(X \leq 2)$, so the answer to the second question is "about 33 percent."

3. NormalDist $(7.5; 5, 2) = 0.89435 = P(X \leq 7.5)$, so the probability that X is greater than 7.5 is $1 - 0.894 = 0.106$, or 10.6 percent.

   NormalDist $(2; 5, 2) = 6.6807 \times 10^{-2} = P(X \leq 2)$, so the answer to the second question is "about 7 percent."

4. For the normal distribution, **Evaluate** gives

$$\int_{-\infty}^{\infty} \text{NormalDen}\,(u;\mu,\sigma)\,u\,du \;=\; \mu$$

$$\int_{-\infty}^{\infty} \text{NormalDen}\,(u;\mu,\sigma)\,(u-\mu)^2\,du \;=\; \sigma^2$$

For the Student's t distribution, with five degrees of freedom, **Evaluate** gives

$$\int_{-\infty}^{\infty} u\,\text{TDen}\,(u;5)\,du = 0$$

for the mean and

$$\int_{-\infty}^{\infty} u^2\,\text{TDen}\,(u;5)\,du = \frac{25}{3}\sqrt{5}$$

for the variance. (When the parameters are symbolic, there are problems in carrying out computations of the integrals that give mean and variance. You should have no difficulty when you specify numerical parameters.)

5. Following is a sample solution.

- For the binomial distribution, **Evaluate** followed by **Simplify** gives the mean:

$$\sum_{x=1}^{\infty} x\binom{n}{x}p^x\,(1-p)^{n-x} = np\,(1-p)^{n-1}\left(-\frac{1}{-1+p}\right)^{n-1} = pn$$

Evaluate followed by **Simplify** and then **Factor** gives the variance:

$$\sum_{x=0}^{n}(x-pn)^2\binom{n}{x}p^x\,(1-p)^{n-x} = pn - p^2 n = (1-p)\,pn$$

(The intermediate expression for the variance is complicated and does not appear here. Also, you need to make the simplifications $(-1)^{2n} = 1$ and $(-1)^{2n+1} = -1$. Note that the symbol $\binom{n}{x}$ is a binomial fraction, rather than a matrix. To enter a binomial fraction, from the **Insert** menu choose **Binomial**, and choose **None** for **Line**.)

- If the probability density function for a distribution is $f(n) = \left(\frac{1}{2}\right)^n$ for $n \geq 0$, then the mean of the distribution is $\sum_{n=1}^{\infty} n\left(\frac{1}{2}\right)^n = 2$, and the variance is $\sum_{n=1}^{\infty}(n-2)^2\left(\frac{1}{2}\right)^n = 2$.

6. The probability of getting a 4 on a single cast is $\frac{1}{6}$, so the probability of getting a different result is $\frac{5}{6}$. The probability of casting the die 5 times without getting a 4 is $\left(\frac{5}{6}\right)^5 = 0.40188$.

7. With 600 calls on average during rush hour, the average number of calls per minute is 10. The probability that the number of connections in a given minute is less than or equal to 20 is the sum $\sum_{k=0}^{20} \text{PoissonDen}(k,10) = \sum_{k=0}^{20} \frac{10^k e^{-10}}{k!} = 0.99841$. Thus, the probability that the board will be overtaxed is $1 - 0.99841 = 0.00159$.

8. $x^2 \leq 4$ when $-2 \leq x \leq 2$. So
$$\begin{aligned}\Pr\left(x^2 \leq 4\right) &= \Pr(x \leq 2) - \Pr(x \leq -2)\\ &= \text{NormalDist}(2;1,1) - \text{NormalDist}(-2;1,1) = 0.84\end{aligned}$$

12 Applied Modern Algebra

In this chapter we describe some of the techniques of applied algebra—including finding integer solutions to equations, solving modular equations, and linear programming—and show how the computational engine in Scientific WorkPlace and Scientific Notebook can be used to help with such computations.

Solving Equations

Many techniques in applied modern algebra are designed to solve equations, from integer equations to polynomial equations to matrix equations. In this section, we describe a few of the methods that can be applied to such problems.

Integer Solutions

The operation Integer on the Solve submenu finds integer solutions to equations and systems of equations.

▶ Solve + Integer

$41x + 421x^2 - 165x^3 - 4x^4 + 4x^5 - 105$, Solution is: $\{-7, 3, 5\}$

Continued Fractions

A simple continued fraction is an expression of the form

$$a_1 + \cfrac{1}{a_2 + \cfrac{1}{a_3 + \cfrac{1}{a_4 + \cfrac{1}{a_5 + \cfrac{1}{1 + \cdots}}}}}$$

The values of $a_1, a_2, a_3, \ldots$ can be either real or complex values. There can be either an infinite or a finite number of terms a_i. A number is rational if and only if it can be expressed as a simple finite continued fraction. You can find rational approximations to irrational numbers by expanding the irrational as a simple continued fraction, then truncating the continued fraction to obtain a rational.

Continued fractions have been utilized within computer algorithms for computing rational approximations to real numbers, as well as solving indeterminate equations. Connections have been established between continued fractions and chaos theory.

Use the MuPAD command **contfrac** to construct a continued fraction of the real numerical expression x to n significant digits.

447

▶ **To define a continued fractions command**

1. From the Definitions submenu, choose Define MuPAD Name.

2. In the MuPAD Name box, enter **numlib::contfrac(x,n)**

3. In the Scientific WorkPlace [Notebook] Name box, enter $r(x, n)$.

4. Check That is built in to MuPAD or is automatically loaded.

5. Choose OK.

Use the continued fractions command to generate continued fractions.

▶ Evaluate

$$r(\pi, 10) = 3 + \cfrac{1}{7 + \cfrac{1}{15 + \cfrac{1}{1 + \cfrac{1}{292 + \cfrac{1}{1 + \cfrac{1}{1 + \ldots}}}}}}$$

The two dots at the bottom indicate an infinite continued fraction. However, you can easily truncate the continued fraction.

▶ **To find a good rational approximation to π**

1. Select the tail $+\cfrac{1}{292 + \cfrac{1}{1 + \cfrac{1}{1 + \ldots}}}$ of the continued fraction and delete it.

2. Evaluate the remaining finite continued fraction.

▶ Evaluate, Evaluate Numerically

$$3 + \cfrac{1}{7 + \cfrac{1}{15 + \cfrac{1}{1}}} = \frac{355}{113} = 3.1416$$

The rational number $\frac{355}{113}$ is good approximation to π. This is to be expected, because continued fractions provide good rational approximations.

Recursive Solutions

Recursion finds solutions to a recursion or a system of recursions. For example, to solve the recursion $y(n + 2) + 3y(n + 1) + 2y(n) = 0$, choose Recursion from the Solve submenu.

▶ Solve + Recursion

$y(n + 2) + 3y(n + 1) + 2y(n) = 0,$

Solution is: $\{y(n) = C_1(-1)^n + C_2(-2)^n\}$

You can also solve recursive equations written in *sequence notation*.

▶ Solve + Recursion

$$x_n + 2x_{n+1} + x_{n+2} = 0, \text{Solution is}: \{x_n = (C_3 + C_4 n)(-1)^n\}$$

You can specify the initial conditions by listing a system of equations in a column matrix.

▶ Solve + Recursion

$$y(n+2) + 3y(n+1) + 2y(n) = 0$$
$$y(0) = -2$$
$$y(1) = 1$$
, Solution is: $\{y(n) = (-2)^n - 3(-1)^n\}$

This closed-form solution now makes it easy to find specific terms. For example, if you define $y(n) = (-2)^n - 3(-1)^n$, then $y(n)$ can be directly evaluated.

▶ Evaluate

$$y(20) = 1048\,573$$

Integers Modulo m

Two integers a and b are *congruent modulo m* if and only if $a - b$ is a multiple of m, in which case we write $a \equiv b \pmod{m}$. Thus, $15 \equiv 33 \pmod{9}$, because $15 - 33 = -18$ is a multiple of 9. Given integers a and m, the *mod function* is given by $a \bmod m = b$ if and only if $a \equiv b \pmod{m}$ and $0 \le b \le m - 1$; hence, $a \bmod m$ is the smallest *nonnegative residue* of a modulo m.

The underlying computer algebra system does not understand the congruence notation $a \equiv b \pmod{m}$, but it does understand the function notation $a \bmod m$. This section shows how to translate problems in algebra and number theory into language that will be handled correctly by the computational engine.

Note that mod is a function of two variables, with the function written between the two variables. This usage is similar to the common usage of $+$, which is also a function of two variables with the function values expressed as $a + b$, rather than the usual functional notation $+(a, b)$.

Traditionally the congruence notation $a \equiv b \pmod{m}$ is written with the $\bmod m$ enclosed inside parentheses since the $\bmod m$ clarifies the expression $a \equiv b$. In this context, the expression $b \pmod{m}$ never appears without the preceding $a \equiv$. On the other hand, the mod function is usually written in the form $a \bmod m$ without parentheses.

▶ **To evaluate the mod function**

1. Leave the insertion point in the expression $a \bmod b$.

2. Choose Evaluate.

▶ Evaluate

$23 \bmod 14 = 9$

$12345678987654 \bmod 9 = 3$

If a is positive, you can also find the smallest nonnegative residue of a modulo m by applying Expand to the quotient $\frac{a}{m}$. This rewrites a fraction as a mixed number.

▶ Expand

$$\frac{23}{14} = 1\frac{9}{14}$$

$$\frac{12345678987654}{9} = 1371\,742\,109\,739\tfrac{1}{3} \text{ (and } \tfrac{1}{3} = \tfrac{3}{9})$$

Since $1\frac{9}{14} = 1 + \frac{9}{14}$, multiplication of $\frac{23}{14} = 1 + \frac{9}{14}$ by 14 gives $23 = 14 \cdot 1 + 9$, which shows that $23 \bmod 14 = 9$. Also, the multiplication of $\frac{12345678987654}{9} = 1371\,742\,109\,739 + \frac{3}{9}$ by 9 shows that $12345678987654 \bmod 9 = 3$.

In terms of the floor function $\lfloor x \rfloor$ (see page 35) the mod function is given by

$$a \bmod m = a - \left\lfloor \frac{a}{m} \right\rfloor m$$

▶ Evaluate

$23 - \left\lfloor \frac{23}{14} \right\rfloor 14 = 9$

$12345678987654 - \left\lfloor \dfrac{12345678987654}{9} \right\rfloor 9 = 3$

Multiplication Tables Modulo m

You can make tables that display the products modulo m of pairs of integers from the set $\{0, 1, 2, \ldots, m-1\}$.

▶ **To get a multiplication table modulo m with $m = 6$**

1. Define the function $g(i, j) = (i - 1)(j - 1)$.

2. From the Matrices submenu, choose Fill Matrix.

3. Select Defined by Function.

4. Enter g in the Enter Function Name box.

5. Select 6 rows and 6 columns.

6. Choose OK.

7. Type mod 6 at the right of the matrix. (Because the insertion point is in mathematics mode; *mod* automatically turns gray.)

8. Choose Evaluate.

▶ Evaluate

$$
\begin{bmatrix}
0 & 0 & 0 & 0 & 0 & 0 \\
0 & 1 & 2 & 3 & 4 & 5 \\
0 & 2 & 4 & 6 & 8 & 10 \\
0 & 3 & 6 & 9 & 12 & 15 \\
0 & 4 & 8 & 12 & 16 & 20 \\
0 & 5 & 10 & 15 & 20 & 25
\end{bmatrix}
\bmod 6 =
\begin{bmatrix}
0 & 0 & 0 & 0 & 0 & 0 \\
0 & 1 & 2 & 3 & 4 & 5 \\
0 & 2 & 4 & 0 & 2 & 4 \\
0 & 3 & 0 & 3 & 0 & 3 \\
0 & 4 & 2 & 0 & 4 & 2 \\
0 & 5 & 4 & 3 & 2 & 1
\end{bmatrix}
$$

A more efficient way to generate the same multiplication table is to define $g(i, j) = (i - 1)(j - 1) \bmod 6$ and follow steps 2-6 above.

You can also find this matrix as the product of a column matrix with a row matrix.

▶ Evaluate

$$
\begin{bmatrix}
0 \\
1 \\
2 \\
3 \\
4 \\
5
\end{bmatrix}
\begin{bmatrix}
0 & 1 & 2 & 3 & 4 & 5
\end{bmatrix}
\bmod 6 =
\begin{bmatrix}
0 & 0 & 0 & 0 & 0 & 0 \\
0 & 1 & 2 & 3 & 4 & 5 \\
0 & 2 & 4 & 0 & 2 & 4 \\
0 & 3 & 0 & 3 & 0 & 3 \\
0 & 4 & 2 & 0 & 4 & 2 \\
0 & 5 & 4 & 3 & 2 & 1
\end{bmatrix}
$$

Make a copy of this last matrix. From the Edit menu, choose Insert Row(s) and add a new row at the top (position 1); choose Insert Column(s).and add a new column at the left (position 1); fill in the blanks and change the new row and column to Bold font, to get the following multiplication table modulo 6:

×	0	1	2	3	4	5
0	0	0	0	0	0	0
1	0	1	2	3	4	5
2	0	2	4	0	2	4
3	0	3	0	3	0	3
4	0	4	2	0	4	2
5	0	5	4	3	2	1

From the table, we see that $2 \cdot 4 \bmod 6 = 2$ and $3 \cdot 3 \bmod 6 = 3$.

A clever approach, that creates this table in one step, is to define

$$g(i, j) = |i - 2| \, |j - 2| \bmod 6$$

Choose Fill Matrix from the Matrices submenu, choose Defined by Function from the dialog box, specify g for the function, and set the matrix size to 7 rows and 7 columns. Then replace the digit 1 in the upper left corner by × and change the first row and column to Bold font, as before.

You can generate an addition table by defining $g(i, j) = i + j - 2 \bmod 6$.

Example If p is a prime, then the integers modulo p form a field, called a *Galois field* and denoted GF_p. For the prime $p = 7$, you can generate the multiplication table by defining $g(i, j) = (i - 1)(j - 1) \bmod 7$ and choosing **Fill Matrix** from the **Matrix** submenu, then selecting **Defined by function** from the dialog box. You can generate the addition table in a similar manner using the function $f(i, j) = i + j - 2 \bmod 7$.

·	0	1	2	3	4	5	6
0	0	0	0	0	0	0	0
1	0	1	2	3	4	5	6
2	0	2	4	6	1	3	5
3	0	3	6	2	5	1	4
4	0	4	1	5	2	6	3
5	0	5	3	1	6	4	2
6	0	6	5	4	3	2	1

+	0	1	2	3	4	5	6
0	0	1	2	3	4	5	6
1	1	2	3	4	5	6	0
2	2	3	4	5	6	0	1
3	3	4	5	6	0	1	2
4	4	5	6	0	1	2	3
5	5	6	0	1	2	3	4
6	6	0	1	2	3	4	5

Inverses Modulo m

If $ab \bmod m = 1$, then b is called an *inverse* of a *modulo* m, and we write $a^{-1} \bmod m$ for the least positive residue of b. The computation engine also recognizes both of the forms $1/a \bmod m$ and $\frac{1}{a} \bmod m$ for the inverse modulo m.

▶ **Evaluate**

$$5^{-1} \bmod 7 = 3 \qquad \tfrac{1}{5} \bmod 7 = 3 \qquad 1/5 \bmod 7 = 3$$

This calculation satisfies the definition of inverse, because $5 \cdot 3 \bmod 7 = 1$.

▶ **Evaluate**

$$23^{-1} \bmod 257 = 190 \qquad \tfrac{1}{5} \bmod 6 = 5$$

The three notations $ab^{-1} \bmod m$, $a/b \bmod m$, and $\frac{a}{b} \bmod m$ are all interpreted as $a(b^{-1} \bmod m) \bmod m$ that is, first find the inverse of b modulo m, multiply the result by a, and then reduce the product modulo m.

▶ **Evaluate**

$$3/23 \bmod 257 = 56 \qquad \tfrac{2}{5} \bmod 6 = 4$$

Note that $a^{-1} \bmod m$ exists if and only if a is *relatively prime* to m; that is, it exists if and only if $\gcd(a, m) = 1$. Thus, modulo 6, only 1 and 5 have inverses. Modulo any prime, every nonzero residue has an inverse. In terms of the multiplication table modulo m, the integer a has an inverse modulo m if and only if 1 appears in row $a \bmod m$ (and 1 appears in column $a \bmod m$).

Solving Congruences Modulo m

To solve a congruence of the form $ax \equiv b \pmod{m}$, multiply both sides by $a^{-1} \bmod m$ to get $x = b/a \bmod m$.

The congruence $17x \equiv 23 \pmod{127}$ has a solution $x = 91$, as the following two evaluations illustrate.

▶ Evaluate

$23/17 \bmod 127 = 91$

Check this result by substitution back into the original congruence.

▶ Evaluate

$17 \cdot 91 \bmod 127 = 23$

Note that, since 91 is a solution to the congruence $17x \equiv 23 \pmod{127}$, additional solutions are given by $91 + 127n$, where n is any integer. In fact, $x \equiv 91 \pmod{127}$ is just another way of writing $x = 91 + 127n$ for some integer n.

Pairs of Linear Congruences

Since linear congruences of the form $ax \equiv b \pmod{m}$ can be reduced to simple congruences of the form $x \equiv c \pmod{m}$, we consider systems of congruences in this latter form.

Example   Consider the system of two congruences
$$x \equiv 45 \pmod{237}$$
$$x \equiv 19 \pmod{419}$$
Checking, $\gcd(237, 419) = 1$, so 237 and 419 are relatively prime. The first congruence can be rewritten in the form $x = 45 + 237k$ for some integer k. Substituting this value into the second congruence, we see that
$$45 + 237k = 19 + 419r$$
for some integer r. This last equation can be rewritten in the form $237k = 19 - 45 \bmod 419$, which has the solution
$$k = (19 - 45)/237 \bmod 419 = 60$$
Hence,
$$x = 45 + 237 \cdot 60 = 14265$$
Checking, $14265 \bmod 237 = 45$ and $14265 \bmod 419 = 19$.
The complete set of solutions is given by
$$x = 14265 + 237 \cdot 419s \equiv 14265 \pmod{99303}$$
Thus, the original pair of congruences has been reduced to a single congruence,
$$x \equiv 14265 \pmod{99303}$$

In general, if m and n are relatively prime, then a solution to the pair

$$x \equiv a \pmod{m}$$
$$x \equiv b \pmod{n}$$

is given by

$$x = a + m\left[(b - a)/m \bmod n\right]$$

A complete set of solutions is given by

$$x = a + m\left[(b - a)/m \bmod n\right] + rmn$$

where r is an arbitrary integer.

Systems of Linear Congruences

You can reduce systems of any number of congruences to a single congruence by solving systems of congruences two at a time. The *Chinese remainder theorem* states that, if the moduli are relatively prime in pairs, then there is a unique solution modulo the product of all the moduli.

Example   Consider the system of three linear congruences

$$x \equiv 45 \pmod{237}$$
$$x \equiv 19 \pmod{419}$$
$$x \equiv 57 \pmod{523}$$

Checking, $\gcd(237 \cdot 419, 523) = 1$; hence this system has a solution. The first two congruences can be replaced by the single congruence $x \equiv 14265 \pmod{99303}$; hence the three congruences can be replaced by the pair

$$x \equiv 14265 \pmod{99303}$$
$$x \equiv 57 \pmod{523}$$

As before, $14265 + 99303k = 57 + 523r$ for some integers k and r. Thus, $k = (57 - 14265)/99303 \bmod 523 = 134$; hence $x = 14265 + 99303 \cdot 134 = 13320867$. This system of three congruences can thus be reduced to the single congruence

$$x \equiv 13320867 \pmod{51935469}$$

Extended Precision Arithmetic

Computer algebra systems support exact sums and products of integers that are hundreds of digits long. One way to do such extended precision arithmetic is to generate a set of mutually relatively prime bases, and to do modular arithmetic modulo all of these bases. For example, consider the vector

$$(997, 999, 1000, 1001, 1003, 1007, 1009)$$

of bases. Factorization shows that the entries are pairwise relatively prime.

▶ Factor

$$
\begin{bmatrix}
997 \\
999 \\
1000 \\
1001 \\
1003 \\
1007 \\
1009
\end{bmatrix}
=
\begin{bmatrix}
997 \\
3^3 37 \\
2^3 5^3 \\
7 \times 11 \times 13 \\
17 \times 59 \\
19 \times 53 \\
1009
\end{bmatrix}
$$

Consider the two numbers 23890864094 and 1883289456. You can represent these numbers by reducing the numbers modulo each of the bases. Thus,

$$
23890864094 \longleftrightarrow
\begin{bmatrix}
23890864094 \bmod 997 \\
23890864094 \bmod 999 \\
23890864094 \bmod 1000 \\
23890864094 \bmod 1001 \\
23890864094 \bmod 1003 \\
23890864094 \bmod 1007 \\
23890864094 \bmod 1009
\end{bmatrix}
=
\begin{bmatrix}
350 \\
872 \\
94 \\
97 \\
879 \\
564 \\
218
\end{bmatrix}
$$

$$
1883289456 \longleftrightarrow
\begin{bmatrix}
1883289456 \bmod 997 \\
1883289456 \bmod 999 \\
1883289456 \bmod 1000 \\
1883289456 \bmod 1001 \\
1883289456 \bmod 1003 \\
1883289456 \bmod 1007 \\
1883289456 \bmod 1009
\end{bmatrix}
=
\begin{bmatrix}
324 \\
630 \\
456 \\
48 \\
488 \\
70 \\
37
\end{bmatrix}
$$

Thus, the product $23890864094 \cdot 1883289456$ is represented by the vector

$$
\begin{bmatrix}
350 \cdot 324 \bmod 997 \\
872 \cdot 630 \bmod 999 \\
94 \cdot 456 \bmod 1000 \\
97 \cdot 48 \bmod 1001 \\
879 \cdot 488 \bmod 1003 \\
564 \cdot 70 \bmod 1007 \\
218 \cdot 37 \bmod 1009
\end{bmatrix}
=
\begin{bmatrix}
739 \\
909 \\
864 \\
652 \\
671 \\
207 \\
1003
\end{bmatrix}
$$

The product $23890864094 \cdot 1883289456$ is now a solution to the system

$$
\begin{aligned}
x &\equiv 739 \pmod{997} \\
x &\equiv 909 \pmod{999} \\
x &\equiv 864 \pmod{1000} \\
x &\equiv 652 \pmod{1001} \\
x &\equiv 671 \pmod{1003} \\
x &\equiv 207 \pmod{1007} \\
x &\equiv 1003 \pmod{1009}
\end{aligned}
$$

Powers Modulo m

▶ **To calculate large powers modulo** m

- Evaluate $a^n \bmod m$.

Example   Define $a = 2789596378267275$, $n = 3848590389047349$, and $m = 2838490563537459$. Applying the command **Evaluate** to $a^n \bmod m$ yields the following:
$$a^n \bmod m = 26220\,18141\,09828$$

Fermat's Little Theorem states that, if p is prime and $0 < a < p$, then
$$a^{p-1} \bmod p = 1$$
The integer 1009 is prime, and the following is no surprise.

▶ Evaluate

$2^{1008} \bmod 1009 = 1$

Generating Large Primes

There is not a built-in function to generate large primes, but the underlying computational systems do have such a function. The following is an example of how to define functions that correspond to existing functions in the underlying computational system. (See page 127 for another example of accessing such a function.)

In this example, $p(x)$ is defined as the Scientific WorkPlace (Notebook) Name for the MuPAD function, **nextprime(x)**, which generates the first prime greater than or equal to x.

▶ **To define** $p(x)$ **as the next-prime function**

1. From the Definitions submenu, choose Define MuPAD Name.

2. Enter nextprime(x) as the MuPAD Name.

3. Enter $p(x)$ as the Scientific WorkPlace (Notebook) Name.

4. Under The MuPAD Name is a Procedure, check That is Built In to MuPAD or is Automatically Loaded.

5. Choose OK.

Test the function using Evaluate.

▶ **Evaluate**

$p(5) = 5$ $\qquad$ $p(500) = 503$ $\qquad$ $p(8298) = 8311$

$p\,(273849728952758923) = 273\,849\,728\,952\,758\,923$

Example The Rivest-Shamir-Adleman (RSA) cipher system is based directly on Euler's theorem and requires a pair of large primes. First, generate a pair of large primes— say,

$$q = p(20934834573) = 20934834647$$

and

$$r = p\,(2593843747347) = 2593843747457$$

(In practice, larger primes are used; such as, $q \approx 10^{100}$ and $r \approx 10^{100}$.) Then

$$
\begin{aligned}
n \;&=\; qr \\
&=\; 20934834647 \cdot 2593843747457 \\
&=\; 543\,01689\,95316\,71217\,42679
\end{aligned}
$$

and the number of positive integers $\leq n$ and relatively prime to n is given by

$$
\begin{aligned}
\varphi(n) \;&=\; (q-1)(r-1) \\
&=\; 20934834646 \cdot 2593843747456 \\
&=\; 543\,01689\,95055\,23431\,60576
\end{aligned}
$$

Let $x = 29384737849576728375$ be plaintext (suitably generated by a short section of English text). Long messages must be broken up into small enough chunks that each plaintext integer x is smaller than the modulus n. Choose E to be a moderately large positive integer that is relatively prime to $\varphi(n)$, for example, $E = 1009$. The ciphertext is given by

$$y = x^E \bmod n = 20636340188476258131729$$

Let

$$D = 1009^{-1} \bmod \varphi(n) = 4251569381658706748945$$

Then friendly colleagues can recover the plaintext by calculating

$$z = y^D \bmod n = 29384737849576728375$$

Other Systems Modulo m

The mod function also works with matrices and with polynomials.

Matrices Modulo m

To reduce a matrix A modulo m, enter the expression $A \bmod m$ and evaluate it.

▶ Evaluate

$$\begin{bmatrix} 5 & 8 \\ 9 & 4 \end{bmatrix} \bmod 3 = \begin{bmatrix} 2 & 2 \\ 0 & 1 \end{bmatrix}$$

$$\begin{pmatrix} 3 & 7 & 5 \\ 5 & 4 & 8 \\ 2 & 0 & 5 \end{pmatrix}^{-1} \bmod 11 = \begin{pmatrix} 9 & 9 & 3 \\ 2 & 5 & 1 \\ 3 & 3 & 10 \end{pmatrix}$$

$$\begin{pmatrix} 3 & 7 & 5 \\ 5 & 4 & 8 \\ 2 & 0 & 5 \end{pmatrix} \begin{pmatrix} 9 & 9 & 3 \\ 2 & 5 & 1 \\ 3 & 3 & 10 \end{pmatrix} \bmod 11 = \begin{pmatrix} 1 & 0 & 0 \\ 0 & 1 & 0 \\ 0 & 0 & 1 \end{pmatrix}$$

Example   The Hamming (7,4) code operates on 4-bit message words (half a byte, and hence a *nibble*) and generates 7-bit code words. Let

$$H = \begin{bmatrix} 1 & 1 & 1 & 0 & 0 & 0 & 0 \\ 1 & 0 & 0 & 1 & 1 & 0 & 0 \\ 0 & 1 & 0 & 1 & 0 & 1 & 0 \\ 1 & 1 & 0 & 1 & 0 & 0 & 1 \end{bmatrix}$$

be the Hamming encoding matrix. Given a message word $\mathbf{m} = \begin{bmatrix} 1 & 0 & 1 & 1 \end{bmatrix}$, the corresponding code word is given by

$$\begin{aligned} \mathbf{c} &= \mathbf{m}H \bmod 2 \\ &= \begin{bmatrix} 1 & 0 & 1 & 1 \end{bmatrix} \begin{bmatrix} 1 & 1 & 1 & 0 & 0 & 0 & 0 \\ 1 & 0 & 0 & 1 & 1 & 0 & 0 \\ 0 & 1 & 0 & 1 & 0 & 1 & 0 \\ 1 & 1 & 0 & 1 & 0 & 0 & 1 \end{bmatrix} \bmod 2 \\ &= \begin{bmatrix} 0 & 1 & 1 & 0 & 0 & 1 & 1 \end{bmatrix} \end{aligned}$$

The message word appears in bits 3, 5, 6, and 7. The remaining 3 bits can be thought of as generalized parity bits. The code word is then transmitted.

At the receiving end, another matrix multiplication is done to check for errors. Let

$$P = \begin{bmatrix} 0 & 0 & 0 & 1 & 1 & 1 & 1 \\ 0 & 1 & 1 & 0 & 0 & 1 & 1 \\ 1 & 0 & 1 & 0 & 1 & 0 & 1 \end{bmatrix}^T$$

be the parity check matrix. If no errors occur, then the matrix product

$$\begin{aligned} \mathbf{c}P \bmod 2 &= \begin{bmatrix} 0 & 1 & 1 & 0 & 0 & 1 & 1 \end{bmatrix} \begin{bmatrix} 0 & 0 & 0 & 1 & 1 & 1 & 1 \\ 0 & 1 & 1 & 0 & 0 & 1 & 1 \\ 1 & 0 & 1 & 0 & 1 & 0 & 1 \end{bmatrix}^T \bmod 2 \\ &= \begin{bmatrix} 0 & 0 & 0 \end{bmatrix} \end{aligned}$$

indicates that no errors were detected and the message word $\mathbf{m}$ is taken to be bits 3, 5, 6, and 7 of the code word.

However, if the code word is corrupted and $c' = \begin{bmatrix} 0 & 0 & 1 & 0 & 0 & 1 & 1 \end{bmatrix}$ is received, then

$$c'P \bmod 2 = \begin{bmatrix} 0 & 0 & 1 & 0 & 0 & 1 & 1 \end{bmatrix} \begin{bmatrix} 0 & 0 & 0 & 1 & 1 & 1 & 1 \\ 0 & 1 & 1 & 0 & 0 & 1 & 1 \\ 1 & 0 & 1 & 0 & 1 & 0 & 1 \end{bmatrix}^T \bmod 2$$

$$= \begin{bmatrix} 0 & 1 & 0 \end{bmatrix}$$

which is the binary equivalent of 2, and hence an error occurred in the second bit. Thus, the correct code word is $\begin{bmatrix} 0 & 1 & 1 & 0 & 0 & 1 & 1 \end{bmatrix}$. The message word from columns 3, 5, 6, and 7 is then $\begin{bmatrix} 1 & 0 & 1 & 1 \end{bmatrix}$.

Example A 2×2 *block cipher* is given by

$$\begin{bmatrix} y_1 \\ y_2 \end{bmatrix} = \begin{bmatrix} a_{11} & a_{12} \\ a_{21} & a_{22} \end{bmatrix} \begin{bmatrix} x_1 \\ x_2 \end{bmatrix} \bmod 26$$

where the x_is represent plaintext, the y_is represent ciphertext, and the matrix entries are integers. For example, $\begin{bmatrix} 5 & 8 \\ 2 & 7 \end{bmatrix} \begin{bmatrix} 4 \\ 11 \end{bmatrix} \bmod 26 = \begin{bmatrix} 4 \\ 7 \end{bmatrix}$ means that the plaintext pair $[E, L]$ (two adjacent letters in the secret message "Elroy was here") gets mapped to the ciphertext pair $[E, H]$, using the correspondence $A \leftrightarrow 0$, $B \leftrightarrow 1$, $C \leftrightarrow 2, \ldots$, $Z \leftrightarrow 25$.

Given the ciphertext, you can recover the plaintext by computing the inverse of the two-by-two matrix modulo 26. For example,

$$\begin{bmatrix} 5 & 8 \\ 2 & 7 \end{bmatrix}^{-1} \bmod 26 = \begin{bmatrix} 25 & 16 \\ 4 & 3 \end{bmatrix}$$

and hence

$$\begin{bmatrix} 25 & 16 \\ 4 & 3 \end{bmatrix} \begin{bmatrix} 4 \\ 7 \end{bmatrix} \bmod 26 = \begin{bmatrix} 4 \\ 11 \end{bmatrix}$$

recovers the original plaintext. You can handle longer messages by replacing the column vector $\begin{bmatrix} E \\ L \end{bmatrix}$ by the matrix $\begin{bmatrix} E & R & Y & A & H & R \\ L & O & W & S & E & E \end{bmatrix}$ and calculating the product

$$\begin{bmatrix} 25 & 16 \\ 4 & 3 \end{bmatrix} \begin{bmatrix} 4 & 17 & 24 & 0 & 7 & 17 \\ 11 & 14 & 22 & 18 & 4 & 4 \end{bmatrix} \bmod 26 = \begin{bmatrix} 16 & 25 & 16 & 2 & 5 & 21 \\ 23 & 6 & 6 & 2 & 14 & 2 \end{bmatrix}$$

Polynomials Modulo m

The mod function can also be combined with polynomials to reduce each of the coefficients modulo m.

▶ Evaluate

$$x^5 + 9x^4 - x^3 + 7x - 2 \bmod 5 = x^5 + 4x^4 + 4x^3 + 2x + 3$$

Given a prime p, the set of polynomials with coefficients reduced modulo p is a ring, denoted by $GF_p[x]$.

▶ **To calculate a product of polynomials** $a(x)$ **and** $b(x)$ **in** $GF_p[x]$

1. Expand the product $a(x)b(x)$.

2. Reduce the product modulo p.

To calculate the product of $4x^5 + 5x + 3$ and $6x^4 + x^3 + 3$ in $GF_7[x]$, do the following two operations.

▶ Expand

$$\left(4x^5 + 5x + 3\right)\left(6x^4 + x^3 + 3\right) = 24x^9 + 4x^8 + 42x^5 + 23x^4 + 3x^3 + 15x + 9$$

▶ Evaluate

$$24x^9 + 4x^8 + 42x^5 + 23x^4 + 3x^3 + 15x + 9 \bmod 7 = 3x^9 + 4x^8 + 2x^4 + 3x^3 + x + 2$$

The sum of $4x^5 + 5x + 3$ and $6x^4 + x^3 + 3$ in $GF_7[x]$ is slightly simpler.

▶ Evaluate

$$\left(4x^5 + 5x + 3\right) + \left(6x^4 + x^3 + 3\right) \bmod 7 = 4x^5 + 6x^4 + x^3 + 5x + 6$$

▶ **To factor a polynomial** $a(x)$ **in** $GF_p[x]$

• Factor $a(x) \bmod p$

To factor $x^{16} + x$ in $GF_2[x]$, apply the command Factor to the expression $x^{16} + x \bmod 2$.

▶ Factor

$$x^{16} + x \bmod 2$$
$$= x\left(x+1\right)\left(x+x^2+1\right)\left(x+x^4+1\right)\left(x^3+x^4+1\right)\left(x+x^2+x^3+x^4+1\right)$$

Notice that $x^{16} + x$ factors as the product of all of the irreducible polynomials of degrees 1, 2, and 4. In particular, $x^2 + x + 1$ is the only irreducible polynomial of degree 2 in $GF_2[x]$.

Polynomials Modulo Polynomials

Two polynomials $f(x)$ and $g(x)$ are *congruent modulo a polynomial* $q(x)$ if and only if $f(x) - g(x)$ is a multiple of $q(x)$, in which case we write

$$f(x) \equiv g(x) \pmod{q(x)}$$

We write
$$g(x) \bmod q(x) = h(x)$$
if $h(x)$ is a polynomial of minimal degree that is congruent to $g(x)$ modulo $q(x)$.

▶ **Evaluate**

$$x^4 + x + 1 \bmod \left(x^2 + 4x + 5\right) = -23x - 54$$

To verify this calculation, note the following computation.

▶ **Polynomials + Divide**

$$\frac{x^4 + x + 1}{x^2 + 4x + 5} = x^2 - 4x + 11 + \frac{-23x - 54}{x^2 + 4x + 5}$$

This result implies that indeed $x^4 + x + 1 \bmod \left(x^2 + 4x + 5\right) = -23x - 54$.

Greatest Common Divisor of Polynomials

The *greatest common divisor of two polynomials* $p(x)$ and $q(x)$ is a polynomial $d(x)$ of highest degree that divides both $p(x)$ and $q(x)$.

Define $p(x) = 18x^7 - 9x^5 + 36x^4 + 4x^3 - 16x^2 + 19x + 12$ and $q(x) = 15x^5 - 9x^4 + 11x^3 + 17x^2 - 10x + 8$, then use **Evaluate** to calculate $\gcd\left(p(x), q(x)\right)$.

▶ **Evaluate**

$$\gcd\left(p(x), q(x)\right) = 3x^3 + x + 4$$

Use the following procedure to verify that $3x^3 + x + 4$ is indeed a common divisor.

▶ **Polynomials + Divide**

$$\frac{18x^7 - 9x^5 + 36x^4 - 5x^3 - 16x^2 + 16x}{3x^3 + x + 4} = 4x - 5x^2 + 6x^4$$

$$\frac{15x^5 - 9x^4 + 11x^3 + 17x^2 - 10x + 8}{3x^3 + x + 4} = 5x^2 - 3x + 2$$

This result demonstrates that $p(x) = \left(6x^4 - 5x^2 + 4x\right)\left(3x^3 + x + 4\right)$ and $q(x) = \left(5x^2 - 3x + 2\right)\left(3x^3 + x + 4\right)$.

Multiplicity of Roots of Polynomials

A root a of a polynomial $f(x)$ has *multiplicity* k if $f(x) = (x - a)^k g(x)$, where $g(a) \neq 0$. If $k > 1$, then $f'(x) = k(x - a)^{k-1}g(x) + (x - a)^k g'(x) = (x - a)^{k-1}(kg(x) + (x - a)g'(x))$, and hence $\gcd(f(x), f'(x)) = (x - a)^{k-1}h(x) \neq 1$. This observation provides a test for multiple roots: If $\gcd(f(x), f'(x))$ is a constant, then $f(x)$ has no multiple roots; otherwise, $f(x)$ has at least one multiple root—in fact, each root of $\gcd(f(x), f'(x))$ is a multiple root of $f(x)$.

The graphs of
$$f(x) = 5537x^5 - 34804x^4 + 60229x^3 - 29267x^2 + 19888x + 54692$$
and
$$g(x) = 5537x^5 - 34797x^4 + 60207x^3 - 29260x^2 + 19873x + 54670$$
appear indistinguishable. Both appear to have a root near 3.1.

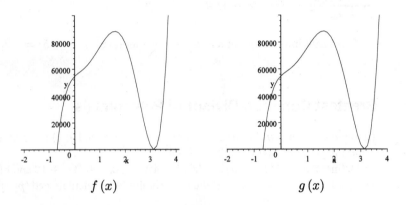

$$f(x) \qquad\qquad\qquad g(x)$$

However, the test for multiple roots gives a different result for the two functions.

▶ Evaluate

$$\gcd(f(x), f'(x)) = 791x - 2486 = 791\left(x - \tfrac{2486}{791}\right) \qquad\qquad \gcd(g(x), g'(x)) = 7$$

Thus, $x = 2486/791 = 22/7$ is a root of $f(x)$ of multiplicity at least two, whereas $g(x)$ has no multiple roots. Solving $f(x) = 0$ and $g(x) = 0$, the real solutions are as follows. We show both symbolic exact and numeric solutions.

▶ Solve + Exact

$$f(x) = 0, \text{ Solution is: } \tfrac{22}{7}, \; \sqrt[3]{\tfrac{1}{108}\sqrt{31}\sqrt{108}} - \tfrac{1}{2} - \frac{1}{3\sqrt[3]{\tfrac{1}{108}\sqrt{31}\sqrt{108}} - \tfrac{1}{2}}$$

$$g(x) = 0, \text{ Solution is: } \tfrac{22}{7}, \tfrac{355}{113}, \; \sqrt[3]{\tfrac{1}{108}\sqrt{31}\sqrt{108}} - \tfrac{1}{2} - \frac{1}{3\sqrt[3]{\tfrac{1}{108}\sqrt{31}\sqrt{108}} - \tfrac{1}{2}}$$

▶ Solve + Numeric

$f(x) = 0$, Solution is: $\{[x = 0.341\,16 - 1.161\,5i], [x = 0.341\,16 + 1.161\,5i],$
  $[x = -0.682\,33], [x = 3.142\,9]\}$

$g(x) = 0$, Solution is: $\{[x = 0.341\,16 + 1.161\,5i], [x = 0.341\,16 - 1.161\,5i],$
  $[x = -0.682\,33], [x = 3.141\,6], [x = 3.142\,9]\}$

Note   If you do not obtain all of these solutions, from the **Tools** menu, choose **Engine Setup**. On the **General** page, verify that the **Solve Options**—Principal Value Only and Ignore Special Cases—are not checked.

To find an approximation to the roots of these two polynomials with multiplicities, from the **Polynomials** submenu, apply **Roots**.

▶ **Polynomials + Roots**

$$f(x),\text{ roots: } \begin{bmatrix} 3.142\,9 \\ 3.142\,9 \\ -0.682\,33 \\ 0.341\,16 + 1.161\,5i \\ 0.341\,16 - 1.161\,5i \end{bmatrix} \qquad g(x),\text{ roots: } \begin{bmatrix} 3.142\,9 \\ 3.141\,6 \\ -0.682\,33 \\ 0.341\,16 - 1.161\,5i \\ 0.341\,16 + 1.161\,5i \end{bmatrix}$$

Thus, g has two distinct real roots that are extremely close, whereas f has a real root of multiplicity two at $\frac{22}{7} \approx 3.142\,9$.

The Galois Field GF_{p^n}

Assume that $q(x)$ is an *irreducible* polynomial of degree n over GF_p; that is, assume that $q(x)$ is of degree n and, whenever $q(x) = a(x)b(x)$ for some $a(x)$ and $b(x)$ in $GF_p[x]$, then either $\deg(a(x)) = 0$ or $\deg(b(x)) = 0$. Given two polynomials $f(x)$ and $g(x)$ in $GF_p[x]$, define the product to be the polynomial $(f(x)g(x) \bmod q(x)) \bmod p$ and the sum to be the polynomial $(f(x) + g(x)) \bmod p$. With these definitions, the set of polynomials in $GF_p[x]$ of degree less than n forms a field called the *Galois field* GF_{p^n}.

The set of polynomials in $GF_2[x]$ of degree less than 2 forms the field $GF_{2^2} = GF_4$. The multiplication and addition tables for GF_2 are given by

×	0	1
0	0	0
1	0	1

+	0	1
0	0	1
1	1	0

The polynomial $q(x) = x^2 + x + 1$ is an irreducible polynomial of degree 2 over GF_2. It is, in fact, the only one. The elements of GF_4 are 0, 1, x, and $1 + x$.

To find the product $x \cdot x$ in GF_4, reduce the product modulo $x^2 + x + 1$, then reduce the result modulo 2.

▶ Evaluate

$$(x^2 \bmod q(x)) \bmod 2 = x + 1$$

Thus, $x^2 = x + 1$ in GF_4. You can generate the entire multiplication table efficiently using matrix and modular arithmetic.

▶ Evaluate

$$
\begin{bmatrix} 0 \\ 1 \\ x \\ x+1 \end{bmatrix}
\begin{bmatrix} 0 & 1 & x & x+1 \end{bmatrix}
=
\begin{bmatrix}
0 & 0 & 0 & 0 \\
0 & 1 & x & x+1 \\
0 & x & x^2 & x(x+1) \\
0 & x+1 & x(x+1) & (x+1)^2
\end{bmatrix}
$$

$$
\begin{bmatrix}
0 & 0 & 0 & 0 \\
0 & 1 & x & x+1 \\
0 & x & x^2 & x(x+1) \\
0 & x+1 & x(x+1) & (x+1)^2
\end{bmatrix}
\bmod q(x)
=
\begin{bmatrix}
0 & 0 & 0 & 0 \\
0 & 1 & x & x+1 \\
0 & x & -x-1 & -1 \\
0 & x+1 & -1 & x
\end{bmatrix}
$$

$$
\begin{bmatrix}
0 & 0 & 0 & 0 \\
0 & 1 & x & x+1 \\
0 & x & -x-1 & -1 \\
0 & x+1 & -1 & x
\end{bmatrix}
\bmod 2
=
\begin{bmatrix}
0 & 0 & 0 & 0 \\
0 & 1 & x & x+1 \\
0 & x & x+1 & 1 \\
0 & x+1 & 1 & x
\end{bmatrix}
$$

Sums require only reduction of polynomial sums modulo 2. The multiplication and addition tables are given by

×	0	1	x	$x+1$
0	0	0	0	0
1	0	1	x	$x+1$
x	0	x	$x+1$	1
$x+1$	0	$x+1$	1	x

+	0	1	x	$x+1$
0	0	1	x	$x+1$
1	1	0	$x+1$	x
x	x	$x+1$	0	1
$x+1$	$x+1$	x	1	0

Given a polynomial $f(x) = ax + b$ with a and b in GF_2, consider the binary representation $(ab)_2$. The binary representations for the multiplication and addition tables for GF_4 are given by

×	00	01	10	11
00	00	00	00	00
01	00	01	10	11
10	00	10	11	01
11	00	11	01	10

+	00	11	10	11
00	00	01	10	11
01	01	00	11	10
10	10	11	00	01
11	11	10	01	00

Converting from binary to decimal, we have $0 = (00)_2$, $1 = (01)_2$, $2 = (10)_2$, and $3 = (11)_2$. Using this shorthand notation for polynomials, the multiplication and addition tables become

×	0	1	2	3
0	0	0	0	0
1	0	1	2	3
2	0	2	3	1
3	0	3	1	2

+	0	1	2	3
0	0	1	2	3
1	1	0	3	2
2	2	3	0	1
3	3	2	1	0

Calculations in larger finite Galois fields can be done without generating addition and multiplication tables. In the following few paragraphs, assume that α is a root of the irreducible polynomial $q(x)$ of degree n used to generate GF_{p^n}. Since every element of GF_{p^n} satisfies the polynomial $x^{p^n} - x$ modulo p, it follows that every nonzero element u of GF_{p^n} satisfies the polynomial $x^{p^n-1} - 1$ modulo p, and hence the inverse of u is given by u^{p^n-2}.

▶ **To calculate the inverse of an element u in GF_{p^n}**

- Evaluate $\left(u^{p^n-2} \bmod q(\alpha)\right) \bmod p$

Let $q(x) = x^4 + x + 1$ and let α be a root of $q(x)$, so that $\alpha^4 + \alpha + 1 = 0$. To calculate the inverse of $\alpha^3 + \alpha^2 + 1$ in GF_{2^4}, carry out the following steps.

▶ Evaluate

$$\left(\left(\alpha^3 + \alpha^2 + 1\right)^{14} \bmod \alpha^4 + \alpha + 1\right) \bmod 2 = \alpha^2$$

▶ **To calculate the product of two elements u and v in GF_{p^n}**

1. Expand the product uv.

2. Evaluate the result modulo $q(\alpha)$.

3. Evaluate the result modulo p.

Let $q(x) = x^4 + x + 1$ and let α be a root of $q(x)$, so that $\alpha^4 + \alpha + 1 = 0$. To calculate the product of $u = \alpha^3 + \alpha^2 + 1$ and $v = \alpha^2$ in GF_{2^4}, carry out the following steps.

▶ Expand

$$\left(\alpha^3 + \alpha^2 + 1\right)\alpha^2 = \alpha^5 + \alpha^4 + \alpha^2$$

▶ Evaluate

$$\alpha^5 + \alpha^4 + \alpha^2 \bmod \alpha^4 + \alpha + 1 = -2\alpha - 1$$
$$-2\alpha - 1 \bmod 2 = 1$$

These steps can also be combined.

▶ Evaluate

$$\left(\left(\alpha^3 + \alpha^2 + 1\right)\alpha^2 \bmod \alpha^4 + \alpha + 1\right) \bmod 2 = 1$$

Example   This setting provides the basis for the Bose-Chaudhuri-Hocquenghem (BCH) Codes. Given the message word $(a_r, a_{r-1}, ..., a_2, a_1, a_0)_2$ as a number in base 2, associate the polynomial

$$a(x) = a_r x^r + a_{r-1} x^{r-1} + \cdots + a_2 x^2 + a_1 x + a_0$$

in $GF_2[x]$. A codeword is then generated by the formula $a(x)q(x) \bmod 2$, where $q(x)$ is a specially selected polynomial.

Consider the Galois field $GF_{2^4} = GF_{16}$. Let α be a primitive element in GF_{16}, so that the nonzero elements of GF_{16} are all powers of α. In particular, this property holds if we take α to be a root of the irreducible polynomial $x^4 + x + 1$. Let $m_i(x)$ be the minimal polynomial of α^i. If

$$q(x) = \mathrm{lcm}[m_1(x), m_2(x), ..., m_{2t}(x)]$$

then the corresponding BCH code corrects at least t errors.

Since $\alpha^4 + \alpha + 1 = 0$, it follows that

$$0^2 = \left(\alpha^4 + \alpha + 1\right)^2 = \left(\alpha^4\right)^2 + \alpha^2 + 1 = \left(\alpha^2\right)^4 + \alpha^2 + 1$$

Hence, $m_1(x) = m_2(x)$. By the same reasoning, $m_2(x) = m_4(x) = m_8(x)$. Likewise,

$$m_3(x) = (x - \alpha^3)(x - \alpha^6)(x - \alpha^{12})(x - \alpha^9) = x^4 + x^3 + x^2 + x + 1$$

Hence, a double error-correcting code is generated by

$$q(x) = \mathrm{lcm}[m_1(x), m_2(x), m_3(x), m_4(x)] \bmod 2 = x^8 + x^7 + x^6 + x^4 + 1$$

Linear Programming

A *linear programming problem* consists of minimizing (or maximizing) a linear function subject to certain conditions or constraints expressible as linear inequalities. The word "programming" is used here in the sense of "planning." The importance of linear programming derives in part from its many applications and in part from the existence of good general-purpose techniques for finding optimal solutions.

The Simplex Algorithm

The basic purpose of the simplex algorithm is to solve linear programming problems. In the following example, the function $f(x, y) = x + y$ is to be maximized subject to the two inequalities shown. The function $f(x, y)$ is the *objective function*, and the set of linear constraints is called the *linear system*.

▶ **To enter a linear programming problem with two constraints**

1. Create a 3×1 matrix.

2. Type the function to be maximized in the first row.

3. Type the linear constraints in the subsequent rows.

4. Leave the insertion point in the matrix.

5. From the Simplex submenu, choose Maximize.

▶ Simplex + Maximize

$$\begin{bmatrix} x + y \\ 4x + 3y \leq 6 \\ 3x + 4y \leq 4 \end{bmatrix}, \text{Maximum is at: } \left\{ x = \tfrac{12}{7}, y = -\tfrac{2}{7} \right\}$$

Of course, these are the same coordinates that minimize $-x - y$. In the following linear programming problem, place the insertion point in the matrix and, from the Simplex submenu, choose Minimize.

▶ Simplex + Minimize

$$\begin{bmatrix} -x - y \\ 4x + 3y \leq 6 \\ 3x + 4y \leq 4 \end{bmatrix}, \text{Minimum is at: } \left\{ y = -\tfrac{2}{7}, x = \tfrac{12}{7} \right\}$$

Feasible Systems

Two things may prevent the existence of a solution. There may be no values of x and y satisfying the constraints. Even if there are such values, there may be none maximizing the objective function. If there are values satisfying the constraints, the system is called *feasible*.

The following example illustrates a set of inequality constraints with no function to be maximized or minimized. You can ask whether the constraints are feasible—that is, whether they define a nonempty set. Just place the insertion point in the matrix and, from the **Simplex** submenu, choose **Feasible**.

▶ Simplex + Feasible?

$$\begin{bmatrix} 4x + 3y \le 6 \\ 3x + 4y \le 4 \\ x \ge 0 \\ y \ge 0 \end{bmatrix}, \text{ Is feasible? true}$$

$$\begin{bmatrix} 4x + 3y \le 6 \\ 4x + 3y \ge 7 \end{bmatrix}, \text{ Is feasible? false}$$

Saying that the system $\begin{matrix} 4x + 3y \le 6 \\ 4x + 3y \ge 7 \end{matrix}$ is not feasible implies, in particular, that there are no values minimizing the objective function in the problem $\begin{matrix} x + y \\ 4x + 3y \le 6 \\ 4x + 3y \ge 7 \end{matrix}$.

Standard Form

A system of linear inequalities is in *standard form* when all the inequalities are of the form $\le$. To convert a system of linear inequalities to a system in standard form, choose **Standardize** from the **Simplex** submenu.

▶ Simplex + Standardize

$$\begin{bmatrix} 4x + 3y \le 6 \\ 3x + 4y \le 4 \\ x \ge 0 \\ y \ge 0 \end{bmatrix}, \text{ System in standard form is: } \begin{bmatrix} -x \le 0 \\ -y \le 0 \\ 3x + 4y \le 4 \\ 4x + 3y \le 6 \end{bmatrix}$$

With a linear function added, you can maximize the resulting linear programming problem.

▶ Simplex + Maximize

$$\begin{bmatrix} x + 3y \\ 3x - y \le 4 \\ 4x + 3y \le 6 \\ -y \le 0 \\ -x \le 0 \end{bmatrix}, \text{ Maximum is at: } \{x = 0, y = 2\}$$

The Dual of a Linear Program

The other item on the Simplex menu is Dual. It computes the dual of a linear program.

▶ Simplex + Dual

$$
\begin{bmatrix}
x + y \\
4x + 3y \le 6 \\
3x + 4y \le 4 \\
x \ge 0 \\
-y \le 0
\end{bmatrix}, \text{Dual system is: }
\begin{bmatrix}
4u_7 + 6u_8 \\
u_5 - 3u_7 - 4u_8 \le -1 \\
u_6 - 4u_7 - 3u_8 \le -1
\end{bmatrix}
$$

Applying the simplex algorithm to these two linear programs yields the following results.

▶ Simplex + Maximize

$$
\begin{bmatrix}
x + y \\
4x + 3y \le 6 \\
3x + 4y \le 4 \\
x \ge 0 \\
-y \le 0
\end{bmatrix}, \text{Maximum is at: } \left\{ y = 0, x = \tfrac{4}{3} \right\}
$$

▶ Simplex + Minimize

$$
\begin{bmatrix}
6s_1 + 4s_2 \\
1 \le 4s_1 + 3s_2 - s_4 \\
1 \le 3s_1 + 4s_2 - s_3 \\
s_1 \ge 0 \\
s_2 \ge 0 \\
s_3 \ge 0 \\
s_4 \ge 0
\end{bmatrix}, \text{Minimum is at: } \left\{ s_4 = 0, s_1 = 0, s_2 = \tfrac{1}{3}, s_3 = \tfrac{1}{3} \right\}
$$

Exercises

1. Give a multiplication table for the integers modulo 11. From the table, find the inverses of 2 and 3. Verify your answers by evaluating $2^{-1} \bmod 11$ and $3^{-1} \bmod 11$.

2. Solve the congruence $5x + 4 \equiv 8 \pmod{13}$. Verify your answer by evaluating $5x + 4 \bmod 13$.

3. A jar is full of jelly beans. If the jelly beans are evenly divided among five children, there are three jelly beans left over; and if the jelly beans are evenly divided among

seven adults, there are five jelly beans left over. How many jelly beans are in the jar? Are other solutions possible? If so, what are they?

4. What is the smallest 100-digit prime?

5. If p is the smallest 100-digit prime, what is $2^{p-1} \bmod p$? What is $2^{(p-1)/2} \bmod p$? What about $2^{(p-1)/4} \bmod p$?

6. The matrix $M = \begin{bmatrix} 1 & 1 & 1 \\ 1 & 2 & 4 \\ 1 & 4 & 9 \end{bmatrix}$ is used as a block cipher modulo 26 to scramble

letters in a message, three letters at a time. Assume $A \leftrightarrow 0$, $B \leftrightarrow 1$, $C \leftrightarrow 2$, and so forth. Descramble the ciphertext $FKBHRTMTU$.

7. Find an irreducible polynomial of degree 3. Use this polynomial to describe how to calculate sums and products in the field GF_{27}.

8. A barge company transports bales of hay and barrels of beer up the Mississippi River. The company charges \$2.30 for each bale of hay and \$3.00 for each barrel of beer. The bales of hay average 75 pounds and take up 5 cubic feet of space; the barrels of beer weigh 100 pounds and take up 4 cubic feet of space. A barge is limited to a payload of 150,000 pounds and 8,000 cubic feet. How much beer and how much hay should a barge transport to maximize the shipping charges?

9. The Riemann Hypothesis states that all of the nontrivial zeros of the Riemann zeta function lie on the line $\mathrm{Re}(s) = \frac{1}{2}$. Visualize the Riemann zeta function along $\mathrm{Re}(s) = \frac{1}{2}$ by drawing a curve in three space.

10. Let $\mathbb{Z}_{30}$ denote the integers modulo 30. Write $\mathbb{Z}_{30}$ as a (disjoint) union of groups.

Solutions

1. Define the function $f(i, j) = ij$. From the **Matrix** submenu, choose **Fill Matrix** with 10 rows and 10 columns, and use the function f to generate a matrix. Then, reduce the matrix mod 11 to get the following:

1	2	3	4	5	6	7	8	9	10
2	4	6	8	10	1	3	5	7	9
3	6	9	1	4	7	10	2	5	8
4	8	1	5	9	2	6	10	3	7
5	10	4	9	3	8	2	7	1	6
6	1	7	2	8	3	9	4	10	5
7	3	10	6	2	9	5	1	8	4
8	5	2	10	7	4	1	9	6	3
9	7	5	3	1	10	8	6	4	2
10	9	8	7	6	5	4	3	2	1

Select the matrix and, from the **Edit** menu, choose **Insert Column(s)**. Add one column at position 1. You have now added a column on the left. Repeat this procedure using Insert Row(s), adding a row at position 1. Fill in the empty boxes with $\times$ and the integers 1 through 10 to generate the final multiplication table,

$\times$	1	2	3	4	5	6	7	8	9	10
1	1	2	3	4	5	6	7	8	9	10
2	2	4	6	8	10	1	3	5	7	9
3	3	6	9	1	4	7	10	2	5	8
4	4	8	1	5	9	2	6	10	3	7
5	5	10	4	9	3	8	2	7	1	6
6	6	1	7	2	8	3	9	4	10	5
7	7	3	10	6	2	9	5	1	8	4
8	8	5	2	10	7	4	1	9	6	3
9	9	7	5	3	1	10	8	6	4	2
10	10	9	8	7	6	5	4	3	2	1

From the table, $2 \cdot 6 = 1$ implies $2^{-1} = 6$, and $3 \cdot 4 = 1$ implies $3^{-1} = 4$. As a check, $2^{-1} \bmod 11 = 6$ and $3^{-1} \bmod 11 = 4$.

2. The solution is given by
$$x = (8 - 4)/5 \bmod 13 = 6$$
As a check,
$$6 \cdot 5 + 4 \bmod 13 = 8$$

3. The problem requires the solution to the system
$$x \equiv 3 \pmod 5$$
$$x \equiv 5 \pmod 7$$
of congruences. The system is equivalent to the equation $x = 3 + 5a = 5 + 7b$, or $3 + 5a \equiv 5 \pmod 7$, which has a solution $a = (5 - 3)/5 \bmod 7 = 6$, which means $x = 3 + 5a = 33$ jelly beans. Other possible solutions are $x = 33 + 35n$, where n is any positive integer.

4. Define the function nextp as indicated in this chapter. Then nextp(10^{99}) produces a number with lots of zeroes that ends in 289. The prime p can be written as $p = 10^{99} + 289$.

5. Note that $2^{p-1} \bmod p = 1$ and $2^{(p-1)/2} \bmod p = 1$, whereas $2^{(p-1)/4} \bmod p$ produces another number with lots of zeroes that ends in 288. More precisely, $2^{(p-1)/4} \equiv -1 \bmod p$. This congruence illustrates the fact that, if p is a prime, then $x^2 \equiv 1 \pmod p$ has only two solutions, $x \equiv 1 \pmod p$ and $x \equiv -1 \pmod p$.

6. We have
$$\begin{bmatrix} 1 & 1 & 1 \\ 1 & 2 & 4 \\ 1 & 4 & 9 \end{bmatrix}^{-1} \bmod 26 = \begin{bmatrix} 24 & 5 & 24 \\ 5 & 18 & 3 \\ 24 & 3 & 25 \end{bmatrix}.$$
The ciphertext F K B H R

T M T U has a numerical equivalent of $[5, 10, 1, 7, 17, 19, 12, 19, 20]$. Picking three at a time, we get
$$\begin{bmatrix} 24 & 5 & 24 \\ 5 & 18 & 3 \\ 24 & 3 & 25 \end{bmatrix} \begin{bmatrix} 5 & 7 & 12 \\ 10 & 17 & 19 \\ 1 & 19 & 20 \end{bmatrix} \bmod 26 = \begin{bmatrix} 12 & 7 & 5 \\ 0 & 8 & 20 \\ 19 & 18 & 13 \end{bmatrix}$$

The vector $[12, 0, 19, 7, 8, 18, 5, 20, 13]$ corresponds to the plaintext M A T H I S F U N, or MATH IS FUN.

7. Defining $g(x) = x^3 + x + 1$, we see that $g(1) \bmod 3 = 0$, and hence $g(x)$ is not irreducible (since it has a root in GF_3). However, if $f(x) = x^3 + 2x + 1$, then $f(0) \bmod 3 = 1$, $f(1) \bmod 3 = 1$, and $f(2) \bmod 3 = 1$, and hence $f(x)$ is irreducible. (If $f(x)$ were reducible, it would have a linear factor, and hence a root.) An element of GF_{27} can be thought of as a polynomial of degree less than 3 with coefficients in GF_3. Given the field elements $2x^2 + x + 2$ and $2x + 1$, the product is $\left((2x^2 + x + 2)(2x + 1) \bmod x^3 + 2x + 1\right) \bmod 3 = x^2 + 1$, and the sum is given by $(2x^2 + x + 2) + (2x + 1) \bmod 3 = 2x^2$.

8. The objective function is $2.3h + 3b$. The constraints are $4h + 5b \le 8000$, $75h + 100b \le 150000$, $b \ge 0$, and $h \ge 0$. Apply **Maximize** from the **Simplex** submenu to the system

$$2.3h + 3b$$
$$5h + 4b \le 8000$$
$$75h + 100b \le 150000$$
$$b \ge 0$$
$$h \ge 0$$

to get the result: Maximum is at: $\{b = 750, h = 1000\}$. Thus the maximum is $2.3(1000) + 3(750) = 4550.0$.

9. Type $\left(t, \operatorname{Re}\left(\zeta\left(\frac{1}{2} + ti\right)\right), \operatorname{Im}\left(\zeta\left(\frac{1}{2} + ti\right)\right)\right)$ and, from the **Plot 3D** submenu, choose **Tube**. Type $(t, 0, 0)$ and drag it to the plot frame. From the **Plot Properties** dialog, choose the **Items Plotted** page. For **Items** 1 and 2, set **Intervals:** 0 to 35, **Points Samples:** 99, **Points per Cross Section:** 7, **Radius:** 0.2 and set the **Surface Style** to **Hidden Line**.

View the curve from several different angles. Note that the intersection points display zeros of the Riemann zeta function.

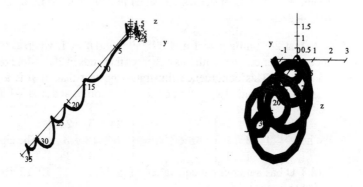

Turn: 45°   Turn: 5°
Tilt: 45°   Tilt: 85°

10. Consider first the positive integers < 30 that are relatively prime to 30. Let $G_1 = \{1, 7, 11, 13, 17, 19, 23, 29\}$ be the group of units modulo 30. In a similar fashion, for each divisor n of 30 define G_n to be the positive integers $a < 30$ such that $\gcd(a, 30) = n$. Thus

$$
\begin{aligned}
G_1 &= \{1, 7, 11, 13, 17, 19, 23, 29\} \\
G_2 &= \{2, 4, 8, 14, 16, 22, 26, 28\} \\
G_3 &= \{3, 9, 21, 27\} \\
G_5 &= \{5, 25\} \\
G_6 &= \{6, 12, 18, 24\} \\
G_{10} &= \{10, 20\} \\
G_{15} &= \{15\} \\
G_{30} &= \{0\}
\end{aligned}
$$

For each of these subsets, create a multiplication table modulo 30 (see page 450) such as the following one for G_2, for which 16 acts as an identity.

$$
\begin{bmatrix} 2 \\ 4 \\ 8 \\ 14 \\ 16 \\ 22 \\ 26 \\ 28 \end{bmatrix}
\begin{bmatrix} 2 & 4 & 8 & 14 & 16 & 22 & 26 & 28 \end{bmatrix} =
$$

$$
\begin{bmatrix}
4 & 8 & 16 & 28 & 32 & 44 & 52 & 56 \\
8 & 16 & 32 & 56 & 64 & 88 & 104 & 112 \\
16 & 32 & 64 & 112 & 128 & 176 & 208 & 224 \\
28 & 56 & 112 & 196 & 224 & 308 & 364 & 392 \\
32 & 64 & 128 & 224 & 256 & 352 & 416 & 448 \\
44 & 88 & 176 & 308 & 352 & 484 & 572 & 616 \\
52 & 104 & 208 & 364 & 416 & 572 & 676 & 728 \\
56 & 112 & 224 & 392 & 448 & 616 & 728 & 784
\end{bmatrix}
$$

and

$$
\begin{bmatrix}
4 & 8 & 16 & 28 & 32 & 44 & 52 & 56 \\
8 & 16 & 32 & 56 & 64 & 88 & 104 & 112 \\
16 & 32 & 64 & 112 & 128 & 176 & 208 & 224 \\
28 & 56 & 112 & 196 & 224 & 308 & 364 & 392 \\
32 & 64 & 128 & 224 & 256 & 352 & 416 & 448 \\
44 & 88 & 176 & 308 & 352 & 484 & 572 & 616 \\
52 & 104 & 208 & 364 & 416 & 572 & 676 & 728 \\
56 & 112 & 224 & 392 & 448 & 616 & 728 & 784
\end{bmatrix}
\bmod 30 =
$$

$$\begin{bmatrix} 4 & 8 & 16 & 28 & 2 & 14 & 22 & 26 \\ 8 & 16 & 2 & 26 & 4 & 28 & 14 & 22 \\ 16 & 2 & 4 & 22 & 8 & 26 & 28 & 14 \\ 28 & 26 & 22 & 16 & 14 & 8 & 4 & 2 \\ 2 & 4 & 8 & 14 & 16 & 22 & 26 & 28 \\ 14 & 28 & 26 & 8 & 22 & 4 & 2 & 16 \\ 22 & 14 & 28 & 4 & 26 & 2 & 16 & 8 \\ 26 & 22 & 14 & 2 & 28 & 16 & 8 & 4 \end{bmatrix}$$

Note that each of these sets is closed under multiplication, and that each element appears once in each row and once in each column. Since multiplication is certainly commutative and associative, it follows that each subset is in fact a group.

Index

Software

 Scientific WorkPlace®

Scientific WorkPlace makes writing, sharing, and doing mathematics easier than you ever imagined possible. This scientific word processor increases your productivity because it is easy to learn and use. You can compose and edit your documents directly on the screen, without being forced to think in a programming language. With a simple click of a button, you can typeset your document in LaTeX. You can also compute and plot solutions with the included computer algebra system. With *Scientific WorkPlace*, both professional and support staff can produce stunning results quickly and easily, without knowing TeX™, LaTeX, or computer algebra syntax. ***Contact us for a free 30-day trial version.***

 Scientific Word®

The Gold Standard for mathematical publishing since 1992, *Scientific Word* makes writing and sharing scientific documents straightforward and easy. With over 100 LaTeX styles included, *Scientific Word* ensures your documents will be beautiful. This means you can concentrate on the content, not the style. It has been estimated that support staff using *Scientific Word* experiences a doubling or tripling of productivity over the use of straight LaTeX. Best of all, MacKichan Software provides free, prompt, and knowledgeable technical support. ***Contact us for a free 30-day trial version.***

 Scientific Notebook®

Scientific Notebook makes word processing and doing mathematics easy. With this complete word processor, you can enter text and mathematics quickly, without having to use an inefficient equation editor. The built-in computer algebra system lets you solve and plot equations without having to learn a special syntax. After creating your scientific documents and exams in *Scientific Notebook*, you can publish them in print and on the World Wide Web. ***Contact us for a free 30-day trial version.***

 MuPAD® **Pro**

MuPAD Pro is a modern, full-featured computer algebra system in an integrated and open environment for symbolic and numeric computing. Its domains and categories are like object-oriented classes that allow overriding and overloading methods and operators, inheritance, and generic algorithms. The *MuPAD* language has a Pascal-like syntax and allows imperative, functional, and object-oriented programming. A comfortable notebook interface includes a graphics tool for visualization, an integrated source-level debugger, a profiler, and hypertext help. ***Contact us for a free 30-day trial version.***

TO ORDER: Visit our webstore, fax, email, or phone us.
Website: www.mackichan.com ♦ Fax: 360-394-6039 ♦ Email: info@mackichan.com ♦ Toll-free: 877-724-9673
19307 8th Avenue NE ♦ Suite C ♦ Poulsbo, WA 98370

Additional Software

MathTalk™/Scientific Notebook®

MathTalk/Scientific Notebook, created by Metroplex Voice Computing, provides voice input for *Scientific Notebook*. With this program, you can enter even the most complex mathematics using voice commands. You can use it in conjunction with the keyboard and mouse to speed the entry of text and mathematics, or to completely replace the keyboard and mouse. *MathTalk/Scientific Notebook* requires *Dragon NaturallySpeaking®*, which is not included. Visit *www.mathtalk.com* for more information.

Duxbury Braille Translator

Duxbury Systems leads the world in software for braille. Their *Duxbury Braille Translator* converts *Scientific Notebook* files into braille. Create your print or large print math, save, then open your *Scientific Notebook* file and go to braille. Used with *MathTalk/Scientific Notebook*, the *Duxbury Braille Translator* provides dramatic new power to visually impaired students and professionals. Visit *www.duxburysystems.com* for more information.

Books

Doing Mathematics with Scientific WorkPlace® and Scientific Notebook®, Version 5

By Darel W. Hardy and Carol L. Walker

Doing Mathematics with Scientific WorkPlace and Scientific Notebook describes how to use the built-in computer algebra system to do a wide range of mathematics, without having to deal directly with the computer algebra syntax.

Creating Documents with Scientific WorkPlace® and Scientific Word®, Version 5

By Susan Bagby

Creating Documents with Scientific WorkPlace and Scientific Word gives you an overview of the process of creating beautiful documents using these two powerful software programs. It covers basic editing and entering of mathematical expressions as well as tables, graphics, lists, indexes, cross-references, tables of contents, and large document management. If you are using *Scientific WorkPlace* or *Scientific Word* to prepare documents for publication, this book is highly recommended.

(Continued)

TO ORDER: Visit our webstore, fax, email, or phone us.
Website: www.mackichan.com ◆ Fax: 360-394-6039 ◆ Email: info@mackichan.com ◆ Toll-free: 877-724-9673
19307 8th Avenue NE ◆ Suite C ◆ Poulsbo, WA 98370

MathTalk™ is a trademark of Metroplex Voice Computing.
Dragon NaturallySpeaking® is a registered trademark of Dragon Systems, Inc.

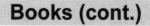

MacKichan
SOFTWARE, INC.

Books (cont.)

Typesetting Documents in Scientific WorkPlace® and Scientific Word®, Second Edition
By Susan Bagby and George Pearson

Typesetting Documents in ScientificWorkPlace and Scientific Word is an aid to choosing and customizing the typeset appearance of your documents. The manual explains how to work from within these two programs to tailor the typeset appearance of a document; how to choose and add document shells; and how document shells use LaTeX document classes, class options, and packages. The manual also documents many of the LaTeX packages that accomplish specific formatting tasks.

A Gallery of Document Shells for Scientific WorkPlace® and Scientific Word®, Version 5
By Susan Bagby and George Pearson (In PDF format on the program CD-ROM)

A Gallery of Document Shells for Scientific WorkPlace and Scientific Word helps you choose document shells that are appropriate for your typesetting purposes. It illustrates and briefly describes the characteristics of almost 200 shells provided with these two programs.

Doing Calculus with Scientific Notebook®
By Darel W. Hardy and Carol L. Walker

Take the mystery out of doing calculus with this must-have companion to the *Scientific Notebook* software. This book provides activities to complete with *Scientific Notebook* that will help develop a clearer understanding of calculus. Think of *Scientific Notebook* as a laboratory for mathematical experimentation and *Doing Calculus with Scientific Notebook* as a lab manual of experiments to perform.

An Interactive Introduction to Mathematical Analysis
By Jonathan Lewin (Cambridge University Press)

This book is a sequel to *An Introduction to Mathematical Analysis*. It includes an on-screen hypertext version that you can read with *Scientific Notebook*. This on-screen version contains alternative approaches to material, more fully explained forms of proofs of theorems, sound movie versions of proofs of theorems, interactive exploration of mathematical concepts using the computing features of *Scientific Notebook*, automatic links to the author's website for solutions to exercises, and more. It can be ordered at any bookstore.

(Continued)

TO ORDER: Visit our webstore, fax, email, or phone us.
Website: www.mackichan.com ♦ Fax: 360-394-6039 ♦ Email: info@mackichan.com ♦ Toll-free: 877-724-9673
19307 8th Avenue NE ♦ Suite C ♦ Poulsbo, WA 98370

Books (cont.)

Precalculus with Scientific Notebook®
By Jonathan Lewin (Kendall Hunt Publishing Company)

This book contains a standard printed version and an on-screen hypertext version designed for interactive reading with *Scientific Notebook*. The on-screen version includes links to solutions to exercises that reside on the author's website. It can be ordered at any bookstore.

Exploring Mathematics with Scientific Notebook®
By Wei-Chi Yang and Jonathan Lewin (Springer-Verlag)

This book is supplied both in printed form and in an on-screen hypertext version for interactive reading with *Scientific Notebook*. It contains a sequence of modules from a variety of mathematical areas and, in each, demonstrates how the editing, Internet, and computing features of *Scientific Notebook* can be combined to deepen the reader's understanding of mathematical concepts. It can be ordered at any bookstore.

MuPAD® Pro Computing Essentials
By Miroslav Majewski (Springer-Verlag)

Intended for teachers of mathematics and their students, *MuPAD Pro Computing Essentials* presents basic information about doing mathematics with *MuPAD Pro*. It includes basic instructions useful in various areas of mathematics that are facilitated by *MuPAD Pro*. Chapters 1 through 7 focus on the basics of using *MuPAD Pro*: Syntax, programming control structures, procedures, libraries, and graphics. Chapters 8 through 13 focus on applications in geometry, algebra, logic, set theory, calculus, and linear algebra. Each chapter includes examples and programming exercises.

TO ORDER: Visit our webstore, fax, email, or phone us.
Website: www.mackichan.com ✦ Fax: 360-394-6039 ✦ Email: info@mackichan.com ✦ Toll-free: 877-724-9673
19307 8th Avenue NE ✦ Suite C ✦ Poulsbo, WA 98370

Workshops

Scientific WorkPlace®, Scientific Word®, and Scientific Notebook®
By Professor Jonathan Lewin, Ph.D.

Seminars, workshops, and training sessions by Jonathan Lewin are available at professional conferences and, by arrangement, at individual campuses of high schools, colleges, and universities.

Scientific Notebook: This presentation introduces the editing, Internet and computing features of *Scientific Notebook* documents. It also covers the publication of mathematical material on websites and how *Scientific Notebook* can be used as an electronic whiteboard in the classroom.

Scientific WorkPlace and Scientific Word: This presentation introduces the editing and typesetting features of *Scientific WorkPlace* and *Scientific Word*. Participants will be trained in the production of documents that will be printed as professional quality hard copy or submitted to an editor for publication.

Each participant will be given a CD containing sound movies that review the material covered in the workshops. Contact Jonathan Lewin for details at *lewins@mindspring.com* or 770-973-5931.

Scientific WorkPlace® and Scientific Notebook®
By Professor Bill Pletsch, Ph.D.

Bill Pletsch has been using computer algebra systems since the early 1980s. He is available for workshops and training sessions on the use of *Scientific WorkPlace* and *Scientific Notebook* as a research and teaching aid.

The workshops and training sessions begin with a demonstration of the capabilities of *Scientific WorkPlace/Notebook*. The introductory demonstration is followed by a thorough nuts and bolts hands-on session on the basics. Participants learn how to compute numerically and symbolically, graph, and manipulate data. Also included are word-processing, Internet techniques, and utilizing other resources. More advanced topics include the use of *Scientific WorkPlace/Notebook* in the classroom and in the preparation of computer classroom lectures and demonstrations.

An overview of computer algebra systems and why every teacher of mathematics should own a copy of *Scientific Notebook* will accompany the presentation. Included in the workshops will be a CD on the electronic delivery of mathematics instruction. No prior computer experience is required. Contact Bill Pletsch for details at *bpletsch@tvi.edu* or 505-224-3672.

(Continued)

Workshops

Scientific Notebook®
By Professor John H. Gresham, Ph.D.

John Gresham has been using *Scientific Notebook* since 1997 to prepare materials for classroom instruction. He is available for workshops and training sessions on *Scientific Notebook*, including the use of *Scientific Notebook*'s Exam Builder to generate both objective and free-response test forms. Workshop participants will receive hands-on step-by-step instruction in using *Scientific Notebook* as a mathematics word processor to prepare class notes, overhead transparencies, and exams. *Scientific Notebook* will also be used to view and prepare web materials, including HTML files. Participants will become familiar with the capability of *Scientific Notebook*'s computer algebra system to perform numeric and symbolic computations, solve equations, and generate graphs and tables. Participants will receive printed materials and a floppy disk with sample files. No prior experience with *Scientific Notebook* is required, but familiarity with basic Windows commands will be helpful. For details, contact John Gresham at jgresham@ranger.cc.tx.us or 254-647-3234.

TO ORDER: Visit our webstore, fax, email, or phone us.
Website: www.mackichan.com ◆ Fax: 360-394-6039 ◆ Email: info@mackichan.com ◆ Toll-free: 877-724-9673
19307 8th Avenue NE ◆ Suite C ◆ Poulsbo, WA 98370